Principles and Practice of Electrical Engineering

Principles and Practice of Electrical Engineering

Contributors :
Yagang Zhang,
Yi Sun, *et al.*

AURIS REFERENCE LTD.
London, UK

Principles & Practice of Electrical Engineering
Contributors : Yi Sun *and* Yagang Zhang, *et al.*

Auris Reference Ltd., UK

www.aurisreference.com

United Kingdom

Copyright 2016

Printed in 2017 for Sale in the Indian Subcontinent

Principles & Practice of Electrical Engineering

ISBN: 978-1-78154-513-3

British Library Cataloguing in Publication Data
A CIP record for this book is available from the British Library

Exclusively distributed by CBS Publishers & Distributors Pvt. Ltd.

Sales & Distribution Rights only for India, Pakistan, Bangladesh, Sri Lanka, Nepal and Bhutan.This book is not to be sold outside these territories.

PREFACE

Electrical engineering is core engineering like civil and mechanical but it has a wide range of subfields. After modernization, many fields of engineering grew out of electrical such as electronics, computer, telecommunication engineering and many more. All the fields of study that directly or indirectly deal with electricity come under electrical engineering.

In a power generation plant where electrical energy is produced, the application of this engineering is huge. All the mechanical and as well as electrical equipment involved in producing electrical energy, such as alternators, boilers, turbines etc. are controlled and protected by electrical signals. All the relays and switches involved in operation of the equipment are either electromechanical or static electronics devices. In the modern age, these devices are digitally controlled by computer software. So in addition to core electrical engineering, these electrical engineering subfields (electronic, computer, software engineering and IT) are also involved in power generation.

When the voltage level of generated power is stepped up, electrical transformers are required. For proper control and protection of these transformers, a sophisticated switchgear system is required. Electrical switchgear includes all circuit breakers, electrical isolators, current transformers, potential transformers, control and protection relay system and many more. Modern electrical engineering deals with production, planning, operation, and maintenance of these systems.

After stepping up the electrical power, it is transmitted to a load center. Huge electrical technology is involved in electrical power transmission systems and networks including large national grids. The problems associated with these systems are solved by electrical engineering including all engineering subfields.

The most familiar form of energy in our daily lives is electrical energy. The branch of engineering which deals with producing, managing and utilizing this energy, is referred as electrical engineering. This field of engineering was introduced in an organized manner in the mid of 19th century. Electrical technology is not as old as civil and mechanical technologies. Interest in this field of study grew after the invention of electricity. It was June 1752 when Benjamin Franklin first tried to catch electricity from clouds during a heavy storm with the help of a flying kite. This was the beginning and to date still we are trying to manage this energy. Managing electrical energy (producing, transmitting, distributing, utilizing) is nothing but electrical engineering. Commercially this became an occupation only after 1950.

The book provides an exhaustive coverage of network theory and analysis, electromagnetic theory and energy conversion, transformers, alternating and direct current machines, and basic instruments.

CONTENTS

Preface   *v*

1. ELECTRICAL ENGINEERING   **1-22**

History of Electrical Engineering   2

Subdisciplines   6

Practicing Engineers   20

2. NETWORK   **23-92**

Network Analysis   23

Equivalent Impedance Transforms   24

Y-Δ Transform   36

Star-Mesh Transform   43

Simple Networks   44

Nodal Analysis   52

Mesh Analysis   55

Choice of Method   58

Transfer Function   58

Non-Linear Networks   79

Introduction to Network Theorems   83

3. MAGNETIC CIRCUITS   **93-116**

Magnetomotive Force   93

Magnetic Flux   94

Hopkinson's Law : The Magnetic Analogy to Ohm's Law   97

Magnetic Reluctance   98

Magnetic Capacitivity   103

Magnetic Capacitance   104

Magnetic Complex Reluctance   104

Magnetic Core   105

Magnetic Effective Resistance   114

Magnetic Impedance   115

Magnetic Inductance   115

Magnetic Tension Force   116

4. THREE-PHASE ELECTRIC POWER   **117-129**

Details   117

Generation and Distribution   118

Transformer Connections   118

Three-Wire and Four-Wire Circuits   119

Balanced Circuits   121

Delta   122

Single-Phase Loads   123

Unbalanced Loads   123

Non-Linear Loads   123

Three-Phase Loads   124

Phase Converters   125

Alternatives to Three-Phase   126

Three-Phase AC Railway Electrification   127

5. TRANSFORMER   **130-185**

Invention   130

Applications   131

Basic Principles   131

Leakage Inductance   135

Equivalent Circuit   141

Types of Transformer   155

6. DC MOTOR   **186-235**

Principle of DC Motor   186

Detailed Description of A DC Motor   187

Working or Operating Principle of DC Motor   189

Construction of DC Motor   193

Torque Equation of DC Motor   196

Types of DC Motor   199

Shunt Wound DC Motor   206

DC Series Motor   210

DC Compound Motor   213

Starting Methods to Limit Starting Current & Torque of DC Motor   216

3 Point Starter 219

4 Point Starter 222

Speed Regulation of DC Motor 224

Speed Control of DC Motor 225

Lap Winding Simplex and Duplex Lap Winding 228

Permanent Magnet DC Motor or PMDC Motor 231

Ward Leonard Method of Speed Control 234

7. SYNCHRONOUS MOTOR 236-242

Type 236

8. INDUCTION MOTOR 243-273

Working Principle of Induction Motor 243

Types Induction Motor 244

Why is Three Phase Induction Motor Self-Starting? 244

Working Principle of Three Phase Induction Motor 244

Single Phase Induction Motor 246

Types of Three Phase Induction Motor 249

Torque Equation of Three Phase Induction Motor 251

Speed Control of Three Phase Induction Motor 255

Losses and Efficiency of Induction Motor 261

Construction of Three Phase Induction Motor 264

Linear Induction Motor 269

9. ELECTROMAGNETISM 274-335

Classical Electromagnetism 277

Electric Field 282

Photoelectric Effect 289

Quantities and Units 300

Electromagnetic Phenomena 300

Electromagnetic Induction 301

Computational Electromagnetics 301

Electromagnetic Wave Equation 309

Electromechanics 316

Electromagnet 317

Relativistic Electromagnetism 327

10. RESEARCH ON THE FAULT COEFFICIENT IN COMPLEX ELECTRICAL ENGINEERING 336-350

Yi Sun, Yagang Zhang and Yinding Wang

Introduction 337

The Fundamental of Feature Extraction about the Fault Coefficient 337

Phasor Measurement Unit and Wide Area Measurement System 340
Fault Coefficient Feature Extraction in Complex Electrical Engineering 341
Conclusions 347
References 348

LIST OF CONTRIBUTORS

Yi Sun

Hebei Electric Power Research Institute, Shijiazhuang, Hebei 050022, China; E-Mail: yisunsjz@163.com

Yagang Zhang

State Key Laboratory of Alternate Electrical Power System with Renewable Energy Sources, North China Electric Power University, Beijing 102206, China
and

Interdisciplinary Mathematics Institute, University of South Carolina, Columbia, SC 29208, USA; E-Mail: dnagct@gmail.com

Yinding Wang

Interdisciplinary Mathematics Institute, University of South Carolina, Columbia, SC 29208, USA; E-Mail: dnagct@gmail.com

This page left intentionally blank.

Chapter 1

ELECTRICAL ENGINEERING

Electrical engineering is a field of engineering that generally deals with the study and application of electricity, electronics, and electromagnetism. This field first became an identifiable occupation in the latter half of the 19th century after commercialization of the electric telegraph, the telephone, and electric power distribution and use. Subsequently, broadcasting and recording media made electronics part of daily life. The invention of the transistor and, subsequently, the integrated circuit brought down the cost of electronics to the point where they can be used in almost any household object. The personal computer and information technology are the most complex electronics yet to be used in everyday life.

Electrical engineering has now sub-divided into a wide range of sub-fields including electronics, digital computers, power engineering, telecommunications, control systems, RF engineering, signal processing, instrumentation, and micro-electronics. The subject of electronic engineering is often treated as its own sub-field but it intersects with all the other sub-fields, including the power electronics of power engineering.

Electrical engineers typically hold a degree in electrical engineering or electronic engineering. Practicing engineers may have professional certification and be members of a professional body. Such bodies include the Institute of Electrical and Electronic Engineers (IEEE) and the Institution of Engineering and Technology (IET).

Electrical engineers work in a very wide range of industries and the skills required are likewise variable. These range from basic circuit theory to the management skills required of project manager. The tools and equipment that an individual engineer may need are similarly variable, ranging from a simple voltmeter to a top end analyzer to sophisticated design and manufacturing software.

HISTORY OF ELECTRICAL ENGINEERING

Ancient Developments

Thales of Miletus, an ancient Greek philosopher, writing at around 600 B.C.E., described a form of static electricity, noting that rubbing fur on various substances, such as amber, would cause a particular attraction between the two. He noted that the amber buttons could attract light objects such as hair and that if they rubbed the amber for long enough they could even get a spark to jump.

At around 450 B.C.E. Democritus, a later Greek philosopher, developed an atomic theory that was remarkably similar to our modern atomic theory. His mentor, Leucippus, is credited with this same theory. The hypothesis of Leucippus and Democritus held everything to be composed of atoms. But these *atoms*, called "atomos", were indivisible, and indestructible. He presciently stated that between atoms lies empty space, and that atoms are constantly in motion. He was incorrect only in stating that atoms come in different sizes and shapes. Each object had its own shaped and sized atom.

An object found in Iraq in 1938, dated to about 250 B.C.E. and called the Baghdad Battery, resembles a galvanic cell and is believed by some to have been used for electroplating in Mesopotamia, although this has not yet been proven.

17th Century Developments

Electricity would remain little more than an intellectual curiosity for millennia until 1500.

When the Italian scientist Girolamo Cardano started a study on electricity in De Subtilitate (1550), distinguishing for the first time the electric strength from the magnetic one. In the 1600 the English scientist, William Gilbert extended the study of Cardano on electricity and magnetism, distinguishing the lodestone effect from static electricity produced by rubbing amber. He coined the New Latin word *electricus* ("of amber" or "like amber", from ἤλεκτρον [*elektron*], the Greek word for "amber") to refer to the property of attracting small objects after being rubbed. This association gave rise to the English words "electric" and "electricity", which made their first appearance in print in Thomas Browne's *Pseudodoxia Epidemica* of 1646.

Further work was conducted by Otto von Guericke who showed electrostatic repulsion. Robert Boyle also published work.

18th Century Developments

By 1705, Hauksbee had discovered that if he placed a small amount of mercury in the glass of his modified version of Otto von Guericke's generator, evacuated the air from it to create a mild vacuum and rubbed the ball in order to build up a charge, a glow was visible if he placed his hand on the outside of the ball. This glow was bright enough to read by. It seemed to be similar to St. Elmo's Fire. This effect later became the basis of the gas-discharge lamp, which led to neon light-ing and mercury vapour lamps. In 1706 he produced an 'Influence machine' to generate this effect. He was elected a Fellow of the Royal Society the same year.

Hauksbee continued to experiment with electricity, making numerous ob-servations and developing machines to generate and demonstrate various electri-cal phenomena. In 1709 he published *Physico-Mechanical Experiments on Various Subjects* which summarized much of his scientific work.

Stephen Gray discovered the importance of insulators and conductors. C. F. du Fay seeing his work, developed a "two-fluid" theory of electricity.

In the 18th century, Benjamin Franklin conducted extensive research in elec-tricity, selling his possessions to fund his work. In June 1752 he is reputed to have attached a metal key to the bottom of a dampened kite string and flown the kite in a storm-threatened sky. A succession of sparks jumping from the key to the back of his hand showed that lightning was indeed electrical in nature. He also explained the apparently paradoxical behaviour of the Leyden jar as a device for storing large amounts of electrical charge, by coming up with the single fluid, two states theory of electricity.

In 1791, Luigi Galvani published his discovery of bioelectricity, demonstrating that electricity was the medium by which nerve cells passed signals to the muscles. Alessandro Volta's battery, or voltaic pile, of 1800, made from alternating layers of zinc and copper, provided scientists with a more reliable source of electrical energy than the electrostatic machines previously used.

19th Century Developments

In the 19th century, the subject of electrical engineering, with the tools of modern research techniques, started to intensify. Notable developments in this century include the work of Georg Ohm, who in 1827 quantified the relationship between the electric current and potential difference in a conductor, Michael Faraday, the discoverer of electromagnetic induction in 1831, and James Clerk Maxwell, who in 1873 published a unified theory of electricity and magnetism in his treatise on *Electricity and Magnetism*. In the 1830s, Georg Ohm also constructed an early electrostatic machine. The homopolar generator was developed first by Michael Faraday during his memorable experiments in 1831. It was the beginning of mod-ern dynamos — that is, electrical generators which operate using a magnetic field. The invention of the industrial generator, which didn't need external magnetic power in 1866 by Werner von Siemens made a large series of other inventions in the wake possible. In 1878, the British inventor James Wimshurst developed an apparatus that had two glass disks mounted on two shafts. It was not till 1883 that the Wimshurst machine was more fully reported to the scientific community.

During the latter part of the 1800s, the study of electricity was largely con-sidered to be a sub-field of physics. It was not until the late 19th century that universities started to offer degrees in electrical engineering. In 1882, Darmstadt University of Technology founded the first chair and the first faculty of electrical engineering worldwide. In the same year, under Professor Charles Cross, the Massachusetts Institute of Technology began offering the first option of Electrical Engineering within a physics department. In 1883, Darmstadt University of Technology and Cornell University introduced the world's first courses of study in electrical engineering and in 1885 the University College London founded the first chair of electrical engineering in the United Kingdom. The University of Missouri subsequently established the first department of electrical engineering in the United States in 1886.

During this period work in the area increased dramatically. In 1882 Edison switched on the world's first large-scale electrical supply network that provided 110 volts direct current to fifty-nine customers in lower Manhattan. In the late 1880s saw the spread of a competing form of power distribution known as alternating current backed by George Westinghouse. The rivalry between the Westinghouse and Edison systems was known as the "War of Currents". AC eventually replaced DC for generation and power distribution, enormously extending the range and improving the safety and efficiency of power distribution.

By the end of the 19th century, figures in the progress of electrical engineering were beginning to emerge. Charles Proteus Steinmetz helped foster the development of alternating current that made possible the expansion of the electric power industry in the United States, formulating mathematical theories for engineers.

Emergence of Radio and Electronics

During the development of radio, many scientists and inventors contributed to radio technology and electronics. In his classic UHF experiments of 1888, Heinrich Hertz demonstrated the existence of airborn electromagnetic waves (radio waves) leading many inventors and scientists to try to adapt them to commercial applications, such as Guglielmo Marconi (1895) and Alexander Popov (1896).

20th Century Developments

John Fleming invented the first radio tube, the diode, in 1904.

Reginald Fessenden recognized that a continuous wave needed to be generated to make speech transmission possible, and he continued the work of Nikola Tesla, John Stone Stone, and Elihu Thomson on this subject. By the end of 1906, Fessenden sent the first radio broadcast of voice. Also in 1906, Robert von Lieben and Lee De Forest independently developed the amplifier tube, called the triode. Edwin Howard Armstrong enabling technology for electronic television, in 1931.

Second World War Years

The second world war saw tremendous advances in the field of electronics; especially in RADAR and with the invention of the magnetron by Randall and Boot at the University of Birmingham in 1940. Radio location, radio communication and radio guidance of aircraft were all developed at this time. An early electronic computing device, Colossus was built by Tommy Flowers of the GPO to decipher the coded messages of the German Lorenz cipher machine. Also developed at this time were advanced clandestine radio transmitters and receivers for use by secret agents.

An American invention at the time was a device to scramble the telephone calls between Winston Churchill and Franklin D. Roosevelt. This was called the Green Hornet system and worked by inserting noise into the signal. The noise was then extracted at the receiving end. This system was never broken by the Germans.

A great amount of work was undertaken in the United States as part of the War Training Program in the areas of radio direction finding, pulsed linear networks, frequency modulation, vacuum tube circuits, transmission line theory and fundamentals of electromagnetic engineering. These studies were published shortly after the war in what became known as the 'Radio Communication Series' published by McGraw-Hill in 1946.

In 1941 Konrad Zuse presented the Z3, the world's first fully functional and programmable computer.

Post War Developments

Prior to the second world war the subject was commonly known as 'radio engineering' and basically was restricted to aspects of communications and RADAR, commercial radio and early television. At this time, study of radio engineering at universities could only be undertaken as part of a physics degree.

Later, in post war years, as consumer devices began to be developed, the field broadened to include modern TV, audio systems, Hi-Fi and latterly computers and microprocessors. In 1946 the ENIAC (Electronic Numerical Integrator and Computer) of John Presper Eckert and John Mauchly followed, beginning the computing era. The arithmetic performance of these machines allowed engineers to develop completely new technologies and achieve new objectives, including the Apollo missions and the NASA moon landing.

The invention of the transistor in 1947 by William B. Shockley, John Bardeen and Walter Brattain opened the door for more compact devices and led to the development of the integrated circuit in 1958 by Jack Kilby and independently in 1959 by Robert Noyce. In the mid to late 1950s, the term radio engineering gradually gave way to the name electronics engineering, which then became a stand alone university degree subject, usually taught alongside electrical engineering with which it had become associated due to some similarities. In 1968 Marcian Hoff invented the first microprocessor at Intel and thus ignited the development of the personal computer. The first realization of the microprocessor was the Intel 4004, a 4-bit processor developed in 1971, but only in 1973 did the Intel 8080, an 8-bit processor, make the building of the first personal computer, the Altair 8800, possible.

SUBDISCIPLINES

Electrical engineering has many subdisciplines. Although there are electrical engineers who focus exclusively on one of these subdisciplines, many deal with a combination of them. Sometimes certain fields, such as electronic engineering and computer engineering, are considered separate disciplines in their own right.

Power Engineering

Power engineering, also called **Power Systems engineering**, is a sub-field of electrical engineering that deals with the generation, transmission, distribution and utilization of electric power as well as the electrical devices connected to such systems including generators, motors and transformers. Although much of the field is concerned with the problems of three-phase AC power - the standard for large-scale power transmission and distribution across the modern world - a significant fraction of the field is concerned with the conversion between AC and DC power as well as the development of specialized power systems such as those used in aircraft or for electric railway networks. It was a sub-field of electrical engineering before the emergence of energy engineering.

History

Electricity became a subject of scientific interest in the late 17th century with the work of William Gilbert. Over the next two centuries a number of important discoveries were made including the incandescent light bulb and the voltaic pile. Probably the greatest discovery with respect to power engineering came from Michael Faraday who in 1831 discovered that a change in magnetic flux induces an electromotive force in a loop of wire—a principle known as electromagnetic induction that helps explain how generators and transformers work.

In 1881 two electricians built the world's first power station at Godalming in England. The station employed two waterwheels to produce an alternating current that was used to supply seven Siemens arc lamps at 250 volts and thirty-four incandescent lamps at 40 volts. However supply was intermittent and in 1882 Thomas Edison and his company, The Edison Electric Light Company, developed the first steam-powered electric power station on Pearl Street in New York City. The Pearl Street Station consisted of several generators and initially powered around 3,000 lamps for 59 customers. The power station used direct current and operated at a single voltage. Since the direct current power could not be easily transformed to the higher voltages necessary to minimise power loss during transmission, the possible distance between the generators and load was limited to around half-a-mile (800 m).

That same year in London Lucien Gaulard and John Dixon Gibbs demonstrated the first transformer suitable for use in a real power system. The practical value of Gaulard and Gibbs' transformer was demonstrated in 1884 at Turin where the transformer was used to light up forty kilometres (25 miles) of railway from a single alternating current generator. Despite the success of the system, the pair made some fundamental mistakes. Perhaps the most serious was connecting the primaries of the transformers in series so that switching one lamp on or off would affect other lamps further down the line. Following the demonstration George Westinghouse, an American entrepreneur, imported a number of the transformers along with a Siemens generator and set his engineers to experimenting with them in the hopes of improving them for use in a commercial power system.

One of Westinghouse's engineers, William Stanley, recognised the problem with connecting transformers in series as opposed to parallel and also realised that making the iron core of a transformer a fully enclosed loop would improve the voltage regulation of the secondary winding. Using this knowledge he built a much improved alternating current power system at Great Barrington, Massachusetts in 1886. In 1885 the Italian physicist and electrical engineer Galileo Ferraris demonstrated an induction motor and in 1887 and 1888 the Serbian-American engineer Nikola Tesla filed a range of patents related to power systems including one for a practical two-phase induction motor which Westinghouse licensed for his AC system.

By 1890 the power industry had flourished and power companies had built literally thousands of power systems (both direct and alternating current) in the United States and Europe-these networks were effectively dedicated to providing electric lighting. During this time a fierce rivalry in the US known as the "War of Currents" emerged between Edison and Westinghouse over which form of transmission (direct or alternating current) was superior. In 1891, Westinghouse installed the first major power system that was designed to drive an electric motor and not just provide electric lighting. The installation powered a 100 horsepower (75 kW) synchronous motor at Telluride, Colorado with the motor being started by a Tesla induction motor. On the other side of the Atlantic, Oskar von Miller built a 20 kV 176 km three-phase transmission line from Lauffen am Neckar to Frankfurt am Main for the Electrical Engineering Exhibition in Frankfurt. In 1895, after a protracted decision-making process, the Adams No. 1 generating station at Niagara Falls began transmitting three-phase alternating current power to Buffalo at 11 kV. Following completion of the Niagara Falls project, new power systems increasingly chose alternating current as opposed to direct current for electrical transmission.

Although the 1880s and 1890s were seminal decades in the field, developments in power engineering continued throughout the 20th and 21st century. In 1936 the first commercial high-voltage direct current (HVDC) line using Mercury-arc valves was built between Schenectady and Mechanicville, New York. HVDC had previously been achieved by installing direct current generators in series (a system known as the Thury system) although this suffered from serious reliability issues. In 1957 Siemens demonstrated the first solid-state rectifier (solid-state rectifiers are now the standard for HVDC systems) however it was not until the early 1970s that this technology was used in commercial power systems. In 1959 Westinghouse demonstrated the first circuit breaker that used SF_6 as the interrupting medium. SF_6 is a far superior dielectric to air and, in recent times, its use has been extended to produce far more compact switching equipment (known as switchgear) and transformers. Many important developments also came from extending innovations in the ICT field to the power engineering field. For example, the development of computers meant load flow studies could be run more efficiently allowing for much better planning of power systems. Advances in information technology and telecommunication also allowed for much better remote control of the power system's switchgear and generators.

Basics of Electric Power

Electric power is the mathematical product of two quantities : current and voltage. These two quantities can vary with respect to time (AC power) or can be kept at constant levels (DC power).

Most refrigerators, air conditioners, pumps and industrial machinery use AC power whereas computers and digital equipment use DC power (the digital devices you plug into the mains typically have an internal or external power adapter to convert from AC to DC power). AC power has the advantage of being easy to transform between voltages and is able to be generated and utilised by brushless machinery. DC power remains the only practical choice in digital systems and can be more economical to transmit over long distances at very high voltages.

The ability to easily transform the voltage of AC power is important for two reasons : Firstly, power can be transmitted over long distances with less loss at higher voltages. So in power networks where generation is distant from the load, it is desirable to step-up the voltage of power at the generation point and then step-down the voltage near the load. Secondly, it is often more economical to install turbines that produce higher voltages than would be used by most appliances, so the ability to easily transform voltages means this mismatch between voltages can be easily managed.

Solid state devices, which are products of the semi-conductor revolution, make it possible to transform DC power to different voltages, build brushless DC machines and convert between AC and DC power. Nevertheless devices utilising solid state technology are often more expensive than their traditional counterparts, so AC power remains in widespread use.

Power

Power Engineering deals with the generation, transmission, distribution and utilization of electricity as well as the design of a range of related devices. These include transformers, electric generators, electric motors and power electronics.

The power grid is an electrical network that connects a variety of electric generators to the users of electric power. Users purchase electricity from the grid so that they do not need to generate their own. Power engineers may work on the design and maintenance of the power grid as well as the power systems that connect to it. Such systems are called on-grid power systems and may supply the grid with additional power, draw power from the grid or do both. The grid is designed and managed using software that performs simulations of power flows.

Power engineers may also work on systems that do not connect to the grid. These systems are called off-grid power systems and may be used in preference to on-grid systems for a variety of reasons. For example, in remote locations it may be cheaper for a mine to generate its own power rather than pay for connection to the grid and in most mobile applications connection to the grid is simply not practical.

Today, most grids adopt three-phase electric power with alternating current. This choice can be partly attributed to the ease with which this type of power can be generated, transformed and used. Often (especially in the USA), the power is split before it reaches residential customers whose low-power appliances rely upon single-phase electric power. However, many larger industries and organizations still prefer to receive the three-phase power directly because it can be used to drive highly efficient electric motors such as three-phase induction motors.

Transformers play an important role in power transmission because they allow power to be converted to and from higher voltages. This is important because higher voltages suffer less power loss during transmission. This is because higher voltages allow for lower current to deliver the same amount of power, as power is the product of the two. Thus, as the voltage steps up, the current steps down. It is the current flowing through the components that result in both the losses and the subsequent heating. These losses, appearing in the form of heat, are equal to the current squared times the electrical resistance through which the current flows, so as the voltage goes up the losses are dramatically reduced.

For these reasons, electrical sub-stations exist throughout power grids to convert power to higher voltages before transmission and to lower voltages suitable for appliances after transmission.

Components

Power engineering is a network of interconnected components which convert different forms of energy to electrical energy. Modern power engineering consists of four main sub-systems : the generation sub-system, the transmission sub-system, the distribution sub-system and the utilization sub-system. In the generation sub-system, the power plant produces the electricity. The transmission sub-system transmits the electricity to the load centers. The distribution sub-system continues to transmit the power to the customers. The utilization system is concerned with the different uses of electrical energy like illumination, refrigeration, traction, electric drives, etc. Utilization is a very recent concept in Power engineering.

Generation

Generation of electrical power is a process whereby energy is transformed into an electrical form. There are several different transformation processes, among which are chemical, photo-voltaic, and electromechanical. Electromechanical energy conversion is used in converting energy from coal, petroleum, natural gas, uranium into electrical energy. Of these, all except the wind energy conversion process take advantage of the synchronous AC generator coupled to a steam, gas or hydro turbine such that the turbine converts steam, gas, or water flow into rotational energy, and the synchronous generator then converts the rotational energy of the turbine into electrical energy. It is the turbine-generator conversion process that is by far most economical and consequently most common in the industry today.

The AC synchronous machine is the most common technology for generating electrical energy. It is called synchronous because the composite magnetic field produced by the three stator windings rotate at the same speed as the magnetic field produced by the field winding on the rotor. A simplified circuit model is used to analyze steady-state operating conditions for a synchronous machine. The phasor diagram is an effective tool for visualizing the relationships between internal voltage, armature current, and terminal voltage. The excitation control system is used on synchronous machines to regulate terminal voltage, and the turbine-governor system is used to regulate the speed of the machine. However, in highly interconnected systems, such as the "Western system", the "Texas system" and the "Eastern system", *one* machine will usually be assigned as the so-called "swing machine", and which generation may be increased or decreased to compensate for small changes in load, thereby maintaining the system frequency at precisely 60 Hz. Should the load dramatically change, as which happens with a system separation, then a combination of "spinning reserve" and the "swing machine" may be used by the system's load dispatcher.

The operating costs of generating electrical energy is determined by the fuel cost and the efficiency of the power station. The efficiency depends on generation level and can be obtained from the heat rate curve. We may also obtain the incremental cost curve from the heat rate curve. *Economic dispatch* is the process of allocating the required load demand between the available generation units such that the cost of operation is minimized. *Emission dispatch* is the process of allocating the required load demand between the available generation units such that air pollution occurring from operation is minimized. In large systems, particularly in the West, a combination of economic and emission dispatch may be used.

Transmission

The electricity is transported to load locations from a power station to a transmission sub-system. Therefore we may think of the transmission system as providing the medium of transportation for electric energy. The transmission system may be sub-divided into the bulk transmission system and the sub-transmission system. The functions of the bulk transmission are to interconnect generators, to interconnect various areas of the network, and to transfer electrical energy from the generators to the major load centers. This portion of the system is called "bulk" because it delivers energy only to so-called bulk loads such as the distribution system of a town, city, or large industrial plant. The function of the sub-transmission system is to interconnect the bulk power system with the distribution system.

Transmission circuits may be built either underground or overhead. Underground cables are used predominantly in urban areas where acquisition of overhead rights of way are costly or not possible. They are also used for transmission under rivers, lakes and bays. Overhead transmission is used otherwise because, for a given voltage level, overhead conductors are much less expensive than underground cables.

The transmission system is a highly integrated system. It is referred to as the sub-station equipment and transmission lines. The sub-station equipment contain the transformers, relays, and circuit breakers. Transformers are important static devices which transfer electrical energy from one circuit to another in the transmission sub-system. Transformers are used to step up the voltage on the transmission line to reduce the power loss which is dissipated on the way. A relay is functionally a level-detector; they perform a switching action when the input voltage (or current) meets or exceeds a specific and adjustable value. A circuit breaker is an automatically operated electrical switch designed to protect an electrical circuit from damage caused by overload or short circuit. A change in the status of any one component can significantly affect the operation of the entire system. Without adequate contact protection, the occurrence of undesired electric arcing causes significant degradation of the contacts, which suffer serious damage. There are three possible causes for power flow limitations to a transmission line. These causes are thermal overload, voltage instability, and rotor angle instability. Thermal overload is caused by excessive current flow in a circuit causing overheating. Voltage instability is said to occur when the power required to maintain voltages at or above acceptable levels exceeds the available power. Rotor angle instability is a dynamic problem that may occur following faults, such as short circuit, in the transmission system. It may also occur tens of seconds after a fault due to poorly damped or undamped oscillatory response of the rotor motion. As long as the *equal area criteria* is maintained, the interconnected system will remain stable. Should the *equal area criteria* be violated, it becomes necessary to separate the unstable component from the remainder of the system.

Distribution

The distribution system transports the power from the transmission system/sub-station to the customer. Distribution feeders can be radial or networked in an open loop configuration with a single or multiple alternate sources. Rural systems tend to be of the former and urban systems the latter. The equipment associated with the distribution system usually begins downstream of the distribution feeder circuit breaker. The transformer and circuit breaker are usually under the jurisdiction of a "sub-stations department". The distribution feeders consist of combinations of overhead and underground conductor, 3 phase and single phase switches with load break and non-loadbreak ability, relayed protective devices, fuses, transformers (to utilization voltage), surge arresters, voltage regulators and capacitors.

More recently, Smart Grid initiatives are being deployed so that 1. Distribution feeder faults are automatically isolated and power restored to unfaulted circuits by automatic hardware/software/communications packages. 2. Capacitors are automatically switched on or off to dynamically control VAR flow and for CVR (Conservation Voltage Reduction)

Utilization

Utilization is the "end result" of the generation, transmission, and distribution of electric power. The energy carried by the transmission and distribution system is turned into useful work, light, heat, or a combination of these items at the utilization point. Understanding and characterizing the utilization of electric power is critical for proper planning and operation of power systems. Improper characterization of utilization can result of over or under building of power system facilities and stressing of system equipment beyond design capabilities. The term load refers to a device or collection of devices that draw energy from the power system. Individual loads (devices) range from small light bulbs to large induction motors to arc furnaces. The term load is often somewhat arbitrarily applied, at times being used to describe a specific device, and other times referring to an entire facility and even being used to describe the lumped power requirements of power system components and connected utilization devices downstream of a specific point in large scale system studies.

A major application of electric energy is in its conversion to mechanical energy. Electromagnetic, or "EM" devices designed for this purpose are commonly called "motors." Actually the machine is the central component of an integrated system consisting of the source, controller, motor, and load. For specialized applications, the system may be, and frequently is, designed as an integrated whole. Many household appliances (*e.g.*, a vacuum cleaner) have in one unit, the controller, the motor, and the load. However, there remain a large number of important stand-alone applications that require the selection of a proper motor and associated control, for a particular load. The reader is cautioned that there is no "magic bullet" to deal with all motor-load applications. Like many engineering problems, there is an artistic, as well as a scientific dimension to its solution. Likewise, each individual application has its own peculiar characteristics, and requires significant experience to manage. Nevertheless, a systematic formulation of the issues can be useful to a beginner in this area of design, and even for experienced engineers faced with a new or unusual application.

Control Engineering

Control engineering or control systems engineering is the engineering discipline that applies control theory to design systems with desired behaviours. The practice uses sensors to measure the output performance of the device being controlled and those measurements can be used to give feedback to the input actuators that can make corrections toward desired performance. When a device is designed to perform without the need of human inputs for correction it is called automatic control (such as cruise control for regulating a car's speed). Multi-disciplinary in nature, control systems engineering activities focus on implementation of control systems mainly derived by mathematical modelling of systems of a diverse range.

Overview

Modern day control engineering (also called control systems engineering) is a relatively new field of study that gained a significant attention during 20th century with the advancement in technology. It can be broadly defined or classified as practical application of control theory. Control engineering has an essential role in a wide range of control systems, from simple household washing machines to high-performance F-16 fighter aircraft. It seeks to understand physical systems, using mathematical modelling, in terms of inputs, outputs and various components with different behaviours; use control systems design tools to develop controllers for those systems; and implement controllers in physical systems employing available technology. A system can be mechanical, electrical, fluid, chemical, financial and even biological, and the mathematical modelling, analysis and controller design uses control theory in one or many of the time, frequency and complex-s domains, depending on the nature of the design problem.

History

Automatic control systems were first developed over two thousand years ago. The first feedback control device on record is thought to be the ancient Ktesibios's water clock in Alexandria, Egypt around the third century B.C. It kept time by regulating the water level in a vessel and, therefore, the water flow from that vessel. This certainly was a successful device as water clocks of similar design were still being made in Baghdad when the Mongols captured the city in 1258 A.D. A variety of automatic devices have been used over the centuries to accomplish useful tasks or simply to just entertain. The latter includes the automata, popular in Europe in the 17th and 18th centuries, featuring dancing figures that would repeat the same task over and over again; these automata are examples of open-loop control. Milestones among feedback, or "closed-loop" automatic control devices, include the temperature regulator of a furnace attributed to Drebbel, circa 1620, and the centrifugal flyball governor used for regulating the speed of steam engines by James Watt in 1788.

In his 1868 paper "On Governors", J. C. Maxwell (who discovered the Maxwell electromagnetic field equations) was able to explain instabilities exhibited by the flyball governor using differential equations to describe the control system. This demonstrated the importance and usefulness of mathematical models and methods in understanding complex phenomena, and signaled the beginning of mathematical control and systems theory. Elements of control theory had appeared earlier but not as dramatically and convincingly as in Maxwell's analysis.

Control theory made significant strides in the next 100 years. New mathematical techniques made it possible to control, more accurately, significantly more complex dynamical systems than the original flyball governor. These techniques include developments in optimal control in the 1950s and 1960s, followed by progress in stochastic, robust, adaptive and optimal control methods in the 1970s and 1980s. Applications of control methodology have helped make possible space travel and communication satellites, safer and more efficient aircraft, cleaner auto engines, cleaner and more efficient chemical processes, to mention but a few.

Before it emerged as a unique discipline, control engineering was practiced as a part of mechanical engineering and control theory was studied as a part of electrical engineering, since electrical circuits can often be easily described using control theory techniques. In the very first control relationships, a current output was represented with a voltage control input. However, not having proper technology to implement electrical control systems, designers left with the option of less efficient and slow responding mechanical systems. A very effective mechanical controller that is still widely used in some hydro plants is the governor. Later on, previous to modern power electronics, process control systems for industrial applications were devised by mechanical engineers using pneumatic and hydraulic control devices, many of which are still in use today.

Control Theory

There are two major divisions in control theory, namely, classical and modern, which have direct implications over the control engineering applications. The scope of classical control theory is limited to single-input and single-output (SISO) system design, except when analyzing for disturbance rejection using a second input. The system analysis is carried out in the time domain using differential equations, in the complex-s domain with the Laplace transform, or in the frequency domain by transforming from the complex-s domain. Many systems may be assumed to have a second order and single variable system response in the time domain. A controller designed using classical theory often requires on-site tuning due to incorrect design approximations. Yet, due to the easier physical implementation of classical controller designs as compared to systems designed using modern control theory, these controllers are preferred in most industrial applications. The most common controllers designed using classical control theory are PID controllers. A less common implementation may include either a Lead or Lag filter and at times both. The ultimate end goal is to meet a requirements set typically provided in the time-domain called the Step response, or at times in the frequency domain called the Open-Loop response. The Step response characteristics applied in a specification are typically percent overshoot, settling time, etc. The Open-Loop response characteristics applied in a specification are typically Gain and Phase margin and bandwidth. These characteristics may be evaluated through simulation including a dynamic model of the system under control coupled with the compensation model.

In contrast, modern control theory is carried out in the state space, and can deal with multi-input and multi-output (MIMO) systems. This overcomes the limitations of classical control theory in more sophisticated design problems, such as fighter aircraft control, with the limitation that no frequency domain analysis is possible. In modern design, a system is represented to the greatest advantage as a set of decoupled first order differential equations defined using state variables. Non-linear, multi-variable, adaptive and robust control theories come under this division. Matrix methods are significantly limited for MIMO systems where linear independence cannot be assured in the relationship between inputs and outputs. Being fairly new, modern control theory has many areas yet to be explored. Scholars like Rudolf E. Kalman and Aleksandr Lyapunov are well-known among the people who have shaped modern control theory.

Control Systems

Control engineering is the engineering discipline that focuses on the modelling of a diverse range of dynamic systems (*e.g.* mechanical systems) and the design of controllers that will cause these systems to behave in the desired manner. Although such controllers need not be electrical many are and hence control engineering is often viewed as a sub-field of electrical engineering. However, the falling price of microprocessors is making the actual implementation of a control system essentially trivial. As a result, focus is shifting back to the mechanical and process engineering discipline, as intimate knowledge of the physical system being controlled is often desired.

Electrical circuits, digital signal processors and microcontrollers can all be used to implement Control systems. Control engineering has a wide range of applications from the flight and propulsion systems of commercial airliners to the cruise control present in many modern automobiles.

In most of the cases, control engineers utilize feedback when designing control systems. This is often accomplished using a PID controller system. For example, in an automobile with cruise control the vehicle's speed is continuously monitored and fed back to the system, which adjusts the motor's torque accordingly. Where there is regular feedback, control theory can be used to determine how the system responds to such feedback. In practically all such systems stability is important and control theory can help ensure stability is achieved.

Although feedback is an important aspect of control engineering, control engineers may also work on the control of systems without feedback. This is known as open loop control. A classic example of open loop control is a washing machine that runs through a pre-determined cycle without the use of sensors.

Control Engineering Education

At many universities, control engineering courses are taught in Electrical and Electronic Engineering, Mechatronics Engineering, Mechanical engineering, and Aerospace engineering; in others it is connected to computer science, as most control techniques today are implemented through computers, often as embedded systems (as in the automotive field). The field of control within chemical engineering is often known as process control. It deals primarily with the control of variables in a chemical process in a plant. It is taught as part of the undergraduate curriculum of any chemical engineering program, and employs many of the same principles in control engineering. Other engineering disciplines also overlap with control engineering, as it can be applied to any system for which a suitable model can be derived. However, specialised control engineering departments do exist, for example, the Department of Automatic Control and Systems Engineering at the University of Sheffield.

Control engineering has diversified applications that include science, finance management, and even human behaviour. Students of control engineering may start with a linear control system course dealing with the time and complex-s domain, which requires a thorough background in elementary mathematics and Laplace transform (called classical control theory). In linear control, the student does frequency and time domain analysis. Digital control and non-linear control courses require z transformation and algebra respectively, and could be said to complete a basic control education. From here onwards there are several sub branches.

Recent Advancement

Originally, control engineering was all about continuous systems. Development of computer control tools posed a requirement of discrete control system engineering because the communications between the computer-based digital controller and the physical system are governed by a computer clock. The equivalent to Laplace transform in the discrete domain is the z-transform. Today many of the control systems are computer controlled and they consist of both digital and analog components.

Therefore, at the design stage either digital components are mapped into the continuous domain and the design is carried out in the continuous domain, or analog components are mapped into discrete domain and design is carried out there. The first of these two methods is more commonly encountered in practice because many industrial systems have many continuous systems components, including mechanical, fluid, biological and analog electrical components, with a few digital controllers.

Similarly, the design technique has progressed from paper-and-ruler based manual design to computer-aided design, and now to computer-automated design (CAutoD), which has been made possible by evolutionary computation. CAutoD can be applied not just to tuning a predefined control scheme, but also to controller structure optimisation, system identification and invention of novel control systems, based purely upon a performance requirement, independent of any specific control scheme.

Electronic Engineering

Electronics engineering, or **electronic engineering**, is an engineering discipline where non-linear and active electrical components such as electron tubes, and semi-conductor devices, especially transistors, diodes and integrated circuits, are utilized to design electronic circuits, devices and systems, typically also including passive electrical components and based on printed circuit boards. The term denotes a broad engineering field that covers important sub-fields such as analog electronics, digital electronics, consumer electronics, embedded systems and power electronics. Electronics engineering deals with implementation of applications, principles and algorithms developed within many related fields, for example, solid-state physics, radio engineering, telecommunications, control systems, signal processing, systems engineering, computer engineering, instrumentation engineering, electric power control, robotics, and many others.

The Institute of Electrical and Electronics Engineers (IEEE) is one of the most important and influential organizations for electronics engineers.

Relationship to Electrical Engineering

Electronics is a sub-field within the wider electrical engineering academic subject. An academic degree with a major in electronics engineering can be acquired from some universities, while other universities use electrical engineering as the subject. The term electrical engineer is still used in the academic world to include electronic engineers. However, some people consider the term 'electrical engineer' should be reserved for those having specialized in power and heavy current or high voltage engineering, while others consider that power is just one subset of electrical engineering and (and indeed the term 'power engineering' is used in that industry) as well as 'electrical distribution engineering'. Again, in recent years there has been a growth of new separate-entry degree courses such as 'information engineering', 'systems engineering' and 'communication systems engineering', often followed by academic departments of similar name, which are typically not considered as sub-fields of electronics engineering but of electrical engineering.

Beginning in the 1980s, the term computer engineer was often used to refer to a sub-field of electronic or information engineers. However, computer engineering is now considered a subset of electronics engineering and computer science and the term is now becoming archaic.

History

Electronic engineering as a profession sprang from technological improvements in the telegraph industry in the late 19th century and the radio and the telephone industries in the early 20th century. People were attracted to radio by the technical fascination it inspired, first in receiving and then in transmitting. Many who went into broadcasting in the 1920s were only 'amateurs' in the period before World War I.

To a large extent, the modern discipline of electronic engineering was born out of telephone, radio, and television equipment development and the large amount of electronic systems development during World War II of radar, sonar, communication systems, and advanced munitions and weapon systems. In the interwar years, the subject was known as radio engineering and it was only in the late 1950s that the term **electronic engineering** started to emerge.

Electronics

In the field of electronic engineering, engineers design and test circuits that use the electromagnetic properties of electrical components such as resistors, capacitors, inductors, diodes and transistors to achieve a particular functionality. The tuner circuit, which allows the user of a radio to filter out all but a single station, is just one example of such a circuit.

In designing an integrated circuit, electronics engineers first construct circuit schematics that specify the electrical components and describe the interconnections between them. When completed, VLSI engineers convert the schematics into actual layouts, which map the layers of various conductor and semi-conductor materials needed to construct the circuit. The conversion from schematics to layouts can be done by software but very often requires human fine-tuning to decrease space and power consumption. Once the layout is complete, it can be sent to a fabrication plant for manufacturing.

Integrated circuits and other electrical components can then be assembled on printed circuit boards to form more complicated circuits. Today, printed circuit boards are found in most electronic devices including televisions, computers and audio players.

Sub-fields

Signal Processing

Signal processing deals with the analysis and manipulation of signals. Signals can be either analog, in which case the signal varies continuously according to the information, or digital, in which case the signal varies according to a series of discrete values representing the information.

For analog signals, signal processing may involve the amplification and filtering of audio signals for audio equipment or the modulation and demodulation of signals for telecommunications. For digital signals, signal processing may involve the compression, error checking and error detection of digital signals.

Telecommunications engineering deals with the transmission of information across a channel such as a co-axial cable, optical fiber or free space.

Transmissions across free space require information to be encoded in a carrier wave in order to shift the information to a carrier frequency suitable for transmission, this is known as modulation. Popular analog modulation techniques include amplitude modulation and frequency modulation. The choice of modulation affects the cost and performance of a system and these two factors must be balanced carefully by the engineer.

Once the transmission characteristics of a system are determined, telecommunication engineers design the transmitters and receivers needed for such systems. These two are sometimes combined to form a two-way communication device known as a transceiver. A key consideration in the design of transmitters is their power consumption as this is closely related to their signal strength. If the signal strength of a transmitter is insufficient the signal's information will be corrupted by noise.

Control engineering has a wide range of applications from the flight and propulsion systems of commercial airplanes to the cruise control present in many modern cars. It also plays an important role in industrial automation.

Control engineers often utilize feedback when designing control systems. For example, in a car with cruise control the vehicle's speed is continuously monitored and fed back to the system which adjusts the engine's power output accordingly. Where there is regular feedback, control theory can be used to determine how the system responds to such feedback.

Instrumentation engineering deals with the design of devices to measure physical quantities such as pressure, flow and temperature. These devices are known as instrumentation.

The design of such instrumentation requires a good understanding of physics that often extends beyond electromagnetic theory. For example, radar guns use the Doppler effect to measure the speed of oncoming vehicles. Similarly, thermocouples use the Peltier-Seebeck effect to measure the temperature difference between two points.

Often instrumentation is not used by itself, but instead as the sensors of larger electrical systems. For example, a thermocouple might be used to help ensure a furnace's temperature remains constant. For this reason, instrumentation engineering is often viewed as the counterpart of control engineering.

Computer engineering deals with the design of computers and computer systems. This may involve the design of new computer hardware, the design of PDAs or the use of computers to control an industrial plant. Development of embedded systems—systems made for specific tasks (*e.g.*, mobile phones)—is also included in this field. This field includes the micro controller and its applications. Computer engineers may also work on a system's software. However, the design of complex software systems is often the domain of software engineering, which is usually considered a separate discipline.

VLSI Design Engineering VLSI stands for very large scale integration. It deals with fabrication of ICs and various electronics components.

Micro-electronics

Micro-electronics is a sub-field of electronics. As the name suggests, micro-electronics relates to the study and manufacture (or microfabrication) of very small electronic designs and components. Usually, but not always, this means micrometre-scale or smaller. These devices are typically made from semi-conductor materials. Many components of normal electronic design are available in a micro-electronic equivalent. These include transistors, capacitors, inductors, resistors, diodes and (naturally) insulators and conductors can all be found in micro-electronic devices. Unique wiring techniques such as wire bonding are also often used in micro-electronics because of the unusually small size of the components, leads and pads. This technique requires specialized equipment and is expensive.

Digital integrated circuits (ICs) consist mostly of transistors. Analog circuits commonly contain resistors and capacitors as well. Inductors are used in some high frequency analog circuits, but tend to occupy large chip area if used at low frequencies; gyrators can replace them in many applications.

As techniques improve, the scale of micro-electronic components continues to decrease. At smaller scales, the relative impact of intrinsic circuit properties such as interconnections may become more significant. These are called **parasitic effects**, and the goal of the micro-electronics design engineer is to find ways to compensate for or to minimize these effects, while always delivering smaller, faster, and cheaper devices.

PRACTICING ENGINEERS

In most countries, a Bachelor's degree in engineering represents the first step towards professional certification and the degree program itself is certified by a professional body. After completing a certified degree program the engineer must satisfy a range of requirements (including work experience requirements) before being certified. Once certified the engineer is designated the title of Professional Engineer (in the United States, Canada and South Africa), Chartered Engineer or Incorporated Engineer (in India, Pakistan, the United Kingdom, Ireland and Zimbabwe), Chartered Professional Engineer (in Australia and New Zealand) or European Engineer (in much of the European Union).

The advantages of certification vary depending upon location. For example, in the United States and Canada "only a licensed engineer may seal engineering work for public and private clients". This requirement is enforced by state and provincial legislation such as Quebec's Engineers Act. In other countries, no such legislation exists. Practically all certifying bodies maintain a code of ethics that they expect all members to abide by or risk expulsion. In this way these organizations play an important role in maintaining ethical standards for the profession. Even in jurisdictions where certification has little or no legal bearing on work, engineers are subject to contract law. In cases where an engineer's work fails he or she may be subject to the tort of negligence and, in extreme cases, the charge of criminal negligence. An engineer's work must also comply with numerous other rules and regulations such as building codes and legislation pertaining to environmental law.

Professional bodies of note for electrical engineers include the Institute of Electrical and Electronics Engineers (IEEE) and the Institution of Engineering and Technology (IET). The IEEE claims to produce 30% of the world's literature in electrical engineering, has over 360,000 members worldwide and holds over 3,000 conferences annually. The IET publishes 21 journals, has a worldwide membership of over 150,000, and claims to be the largest professional engineering society in

Europe. Obsolescence of technical skills is a serious concern for electrical engineers. Membership and participation in technical societies, regular reviews of periodicals in the field and a habit of continued learning are therefore essential to maintaining proficiency. MIET(Member of the Institution of Engineering and Technology) is recognised in Europe as Electrical and computer (technology) engineer.

In Australia, Canada and the United States electrical engineers make up around 0.25% of the labour force. Outside of Europe and North America, engineering graduates per-capita, and hence probably electrical engineering graduates also, are most numerous in Taiwan, Japan, and South Korea.

Tools and Work

From the Global Positioning System to electric power generation, electrical engineers have contributed to the development of a wide range of technologies. They design, develop, test and supervise the deployment of electrical systems and electronic devices. For example, they may work on the design of telecommunication systems, the operation of electric power stations, the lighting and wiring of buildings, the design of household appliances or the electrical control of industrial machinery.

Fundamental to the discipline are the sciences of physics and mathematics as these help to obtain both a qualitative and quantitative description of how such systems will work. Today most engineering work involves the use of computers and it is commonplace to use computer-aided design programs when designing electrical systems. Nevertheless, the ability to sketch ideas is still invaluable for quickly communicating with others.

Although most electrical engineers will understand basic circuit theory (that is the interactions of elements such as resistors, capacitors, diodes, transistors and inductors in a circuit), the theories employed by engineers generally depend upon the work they do. For example, quantum mechanics and solid state physics might be relevant to an engineer working on VLSI (the design of integrated circuits), but are largely irrelevant to engineers working with macroscopic electrical systems. Even circuit theory may not be relevant to a person designing telecommunication systems that use off-the-shelf components. Perhaps the most important technical skills for electrical engineers are reflected in university programs, which emphasize strong numerical skills, computer literacy and the ability to understand the technical language and concepts that relate to electrical engineering.

A wide range of instrumentation is used by electrical engineers. For simple control circuits and alarms, a basic multi-meter measuring voltage, current and resistance may suffice. Where time-varying signals need to be studied, the oscilloscope is also an ubiquitous instrument. In RF engineering and high frequency telecommunications spectrum analyzers and network analyzers are used. In some disciplines safety can be a particular concern with instrumentation. For instance, medical electronics designers must take into account that much lower voltages than normal can be dangerous when electrodes are directly in contact with internal body fluids. Power transmission engineering also has great safety concerns due to the high voltages used; although voltmeters may in principle be similar to their low voltage equivalents, safety and calibration issues make them very different. Many disciplines of electrical engineering use tests specific to their discipline. Audio electronics engineers use audio test sets consisting of a signal generator and a meter, principally to measure level but also other parameters such as harmonic distortion and noise. Likewise information technology have their own test sets, often specific to a particular data format, and the same is true of television broadcasting.

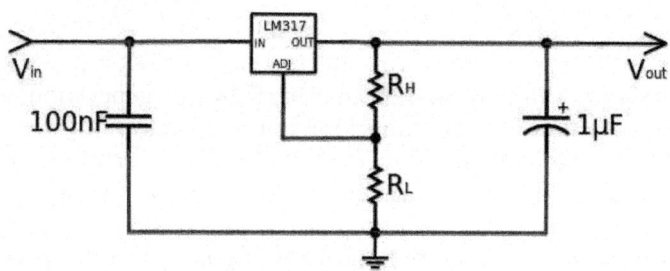

Fig. : Typical electrical engineering diagram used as a troubleshooting tool.

For many engineers, technical work accounts for only a fraction of the work they do. A lot of time may also be spent on tasks such as discussing proposals with clients, preparing budgets and determining project schedules. Many senior engineers manage a team of technicians or other engineers and for this reason project management skills are important. Most engineering projects involve some form of documentation and strong written communication skills are therefore very important.

The workplaces of electrical engineers are just as varied as the types of work they do. Electrical engineers may be found in the pristine lab environment of a fabrication plant, the offices of a consulting firm or on site at a mine. During their working life, electrical engineers may find themselves supervising a wide range of individuals including scientists, electricians, computer programmers and other engineers.

Electrical engineering has an intimate relationship with the physical sciences. For instance the physicist Lord Kelvin played a major role in the engineering of the first transatlantic telegraph cable. Conversely, the engineer Oliver Heaviside produced major work on the mathematics of transmission on telegraph cables. Electrical engineers are often required on major science projects. For instance, large particle accelerators such as CERN need electrical engineers to deal with many aspects of the project : from the power distribution, to the instrumentation, to the manufacture and installation of the superconducting electromagnets.

Chapter 2

NETWORK

NETWORK ANALYSIS

A network, in the context of electronics, is a collection of interconnected components. **Network analysis** is the process of finding the voltages across, and the currents through, every component in the network. There are many different techniques for calculating these values. However, for the most part, the applied technique assumes that the components of the network are all linear.

Definitions

Component	A device with two or more terminals into which, or out of which, charge may flow.
Node	A point at which terminals of more than two components are joined. A conductor with a substantially zero resistance is considered to be a node for the purpose of analysis.
Branch	The component(s) joining two nodes.
Mesh	A group of branches within a network joined so as to form a complete loop.
Port	Two terminals where the current into one is identical to the current out of the other.
Circuit	A current from one terminal of a generator, through load component(s) and back into the other terminal. A circuit is, in this sense, a one-port network and is a trivial case to analyse. If there is any connection to any other circuits then a non-trivial network has been formed and at least two ports must exist. Often, "circuit" and "network" are used interchangeably, but many analysts reserve "network" to mean an idealised model consisting of ideal components.
Transfer function	The relationship of the currents and/or voltages between two ports. Most often, an input port and an output port are discussed and the transfer function is described as gain or attenuation.

Component transfer function	For a two-terminal component (*i.e.* one-port component), the current and voltage are taken as the input and output and the transfer function will have units of impedance or admittance (it is usual ly a matter of arbitrary convenience whether voltage or current is considered the input). A three (or more) terminal component effectively has two (or more) ports and the transfer function cannot be expressed as a single impedance. The usual approach is to express the transfer function as a matrix of parameters.

EQUIVALENT IMPEDANCE TRANSFORMS

An **equivalent impedance** is an equivalent circuit of an electrical network of impedance elements which presents the same impedance between all pairs of terminals as did the given network.

There are a number of very well known and often used equivalent circuits in linear network analysis. These include resistors in series, resistors in parallel and the extension to series and parallel circuits for capacitors, inductors and general impedances. Also well known are the Norton and Thévenin equivalent current generator and voltage generator circuits respectively, as is the Y-Δ transform. None of these are discussed in detail here; the individual linked articles should be consulted.

The number of equivalent circuits that a linear network can be transformed into is unbounded. Even in the most trivial cases this can be seen to be true, for instance, by asking how many different combinations of resistors in parallel are equivalent to a given combined resistor. The number of series and parallel combinations that can be formed grows exponentially with the number of resistors, n. For large n the size of the set has been found by numerical techniques to be approximately 2.53^n and analytically strict bounds are given by a Farey sequence of Fibonacci numbers. Wilhelm Cauer found a transformation that could generate all possible equivalents of a given rational, passive, linear one-port, or in other words, any given two-terminal impedance. Transformations of 4-terminal, especially 2-port, networks are also commonly found and transformations of yet more complex networks are possible.

The vast scale of the topic of equivalent circuits is underscored in a story told by Sidney Darlington. According to Darlington, a large number of equivalent circuits were found by Ronald Foster, following his and George Campbell's 1920 paper on non-dissipative four-ports. In the course of this work they looked at the ways four ports could be interconnected with ideal transformers and maximum power transfer. They found a number of combinations which might have practical applications and asked the AT&T patent department to have them patented. The patent department replied that it was pointless just patenting some of the circuits if a competitor could use an equivalent circuit to get around the patent; they should patent all of them or not bother. Foster therefore set to work calculating every last one of them. He arrived at an enormous total of 83,539 equivalents (577,722 if different output ratios are included). This was too many to patent, so instead the information was released into the public domain in order to prevent any of AT&T's competitors from patenting them in the future.

2-Terminal, 2-Element-kind Networks

A single impedance has two terminals to connect to the outside world, hence can be described as a 2-terminal, or a one-port, network. Despite the simple description, there is no limit to the number of meshes, and hence complexity and number of elements, that the impedance network may have. 2-element-kind networks are common in circuit design; filters, for instance, are often LC-kind networks and printed circuit designers favour RC-kind networks because inductors are less easy to manufacture. Transformations are simpler and easier to find than for 3-element-kind networks. One-element-kind networks can be thought of as a special case of two-element-kind. It is possible to use the transformations in this section on a certain few 3-element-kind networks by substituting a network of elements for element Z_n. However, this is limited to a maximum of two impedances being substituted; the remainder will not be a free choice.

3-Element Networks

One-element networks are trivial and two-element, two-terminal networks are either two elements in series or two elements in parallel, also trivial. The smallest number of elements that is non-trivial is three, and there are two 2-element-kind non-trivial transformations possible, one being both the reverse transformation and the topological dual, of the other.

Description	Network	Transform equations	Transformed network
Transform 1.1 Transform 1.2 is the reverse of this transform.	Z_1 $m_1 Z_1$ Z_2	$p_1 = 1 + m_1,$ $p_2 = m_1(1 + m_1)$ $p_3 = (1 + m_1)^2.$	$p_1 Z_1$ $p_2 Z_1$ $p_3 Z_2$
Transform 1.2 The reverse transform, and topological dual, of Transform 1.1.	$m_1 Z_1$ Z_1 Z_2	$p_1 = \dfrac{m_1{}^2}{1 + m_1},$ $p_2 = \dfrac{m_1}{1 + m_1},$ $p_3 + \left(\dfrac{m_1}{1 + m_1}\right)^2.$	$p_1 Z_1$ $p_2 Z_1$ $p_3 Z_2$

Description	Network	Transform equations	Transformed network
Example 1. An example of Transform 1.2. The reduced size of the inductor has practical advantages.	$2C_1$ / C_1 L_2	$m_1 = 0.5,$ $p_1 = \dfrac{1}{6},$ $p_2 = \dfrac{1}{3},$ $p_3 = \dfrac{1}{9}.$	$6C_1$ $3C_1$ / $\frac{1}{9}L_2$

4-Element Networks

There are four non-trivial 4-element transformations for 2-element-kind networks. Two of these are the reverse transformations of the other two and two are the dual of a different two. Further transformations are possible in the special case of Z_2 being made the same element kind as Z_1, that is, when the network is reduced to one-element-kind. The number of possible networks continues to grow as the number of elements is increased :

$$q_1 := 1 + m_1 + m_2,$$

$$q_2 := \sqrt{q_1^2 - 4m_1 m_2},$$

$$q_3 := \frac{(1 + m_1)(1 + m_2)}{(m_1 - m_2)^2},$$

$$q_4 := \frac{q_2 - q_1 + 2m_2}{2q_2},$$

$$q_5 := \frac{q_2 + q_1 - 2m_2}{2q_2}.$$

Description	Network	Transform equations	Transformed network
Transform 2.1 Transform 2.2 is the reverse of this transform. Transform 2.3 is the topological dual of this transform.	Z_1 $m_1 Z_1$ $m_2 Z_2$ / Z_2	$p_1 = \dfrac{q_1 + q_2}{2q_5},$ $p_2 = \dfrac{q_1 - q_2}{2q_4},$ $p_3 = \dfrac{m_2}{q_5}$ $p_4 = \dfrac{m_2}{q_4}.$	$p_1 Z_1$ $p_3 Z_2$ / $p_2 Z_1$ $p_4 Z_2$

Description	Network	Transform equations	Transformed network
Transform 2.2 Transform 2.1 is the reverse of this transform. Transform 2.4 is the topological dual of this transform.		$$p_1 = \frac{1}{q_3\left(1+m_2\right)},$$ $$p_2 = \frac{m_1}{1+m_1},$$ $$p_3 = \frac{1}{q_3\left(1+m_1\right)},$$ $$p_4 = \frac{m_2}{1+m_2}.$$	
Transform 2.3 Transform 2.4 is the reverse of this transform. Transform 2.1 is the topological dual of this transform.		$$p_1 = \frac{q_4\left(q_1+q_2\right)}{2m_2},$$ $$p_2 = \frac{q_5\left(q_1-q_2\right)}{2m_2},$$ $$p_3 = q_4.$$ $$p_4 = q_5.$$	
Transform 2.4 Transform 2.3 is the reverse of this transform. Transform 2.2 is the topological dual of this transform.		$$p_1 = 1+m_1,$$ $$p_2 = m_1 q_3\left(1+m_1\right),$$ $$p_3 = 1+m_2,$$ $$p_4 = m_1 q_3\left(1+m_2\right),$$	

Description	Network	Transform equations	Transformed network
Example 2. An example of Transform 2.2.	$3R_1$ C_2 R_1 C_2	$m_1 = 3,$ $m_2 = 1,$ $q_3 = 2,$ $p1 = \dfrac{1}{4},$ $p_2 = \dfrac{3}{4},$ $p_3 = \dfrac{1}{8},$ $p_4 = \dfrac{1}{2},$	$\frac{1}{4}R_1$ $\frac{3}{4}R_1$ $2C_2$ $8C_2$

2-Terminal, N-Element, 3-Element-kind Networks

Simple networks with just a few elements can be dealt with by formulating the network equations "by hand" with the application of simple network theorems such as Kirchhoff's laws. Equivalence is proved between two networks by directly comparing the two sets of equations and equating coefficients. For large networks more powerful techniques are required. A common approach is to start by expressing the network of impedances as a matrix. This approach is only good for rational networks. Any network that includes distributed elements, such as a transmission line, cannot be represented by a finite matrix. Generally, an n-mesh network requires an $n \times n$ matrix to represent it. For instance the matrix for a 3-mesh network might look like;

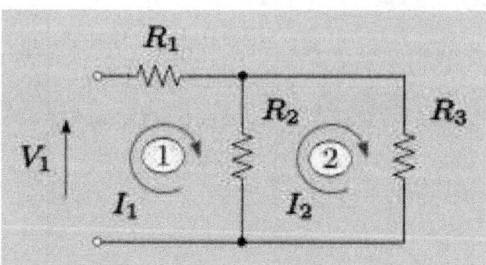

Fig. : Simple example of a network of impedances using resistors only for clarity. However, analysis of networks with other impedance elements proceed by the same principles. Two meshes are shown, with numbers in circles. The sum of impedances around each mesh, p, will form the diagonal of the entries of the matrix, Z_{pp}. The impedance of branches shared by two meshes, p and q, will form the entries $-Z_{pq}$. Z_{pq}, p≠q, will always have a minus sign provided that the convention of loop currents are defined in the same (conventionally counterclockwise) direction and the mesh contains no ideal transformers or mutual inductors.

$$[Z] = \begin{bmatrix} Z_{11} & Z_{12} & Z_{13} \\ Z_{21} & Z_{22} & Z_{23} \\ Z_{31} & Z_{32} & Z_{33} \end{bmatrix}$$

The entries of the matrix are chosen so that the matrix forms a system of linear equations in the mesh voltages and currents (as defined for mesh analysis);

$$[V] = [Z] [I]$$

The example diagram, for instance, can be represented as an impedance matrix by;

$$[Z] = \begin{bmatrix} R_1 + R_2 & -R_2 \\ -R_2 & R_2 + R_3 \end{bmatrix}$$

and the associated system of linear equations are,

$$\begin{bmatrix} V_1 \\ 0 \end{bmatrix} = \begin{bmatrix} R_1 + R_2 & -R_2 \\ -R_2 & R_2 + R_3 \end{bmatrix} \begin{bmatrix} I_1 \\ I_2 \end{bmatrix}$$

In the most general case, each branch, Z_p, of the network may be made up of three elements so that,

$$Z_p = sL_p + R_p + \frac{1}{sC_p}$$

where L, R and C represent inductance, resistance, and capacitance respectively and s is the complex frequency operator $s = \sigma + i\omega$.

This is the conventional way of representing a general impedance but for the purposes of this, it is mathematically more convenient to deal with elastance, D, the inverse of capacitance, C. In those terms the general branch impedance can be represented by,

$$sZ_p = s^2 L_p + sR_p + D_p$$

Likewise, each entry of the impedance matrix can consist of the sum of three elements. Consequently, the matrix can be decomposed into three nxn matrices, one for each of the three element kinds;

$$s[Z] = s^2 [L] + s[R] + [D]$$

It is desired that the matrix [Z] represent an impedance, $Z(s)$. For this purpose, the loop of one of the meshes is cut and $Z(s)$ is the impedance measured between the points so cut. It is conventional to assume the external connection port is in mesh 1, and is therefore connected across matrix entry Z_{11}, although it would be perfectly possible to formulate this with connections to any desired nodes. In the following discussion $Z(s)$ taken across Z_{11} is assumed. $Z(s)$ may be calculated from [Z] by;

$$Z(s) = \frac{|\mathbf{Z}|}{z_{11}}$$

where z_{11} is the complement of Z_{11} and $|\mathbf{Z}|$ is the determinant of $[\mathbf{Z}]$.

For the example network above;

$$|\mathbf{Z}| = (R_1 + R_2)(R_2 + R_3) - R_2{}^2 = R_1R_2 + R_1R_3 + R_2R_3,$$
$$z_{11} = Z_{22} = R_2 + R_3.$$

and,

$$Z(s) = R_1 + \frac{R_2\,R_3}{R_2 + R_3}.$$

This result is easily verified to be correct by the more direct method of resistors in series and parallel. However, such methods rapidly become tedious and cumbersome with the growth of the size and complexity of the network under analysis.

The entries of $[\mathbf{R}]$, $[\mathbf{L}]$ and $[\mathbf{D}]$ cannot be set arbitrarily. For $[\mathbf{Z}]$ to be able to realise the impedance $Z(s)$ then $[\mathbf{R}],[\mathbf{L}]$ and $[\mathbf{D}]$ must all be positive-definite matrices. Even then, the realisation of $Z(s)$ will, in general, contain ideal transformers within the network. Finding only those transforms that do not require mutual inductances or ideal transformers is a more difficult task. Similarly, if starting from the "other end" and specifying an expression for $Z(s)$, this again cannot be done arbitrarily. To be realisable as a rational impedance, $Z(s)$ must be positive-real. The positive-real (PR) condition is both necessary and sufficient but there may be practical reasons for rejecting some topologies.

A general impedance transform for finding equivalent rational one-ports from a given instance of $[\mathbf{Z}]$ is due to Wilhelm Cauer. The group of real affine transformations,

$$[\mathbf{Z'}] = [\mathbf{T}]^T[\mathbf{Z}][\mathbf{T}]$$

where,

$$[\mathbf{T}] = \begin{bmatrix} 1 & 0\ldots0 \\ T_{21} & T_{22}\ldots T_{2n} \\ T_{n1} & T_{n2}\ldots T_{nn} \end{bmatrix}$$

is invariant in $Z(s)$. That is, all the transformed networks are equivalents according to the definition given here. If the $Z(s)$ for the initial given matrix is realisable, that is, it meets the PR condition, then all the transformed networks produced by this transformation will also meet the PR condition.

3 and 4-Terminal Networks

When discussing 4-terminal networks, network analysis often proceeds in terms of 2-port networks, which covers a vast array of practically useful circuits. "2-port", in essence, refers to the way the network has been connected to the outside world: that the terminals have been connected in pairs to a source or load. It is possible to take exactly the same network and connect it to external circuitry in such a way that it is no longer behaving as a 2-port.

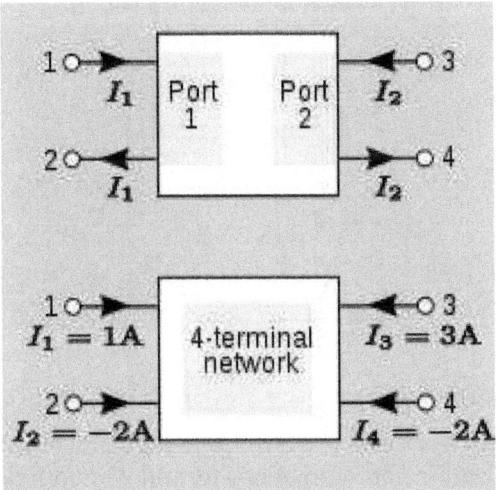

Fig. : A 4-terminal network connected by ports (top) has equal and opposite currents in each pair of terminals. The bottom network does not meet the port condition and cannot be treated as a 2-port. It could, however, be treated as an unbalanced 3-port by splitting one of the terminals into three common terminals shared between the ports.

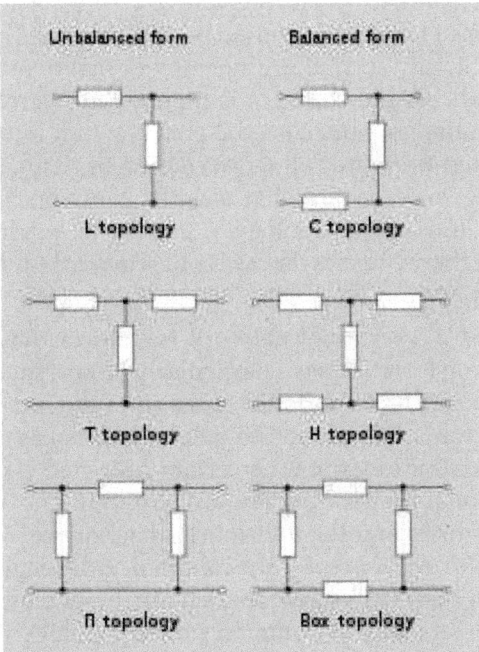

Fig. : Equivalent unbalanced and balanced networks. The impedance of the series elements in the balanced version is half the corresponding impedance of the unbalanced version.

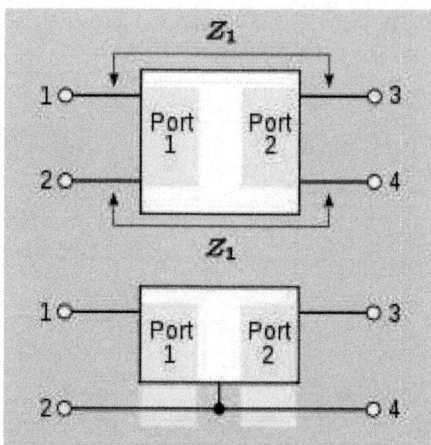

Fig. : To be balanced, a network must have the same impedance in each "leg" of the circuit.

A 3-terminal network can also be used as a 2-port. To achieve this, one of the terminals is connected in common to one terminal of both ports. In other words, one terminal has been split into two terminals and the network has effectively been converted to a 4-terminal network. This topology is known as unbalanced topology and is opposed to balanced topology. Balanced topology requires, that the impedance measured between terminals 1 and 3 is equal to the impedance measured between 2 and 4. This is the pairs of terminals *not* forming ports : the case where the pairs of terminals forming ports have equal impedance is referred to as symmetrical. Strictly speaking, any network that does not meet the balance condition is unbalanced, but the term is most often referring to the 3-terminal topology. Transforming an unbalanced 2-port network into a balanced network is usually quite straightforward : all series connected elements are divided in half with one half being relocated in what was the common branch. Transforming from balanced to unbalanced topology will often be possible with the reverse transformation but there are certain cases of certain topologies which cannot be transformed in this way.

An example of a 3-terminal network transform that is not restricted to 2-ports is the Y-Δ transform. This is a particularly important transform for finding equivalent impedances. Its importance arises from the fact that the total imped-ance between two terminals cannot be determined solely by calculating series and parallel combinations except for a certain restricted class of network. In the general case additional transformations are required. The Y-Δ transform, its in-verse the Δ-Y transform, and the *n*-terminal analogues of these two transforms (star-polygon transforms) represent the minimal additional transforms required to solve the general case. Series and parallel are, in fact, the 2-terminal versions of star and polygon topology. A common simple topology that cannot be solved by series and parallel combinations is the input impedance to a bridge network (except in the special case when the bridge is in balance).

Lattice Transforms

Symmetric 2-port networks can be transformed into lattice networks using Bartlett's bisection theorem. The method is limited to symmetric networks but this includes many topologies commonly found in filters, attenuators and equalisers. The lattice topology is intrinsically balanced, there is no unbalanced counterpart to the lattice and it will usually require more components than the transformed network.

Some common networks transformed to lattices (X-networks)			
Description	Network	Transform equations	Transformed network
Transform 3.1 Transform of T network to lattice network.	Z_1 Z_1 Z_2	$Z_A = Z_1$, $Z_B = Z_1 + 2Z_2$.	Z_A Z_B Z_B Z_A
Transform 3.2 Transform of Π network to lattice network.	Z_1 Z_2 Z_2	$Z_A = \dfrac{Z_1 Z_2}{Z_1 + 2Z_2}$, $Z_B = Z_2$.	Z_A Z_B Z_B Z_A
Transform 3.3 Transform of Bridged-T network to lattice network.	Z_1 Z_0 Z_0 Z_2	$Z_A = \dfrac{Z_1 Z_0}{Z_1 + 2Z_0}$, $Z_B = Z_0 + 2Z_2$.	Z_A Z_B Z_B Z_A

Reverse transformations from a lattice to an unbalanced topology are not always possible in terms of passive components. For instance, this transform,

Description	Network	Transformed network
Transform 3.4 Transform of a lattice phase equaliser to a T network.	L C C L	L L $-\tfrac{1}{2}L$ $2C$

cannot be realised with passive components because of the negative values arising in the transformed circuit. It can however be realised if mutual inductances and ideal transformers are permitted, for instance, in this circuit. Another possibility is to permit the use of active components which would enable negative impedances to be directly realised as circuit components.

It can sometimes be useful to make such a transformation, not for the purposes of actually building the transformed circuit, but rather, for the purposes of aiding understanding of how the original circuit is working. The following circuit in bridged-T topology is a modification of a mid-series m-derived filter T-section. The circuit is due to Hendrik Bode who claims that the addition of the bridging resistor of a suitable value will cancel the parasitic resistance of the shunt inductor. The action of this circuit is clear if it is transformed into T topology - in this form there is a negative resistance in the shunt branch which can be made to be exactly equal to the positive parasitic resistance of the inductor.

Description	Network	Transformed network
Transform 3.5 Transform of a bridged-T low-pass filter section to a T-section.		

Any symmetrical network can be transformed into any other symmetrical network by the same method, that is, by first transforming into the intermediate lattice form and from the lattice form into the required target form. As with the example, this will generally result in negative elements except in special cases.

Eliminating Resistors

A theorem due to Sidney Darlington states that any PR function $Z(s)$ can be realised as a lossless two-port terminated in a positive resistor R. That is, regardless of how many resistors feature in the matrix $[\mathbf{Z}]$ representing the impedance network, a transform can be found that will realise the network entirely as an LC-kind network with just one resistor across the output port (which would normally represent the load). No resistors within the network are necessary in order to realise the specified response. Consequently, it is always possible to reduce 3-element-kind 2-port networks to 2-element-kind (LC) 2-port networks provided the output port is terminated in a resistance of the required value.

Eliminating Ideal Transformers

An elementary transformation that can be done with ideal transformers and some other impedance element is to shift the impedance to the other side of the transformer. In all the following transforms, r is the turns ratio of the transformer.

Description	Network	Transformed network
Transform 4.1 Series impedance through a step-down transformer.		
Transform 4.2 Shunt impedance through a step-down transformer.		
Transform 4.3 Shunt and series impedance network through a step-up transformer.		

These transforms do not just apply to single elements; entire networks can be passed through the transformer. In this manner, the transformer can be shifted around the network to a more convenient location.

Darlington gives an equivalent transform that can eliminate an ideal transformer altogether. This technique requires that the transformer is next to (or capable of being moved next to) an "L" network of same-kind impedances. The transform in all variants results in the "L" network facing the opposite way, that is, topologically mirrored.

Description	Network	Transformed network
Transform 5.1 Elimination of a step-down transformer.		

Description	Network	Transformed network
Transform 5.2 Elimination of a step-up transformer.	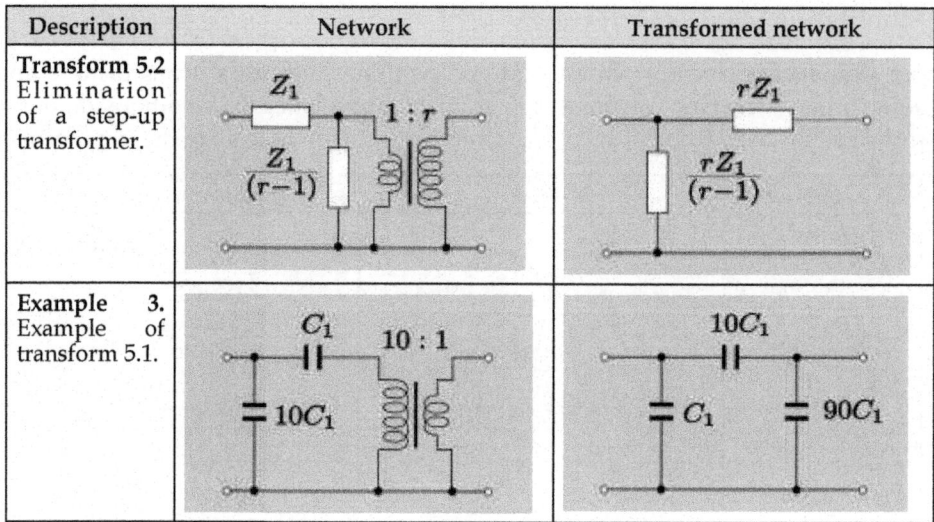	
Example 3. Example of transform 5.1.		

Example 3 shows the result is a Π-network rather than an L-network. The reason for this is that the shunt element has more capacitance than is required by the transform so some is still left over after applying the transform. If the excess were instead, in the element nearest the transformer, this could be dealt with by first shifting the excess to the other side of the transformer before carrying out the transform.

Y-Δ TRANSFORM

The **Y-Δ transform**, also written **wye-delta** and also known by many other names, is a mathematical technique to simplify the analysis of an electrical network. The name derives from the shapes of the circuit diagrams, which look respectively like the letter Y and the Greek capital letter Δ. This circuit transformation theory was published by Arthur Edwin Kennelly in 1899. It is widely used in analysis of three-phase electric power circuits.

The Y-Δ transform can be considered a special case of the star-mesh transform for three resistors.

Names

The **Y-Δ transform** is known by a variety of other names, mostly based upon the two shapes involved, listed in either order. The **Y**, spelled out as **wye**, can also be called **T** or **star**; the Δ, spelled out as **delta**, can also be called **triangle**, **Π** (spelled out as **pi**), or **mesh**. Thus, common names for the transformation include **wye-delta** or **delta-wye**, **star-delta**, **star-mesh**, or **T-Π**.

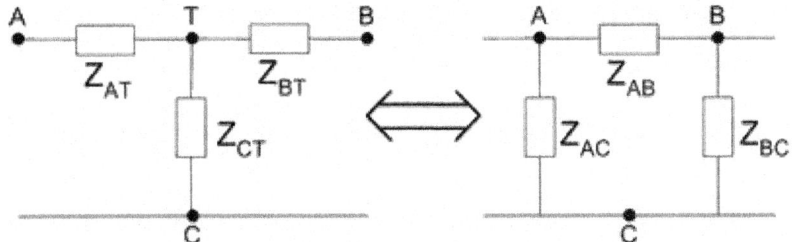

Fig. : Illustration of the transform in its T-Π representation.

Basic Y-Δ Transformation

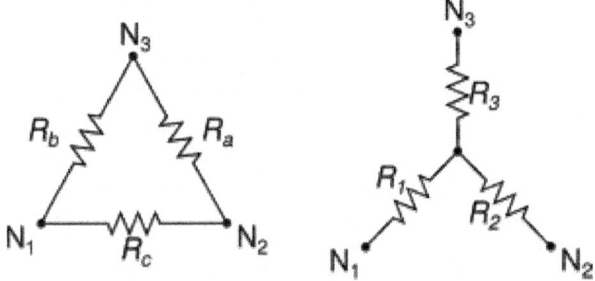

Fig. : Δ and Y circuits with the labels which are used.

The transformation is used to establish equivalence for networks with three terminals. Where three elements terminate at a common node and none are sources, the node is eliminated by transforming the impedances. For equivalence, the impedance between any pair of terminals must be the same for both networks. The equations given here are valid for complex as well as real impedances.

Equations for the Transformation from Δ-Load to Y-Load 3-phase Circuit

The general idea is to compute the impedance R_y at a terminal node of the Y circuit with impedances R', R'' to adjacent node in the Δ circuit by

$$R_y = \frac{R' R''}{\Sigma R_\Delta}$$

where $R\Delta$ are all impedances in the Δ circuit. This yields the specific formulae

$$R_1 = \frac{R_b R_c}{R_a + R_b + R_c}$$

$$R_2 = \frac{R_a R_c}{R_a + R_b + R_c}$$

$$R_3 = \frac{R_a R_b}{R_a + R_b + R_c}$$

Equations for the Transformation from Y-Load to Δ-Load 3-Phase Circuit

The general idea is to compute an impedance $R\Delta$ in the Δ circuit by

$$R\Delta = \frac{R_p}{R_{oppsite}}$$

where $R_p = R_1 R_2 + R_2 R_3 + R_3 R_1$ is the sum of the products of all pairs of impedances in the Y circuit and $R_{opposite}$ is the impedance of the node in the Y circuit which is opposite the edge with $R\Delta$. The formula for the individual edges are thus

$$R_a = \frac{R_1 R_2 + R_2 R_3 + R_3 R_1}{R_1}$$

$$R_b = \frac{R_1 R_2 + R_2 R_3 + R_3 R_1}{R_2}$$

$$R_c = \frac{R_1 R_2 + R_2 R_3 + R_3 R_1}{R_3}$$

Circuit Analysis : Techniques for Solving Δ-Load io Y-Load in 3 Phase Circuits

A given three phase circuit that has a combination of Δ-loads and Y-loads should be converted to the Y configuration. By converting from Δ to Y, each circuit element/phase can be analyzed separately. Converting from Δ to Y is an technique aimed to simplify circuit analysis. (Note : harmonic behaviour from the original circuit remained unchanged). The conversion from the Δ notation to Y notation is as follows.

$$V_{LL} = \sqrt{3} V_{LN} \angle 30$$

$$I_{LL} = \sqrt{3}\, I_{LN} \angle -30$$

$$Z\Delta/3 = Z_Y$$

$$S_3\Phi = |\, S_3\Phi \,|\, \sqrt{3} V_{LL}\, I_L = 3 V_{LN}\, I_L$$

A Proof of the Existence and Uniqueness of the Transformation

The feasibility of the transformation can be shown as a consequence of superposition theorem in electric circuit. A short proof, rather than derived as a corollary of the more general star-mesh transform, can be given as follows. The equivalence lies in the statement that for any external voltages (V_1, V_2 and V_3) applying at the three nodes (N_1, N_2 and N_3), the corresponding currents (I_1, I_2 and I_3) are exactly the same for both the Y and Δ circuit, and *vice versa*. In this proof, we start with given external currents at the nodes. According to superposition theorem, the voltages can be obtained by studying the linear summation of the resulting voltages at the nodes of following three problems : apply at the three nodes with current (1) $(I_1 - I_2)/3, - (I_1 - I_2)/3, 0$, (2) $0, (I_2 - I_3)/3, - (I_2 - I_3)/3$ and (3) $- (I_3 - I_1)/3, 0, (I_3 - I_1)/3$. It can be readily shown that due to Kirchhoff's circuit laws, one has $I_1 + I_2 + I_3 = 0$. One notes that now each problem is relatively simple, since it only involves one single ideal current source. To obtain exactly the same outcome voltages at the nodes for each problem, the equivalent resistances in two circuits must be the same, this can be easily found by using the basic rules of series and parallel circuits :

$$R_3 + R_1 = \frac{(R_c + R_a) R_b}{R_a + R_b + R_c}, \frac{R_3}{R_1} = \frac{R_a}{R_c}.$$

Though usually six equations are more than enough to express three variables (R_1, R_2, R_3) in term of the other three variables (R_a, R_b, R_c), here it is straightforward to show that these equations indeed lead to the above designed expressions. In fact, the superposition theorem not only establishes the relation between the values of the resistances, but also guarantees the uniqueness of such solution.

Simplification of Networks

Resistive networks between two terminals can theoretically be simplified to a single equivalent resistor (more generally, the same is true of impedance). Series and parallel transforms are basic tools for doing so, but for complex networks such as the bridge illustrated here, they do not suffice.

The Y-Δ transform can be used to eliminate one node at a time and produce a network that can be further simplified, as shown.

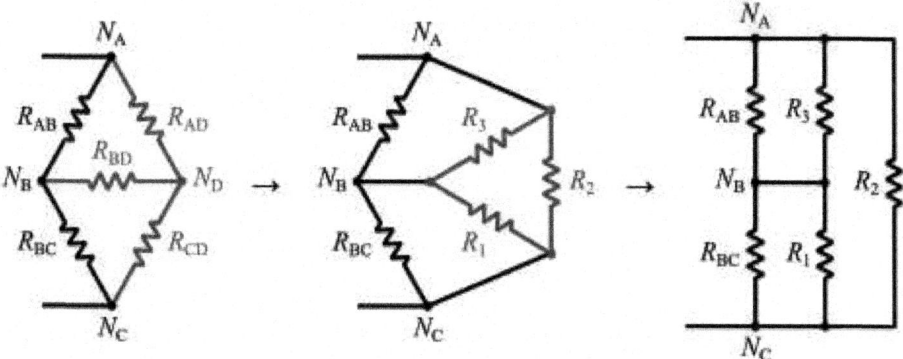

Fig. : Transformation of a bridge resistor network, using the Y-Δ transform to eliminate node D, yields an equivalent network that may readily be simplified further.

The reverse transformation, Δ-Y, which adds a node, is often handy to pave the way for further simplification as well.

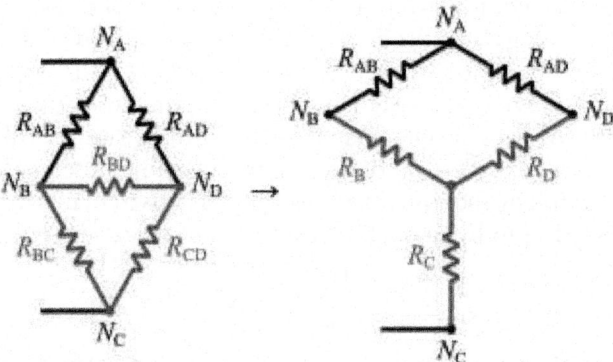

Fig. : Transformation of a bridge resistor network, using the Δ-Y transform, also yields an equivalent network that may readily be simplified further.

Graph Theory

In graph theory, the Y-Δ transform means replacing a Y sub-graph of a graph with the equivalent Δ sub-graph. The transform preserves the number of edges in a graph, but not the number of vertices or the number of cycles. Two graphs are said to be **Y-Δ equivalent** if one can be obtained from the other by a series of Y-Δ transforms in either direction. For example, the Petersen family is a Y-Δ equivalence class.

Demonstration

Δ-Load to Y-Load Transformation Equations

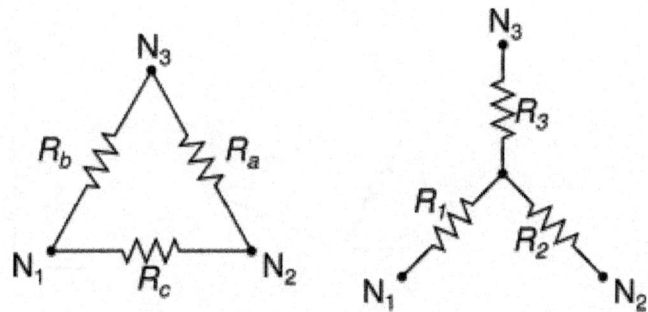

Fig. : Δ and Y circuits with the labels that are used.

To relate $\{R_a, R_b, R_c\}$ from Δ to $\{R_1, R_2, R_3\}$ from Y, the impedance between two corresponding nodes is compared. The impedance in either configuration is determined as if one of the nodes is disconnected from the circuit.

The impedance between N_1 and N_2 with N_3 disconnected in Δ :

$$R\Delta(N_1, N_2) = R_c \,||\, (R_a + R_b)$$

$$= \cfrac{1}{\cfrac{1}{R_c} + \cfrac{1}{R_a + R_b}}$$

$$= \frac{R_c \, (R_a + R_b)}{R_a + R_b + R_c}$$

To simplify, let R_T be the sum of $\{R_a + R_b + R_c\}$.

$$R_T = R_a + R_b + R_c$$

Thus,

$$R\Delta(N_1, N_2) = \frac{R_c(R_a + R_b)}{R_T}$$

The corresponding impedance between N_1 and N_2 in Y is simple :

$$R_Y(N_1, N_2) = R_1 + R_2$$

hence :

$$R_1 + R_2 = \frac{R_c \, (R_a + R_b)}{R_T} \tag{1}$$

Repeating for $R(N_2, N_3)$:

$$R_2 + R_3 = \frac{R_a(R_b + R_c)}{R_T} \tag{2}$$

and for $R(N_1, N_3)$:

$$R_1 + R_3 = \frac{R_b \, (R_a + R_c)}{R_T} \tag{3}$$

From here, the values of $\{R_1, R_2, R_3\}$ can be determined by linear combination (addition and/or subtraction).

For example, adding (1) and (3), then subtracting (2) yields

$$R_1 + R_2 + R_1 + R_3 - R_2 - R_3 = \frac{R_c \, (R_a + R_b)}{R_T} + \frac{R_b \, (R_a + R_c)}{R_T} - \frac{R_a \, (R_b + R_c)}{R_T}$$

$$2R_1 = \frac{2R_b \, R_c}{R_T}$$

thus,

$$R_1 = \frac{R_b \, R_c}{R_T}.$$

where : $R_T = R_a + R_b + R_c$

For completeness :

$$R_1 = \frac{R_b R_c}{R_T} \tag{4}$$

$$R_2 = \frac{R_a R_c}{R_T} \tag{5}$$

$$R_3 = \frac{R_a R_b}{R_T} \tag{6}$$

Y-Load to Δ-Load Transformation Equations

Let

$$R_T = R_a + R_b + R_c.$$

We can write the Δ to Y equations as

$$R_1 = \frac{R_b R_c}{R_T} \tag{1}$$

$$R_2 = \frac{R_a R_c}{R_T} \tag{2}$$

$$R_3 = \frac{R_a R_b}{R_T} \tag{3}$$

Multiplying the pairs of equations yields

$$R_1 R_2 = \frac{R_a R_b R_c^2}{R_T^2} \tag{4}$$

$$R_1 R_3 = \frac{R_a R_b^2 R_c}{R_T^2} \tag{5}$$

$$R_2 R_3 = \frac{R_a^2 R_b R_c}{R_T^2} \tag{6}$$

and the sum of these equations is

$$R_1 R_2 + R_1 R_3 + R_2 R_3 = \frac{R_a R_b R_c^2 + R_a R_b^2 R_c + R_a^2 R_b R_c}{R_T^2} \tag{7}$$

Factor $R_a R_b R_c$ from the right side, leaving R_T in the numerator, canceling with an R_T in the denominator.

$$R_1 R_2 + R_1 R_3 + R_2 R_3 = \frac{(R_a R_b R_c)(R_a + R_b + R_c)}{R_T^2}$$

$$R_1 R_2 + R_1 R_3 + R_2 R_3 = \frac{R_a R_b R_c}{R_T} \tag{8}$$

Note the similarity between (8) and {(1),(2),(3)}

Divide (8) by (1)

$$\frac{R_1R_2 + R_1R_3 + R_2R_3}{R_1} = \frac{R_aR_bR_c}{R_T}\frac{R_T}{R_bR_c},$$

$$\frac{R_1R_2 + R_1R_3 + R_2R_3}{R_1} = R_a,$$

which is the equation for R_a. Dividing (8) by (2) or (3) (expressions for R_2 or R_3) gives the remaining equations.

STAR-MESH TRANSFORM

The **star-mesh transform** (or **star-polygon transform**) is a mathematical circuit analysis technique to transform a resistive network into an equivalent network with one less node. The equivalence follows from the Schur complement identity applied to the Kirchhoff matrix of the network.

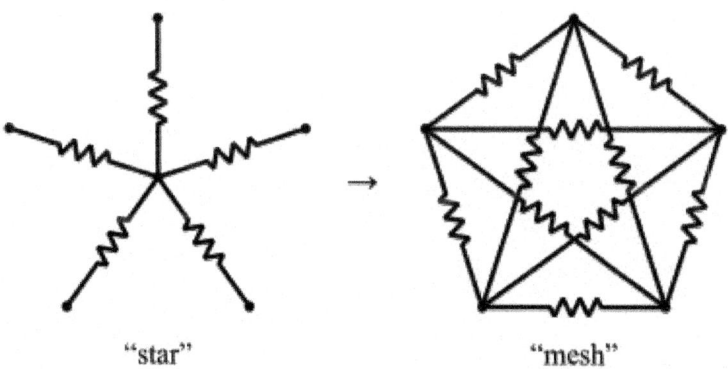

"star"   "mesh"

Given the impedances between the star node (to be eliminated) and N other nodes, the transform yields equivalent impedances between these other nodes :

$$z_{AB} = z_A z_B \sum \frac{1}{z}$$

The transform replaces N resistors with ${}_NC_2$ resistors. For $N > 3$, the result is an increase in the number of resistors, so the transform has no general inverse without additional constraints.

It is possible, though not necessarily efficient, to transform an arbitrarily complex two-terminal resistive network into a single equivalent resistor by repeatedly applying the star-mesh transform to eliminate each non-terminal node.

Special Cases

- $N = 1$: For a single dangling resistor, the transform eliminates the resistor.
- $N = 2$: For two resistors, the "star" is simply the two resistors in series, and the transform yields a single equivalent resistor.
- $N = 3$: The special case of three resistors is better known as the Y-Δ transform. Since the result also has three resistors, this transform has an inverse Δ-Y transform.

SIMPLE NETWORKS

Some very simple networks can be analysed without the need to apply the more systematic approaches.

Voltage Divider

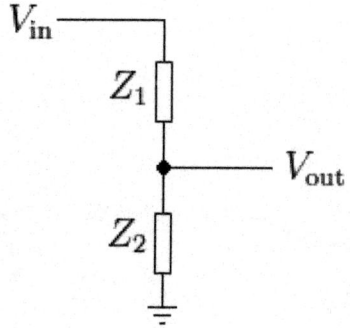

Fig. : Voltage divider.

In electronics or EET, a **voltage divider** (also known as a **potential divider**) is a linear circuit that produces an output voltage (V_{out}) that is a fraction of its input voltage (V_{in}). **Voltage division** refers to the partitioning of a voltage among the components of the divider.

An example of a voltage divider consists of two resistors in series or a potentiometer. It is commonly used to create a reference voltage, or to get a low voltage signal proportional to the voltage to be measured, and may also be used as a signal attenuator at low frequencies. For direct current and relatively low frequencies, a voltage divider may be sufficiently accurate if made only of resistors; where frequency response over a wide range is required (such as in an oscilloscope probe), the voltage divider may have capacitive elements added to allow compensation for load capacitance. In electric power transmission, a capacitive voltage divider is used for measurement of high voltage.

General Case

A voltage divider referenced to ground is created by connecting two electrical impedances in series. The input voltage is applied across the series impedances

Z_1 and Z_2 and the output is the voltage across Z_2. Z_1 and Z_2 may be composed of any combination of elements such as resistors, inductors and capacitors.

Applying Ohm's Law, the relationship between the input voltage, V_{in}, and the output voltage, V_{out}, can be found :

$$V_{out} = \frac{Z_2}{Z_1 + Z_2} \cdot V_{in}$$

Proof :

$$V_{in} = I \cdot (Z_1 + Z_2)$$
$$V_{out} = I \cdot Z_2$$

$$I = \frac{V_{in}}{Z_1 + Z_2}$$

$$V_{out} = V_{in} \cdot \frac{Z_2}{Z_1 + Z_2}$$

The transfer function (also known as the divider's **voltage ratio**) of this circuit is simply :

$$H = \frac{V_{out}}{V_{in}} = \frac{Z_2}{Z_1 + Z_2}$$

In general this transfer function is a complex, rational function of frequency.

Examples

Resistive Divider

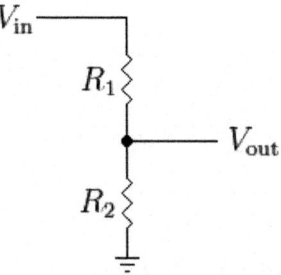

Fig. : Simple resistive voltage divider.

A resistive divider is the case where both impedances, Z_1 and Z_2, are purely resistive.

Substituting $Z_1 = R_1$ and $Z_2 = R_2$ into the previous expression gives :

$$V_{out} = \frac{R_2}{R_1 + R_2} \cdot V_{in}$$

If $R_1 = R_2$ then

$$V_{out} = \frac{1}{2} \cdot V_{in}$$

If V_{out}=6V and V_{in}= 9V (both commonly used voltages), then :

$$\frac{V_{out}}{V_{in}} = \frac{R_2}{R_1 + R_2} = \frac{6}{9} = \frac{2}{3}$$

and by solving using algebra, R_2 must be twice the value of R_1.

To solve for R_1 :

$$R_1 = \frac{R_2 \cdot V_{in}}{V_{out}} - R_2 = R_2 \cdot \left(\frac{V_{in}}{V_{out}} - 1 \right)$$

To solve for R_2 :

$$R_2 = R_1 \cdot \frac{1}{\left(\dfrac{V_{in}}{V_{out}} - 1 \right)}$$

Any ratio V_{out}/V_{in} greater than 1 is not possible. That is, using resistors alone it is not possible to either invert the voltage or increase V_{out} above V_{in}.

Low-pass RC filter

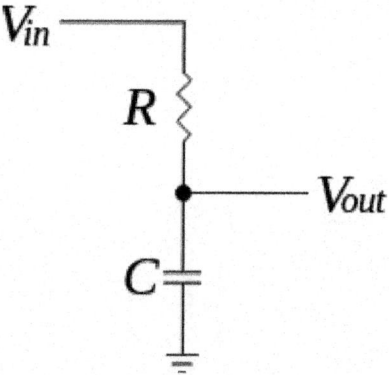

Fig. : Resistor/capacitor voltage divider.

Consider a divider consisting of a resistor and capacitor.

Comparing with the general case, we see Z_1 = R and Z_2 is the impedance of the capacitor, given by

$$Z_2 = j\,X_C = \frac{1}{j\omega C},$$

where X_C is the reactance of the capacitor, C is the capacitance of the capacitor, j is the imaginary unit, and ω (omega) is the radian frequency of the input voltage.

This divider will then have the voltage ratio :

$$\frac{V_{out}}{V_{in}} = \frac{Z_2}{Z_1 + Z_2} = \frac{\dfrac{1}{j\omega C}}{\dfrac{1}{j\omega C} + R} = \frac{1}{1 + j\omega\,RC} \cdot$$

The product τ *(tau)* = RC is called the *time constant* of the circuit.

The ratio then depends on frequency, in this case decreasing as frequency increases. This circuit is, in fact, a basic (first-order) lowpass filter. The ratio contains an imaginary number, and actually contains both the amplitude and phase shift information of the filter. To extract just the amplitude ratio, calculate the magnitude of the ratio, that is :

$$\left| \frac{V_{out}}{V_{in}} \right| = \frac{1}{\sqrt{1 + (\omega RC)^2}}$$

Inductive Divider

Inductive dividers split AC input according to inductance :

$$V_{out} = V_{in} \cdot \frac{L_2}{L_1 + L_2}$$

The above equation is for non-interacting inductors; mutual inductance (as in an autotransformer) will alter the results.

Inductive dividers split DC input according to the resistance of the elements as for the resistive divider above.

Capacitive Divider

Capacitive dividers do not pass DC input.

For an AC input a simple capacitive equation is :

$$V_{out} = V_{in} \cdot \frac{C_1}{C_1 + C_2}$$

Any leakage current in the capactive elements requires use of the generalized expression with two impedances. By selection of parallel R and C elements in the proper proportions, the same division ratio can be maintained over a useful range of frequencies. This is the principle applied in compensated oscilloscope probes to increase measurement bandwidth.

Loading Effect

The voltage output of a voltage divider is not fixed but varies according to the load. To obtain a reasonably stable output voltage the output current should be a small fraction of the input current. The drawback of this is that most of the input current is wasted as heat in the divider. An alternative is to use a voltage regulator.

Applications

Voltage dividers are used for adjusting the level of a signal, for bias of active devices in amplifiers, and for measurement of voltages. A Wheatstone bridge and a multimeter both include voltage dividers. A potentiometer is used as a variable voltage divider in the volume control of a radio. Voltage dividers can also be used to allow a micro-controller to measure the resistance of a sensor. The sensor is hooked up in a voltage divider alongside a known resistor, and a known input voltage is given, and the output voltage is measured, and then used to determine the resistance of the sensor.

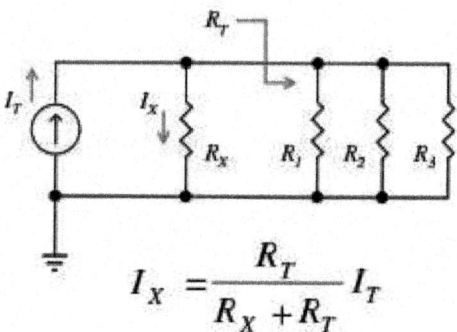

$$I_X = \frac{R_T}{R_X + R_T} I_T$$

Fig. : Schematic of an electrical circuit illustrating current division. Notation R_T refers to the *total* resistance of the circuit to the right of resistor R_X.

Current Divider

In electronics, a **current divider** is a simple linear circuit that produces an output current (I_X) that is a fraction of its input current (I_T). **Current division** refers to the splitting of current between the branches of the divider. The currents in the various branches of such a circuit will always divide in such a way as to minimize the total energy expended.

The formula describing a current divider is similar in form to that for the voltage divider. However, the ratio describing current division places the impedance of the unconsidered branches in the numerator, unlike voltage division where the considered impedance is in the numerator. This is because in current dividers, total energy expended is minimized, resulting in currents that go through paths of least impedance, therefore the inverse relationship with impedance. On the other hand, voltage divider is used to satisfy Kirchhoff's Voltage Law. The voltage around a loop must sum up to zero, so the voltage drops must be divided evenly in a direct relationship with the impedance.

To be specific, if two or more impedances are in parallel, the current that enters the combination will be split between them in inverse proportion to their impedances (according to Ohm's law). It also follows that if the impedances have the same value the current is split equally.

Current Divider

A general formula for the current I_X in a resistor R_X that is in parallel with a combination of other resistors of total resistance R_T is :

$$I_x = \frac{R_T}{(R_x + R_T)} I_T$$

where I_T is the total current entering the combined network of R_X in parallel with R_T. Notice that when R_T is composed of a parallel combination of resistors, say $R_1, R_2, ...$ etc., then the reciprocal of each resistor must be added to find the total resistance R_T :

$$\frac{1}{R_T} = \frac{1}{R_1} + \frac{1}{R_2} + \frac{1}{R_3} + ...$$

General Case

Although the resistive divider is most common, the current divider may be made of frequency dependent impedances. In the general case the current I_X is given by :

$$I_X = \frac{Z_T}{Z_X + Z_T} I_T ,$$

Using Admittance

Instead of using impedances, the current divider rule can be applied just like the voltage divider rule if admittance (the inverse of impedance) is used.

$$I_X = \frac{Y_X}{Y_{total}} I_T$$

Take care to note that Y_{Total} is a straightforward addition, not the sum of the inverses inverted (as you would do for a standard parallel resistive network). For Figure 1, the current I_X would be

$$I_X = \frac{Y_X}{Y_{Total}} I_T = \frac{\frac{1}{R_X}}{\frac{1}{R_X} + \frac{1}{R_1} + \frac{1}{R_2} + \frac{1}{R_3}} I_T$$

Example : RC Combination

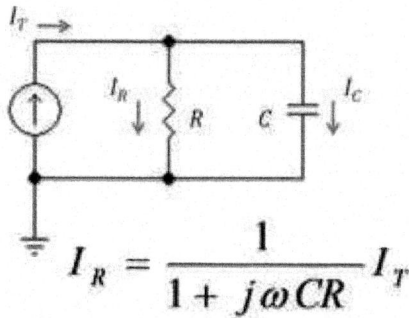

$$I_R = \frac{1}{1 + j\omega CR} I_T$$

Fig. : A low pass RC current divider.

Figure shows a simple current divider made up of a capacitor and a resistor. Using the formula above, the current in the resistor is given by :

$$I_R = \frac{\dfrac{1}{j\omega C}}{R + \dfrac{1}{j\omega C}} I_T$$

$$= \frac{1}{1 + j\omega CR} I_T ,$$

where $Z_C = 1/(j\omega C)$ is the impedance of the capacitor and j is the imaginary unit.

The product $\tau = CR$ is known as the time constant of the circuit, and the frequency for which $\omega CR = 1$ is called the corner frequency of the circuit. Because the capacitor has zero impedance at high frequencies and infinite impedance at low frequencies, the current in the resistor remains at its DC value I_T for frequencies up to the corner frequency, whereupon it drops toward zero for higher frequencies as the capacitor effectively short-circuits the resistor. In other words, the current divider is a low pass filter for current in the resistor.

Loading Effect

The gain of an amplifier generally depends on its source and load terminations. Current amplifiers and transconductance amplifiers are characterized by a short-circuit output condition, and current amplifiers and transresistance amplifiers are characterized using ideal infinite impedance current sources. When an amplifier is terminated by a finite, non-zero termination, and/or driven by a non-ideal source, the effective gain is reduced due to the **loading effect** at the output and/or the input, which can be understood in terms of current division.

Figure shows a current amplifier example. The amplifier (gray box) has input resistance R_{in} and output resistance R_{out} and an ideal current gain A_i. With an ideal current driver (infinite Norton resistance) all the source current i_s becomes input current to the amplifier. However, for a Norton driver a current divider is formed at the input that reduces the input current to

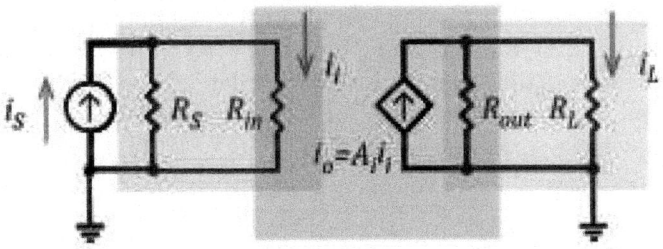

Fig. : A current amplifier (gray box) driven by a Norton source (i_S, R_S) and with a resistor load R_L. Current divider in blue box at input (R_S, R_{in}) reduces the current gain, as does the current divider in green box at the output (R_{out}, R_L).

$$i_i = \frac{R_S}{R_S + R_{in}} i_s,$$

which clearly is less than i_s. Likewise, for a short circuit at the output, the amplifier delivers an output current $i_o = A_i \, i_i$ to the short-circuit. However, when the load is a non-zero resistor R_L, the current delivered to the load is reduced by current division to the value :

$$i_L = \frac{R_{out}}{R_{out} + R_L} A_i i_i.$$

Combining these results, the ideal current gain A_i realized with an ideal driver and a short-circuit load is reduced to the **loaded gain** A_{loaded} :

$$A_{loaded} = \frac{i_L}{i_s} = \frac{R_s}{R_s + R_{in}} \frac{R_{out}}{R_{out} + R_L} A_i.$$

The resistor ratios in the above expression are called the **loading factors**. For more discussion of loading in other amplifier types.

Unilateral versus Bilateral Amplifiers

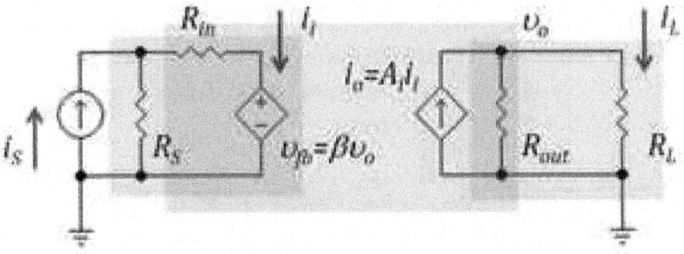

Fig. : Current amplifier as a bilateral two-port network; feedback through dependent voltage source of gain β V/V.

Figure and the associated discussion refers to a unilateral amplifier. In a more general case where the amplifier is represented by a two port, the input resistance of the amplifier depends on its load, and the output resistance on the source impedance. The loading factors in these cases must employ the true amplifier impedances including these bilateral effects. For example, taking the unilateral current amplifier of Figure, the corresponding bilateral two-port network is shown in Figure based upon h-parameters. Carrying out the analysis for this circuit, the current gain with feedback A_{fb} is found to be

$$A_{fb} = \frac{i_L}{i_S} = \frac{A_{loaded}}{1 + \beta(R_L / R_S)A_{loaded}}.$$

That is, the ideal current gain A_i is reduced not only by the loading factors, but due to the bilateral nature of the two-port by an additional factor $(1 + \beta(R_L / R_S) A_{loaded})$, which is typical of negative feedback amplifier circuits. The factor $\beta (R_L / R_S)$ is the current feedback provided by the voltage feedback source of voltage gain β V/V. For instance, for an ideal current source with $R_S = \infty\ \Omega$, the voltage feedback has no influence, and for $R_L = 0\ \Omega$, there is zero load voltage, again disabling the feedback.

NODAL ANALYSIS

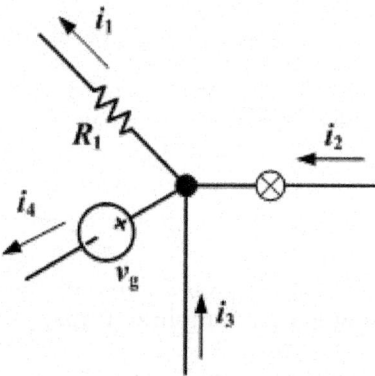

Fig. : Kirchhoff's current law is the basis of nodal analysis.

In electric circuits analysis, **nodal analysis, node-voltage analysis,** or the **branch current method** is a method of determining the voltage (potential difference) between "nodes" (points where elements or branches connect) in an electrical circuit in terms of the branch currents.

In analyzing a circuit using Kirchhoff's circuit laws, one can either do nodal analysis using Kirchhoff's current law (KCL) or mesh analysis using Kirchhoff's voltage law (KVL). Nodal analysis writes an equation at each electrical node, requiring that the branch currents incident at a node must sum to zero. The branch currents are written in terms of the circuit node voltages. As a consequence, each branch constitutive relation must give current as a function of voltage; an admittance representation. For instance, for a resistor, $I_{branch} = V_{branch} * G$, where G (=1/R) is the admittance (conductance) of the resistor.

Nodal analysis is possible when all the circuit elements' branch constitutive relations have an admittance representation. Nodal analysis produces a compact set of equations for the network, which can be solved by hand if small, or can be quickly solved using linear algebra by computer. Because of the compact system of equations, many circuit simulation programs (*e.g.* SPICE) use nodal analysis as a basis. When elements do not have admittance representations, a more general extension of nodal analysis, modified nodal analysis, can be used.

While simple examples of nodal analysis focus on linear elements, more complex non-linear networks can also be solved with nodal analysis by using Newton's method to turn the non-linear problem into a sequence of linear problems.

Method

1. Note all connected wire segments in the circuit. These are the *nodes* of nodal analysis.
2. Select one node as the ground reference. The choice does not affect the result and is just a matter of convention. Choosing the node with the most connections can simplify the analysis.
3. Assign a variable for each node whose voltage is unknown. If the voltage is already known, it is not necessary to assign a variable.
4. For each unknown voltage, form an equation based on Kirchhoff's current law. Basically, add together all currents leaving from the node and mark the sum equal to zero. Finding the current between two nodes is nothing more than "the node you're on, minus the node you're going to, divided by the resistance between the two nodes."
5. If there are voltage sources between two unknown voltages, join the two nodes as a supernode. The currents of the two nodes are combined in a single equation, and a new equation for the voltages is formed.
6. Solve the system of simultaneous equations for each unknown voltage.

Examples

Basic Case

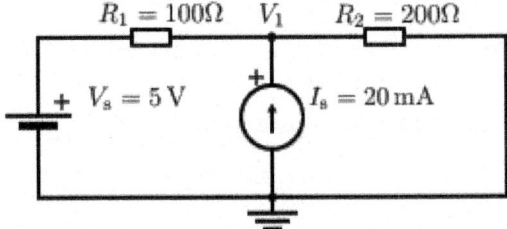

Fig. : Basic example circuit with one unknown voltage, V_1.

The only unknown voltage in this circuit is V_1. There are three connections to this node and consequently three currents to consider. The direction of the currents in calculations is chosen to be away from the node.

1. Current through resistor R_1 : $(V_1 - V_S) / R_1$
2. Current through resistor R_2 : V_1 / R_2
3. Current through current source I_S : $-I_S$

With Kirchhoff's current law, we get :

$$\frac{V_1 - V_S}{R_1} + \frac{V_1}{R_2} - I_S = 0$$

This equation can be solved in respect to V_1 :

$$V_1 = \frac{\left(\dfrac{V_s}{R_1} + I_S\right)}{\left(\dfrac{1}{R_1} + \dfrac{1}{R_2}\right)}$$

Finally, the unknown voltage can be solved by substituting numerical values for the symbols. Any unknown currents are easy to calculate after all the voltages in the circuit are known.

$$V_1 = \frac{\left(\dfrac{5\,V}{100\,\Omega} + 20\,mA\right)}{\dfrac{1}{100\,\Omega} + \dfrac{1}{200\,\Omega}} \approx 4.667\ V$$

Supernodes

In this circuit, we initially have two unknown voltages, V_1 and V_2. The voltage at V_3 is already known to be V_B because the other terminal of the voltage source is at ground potential.

The current going through voltage source V_A cannot be directly calculated. Therefore we can not write the current equations for either V_1 or V_2. However, we know that the same current leaving node V_2 must enter node V_1. Even though the nodes can not be individually solved, we know that the combined current of these two nodes is zero. This combining of the two nodes is called the supernode technique, and it requires one additional equation : $V_1 = V_2 + V_A$.

The complete set of equations for this circuit is :

$$\begin{cases} \dfrac{V_1 - V_B}{R_1} + \dfrac{V_2 - V_B}{R_2} + \dfrac{V_2}{R_3} = 0 \\ V_1 = V_2 + V_A \end{cases}$$

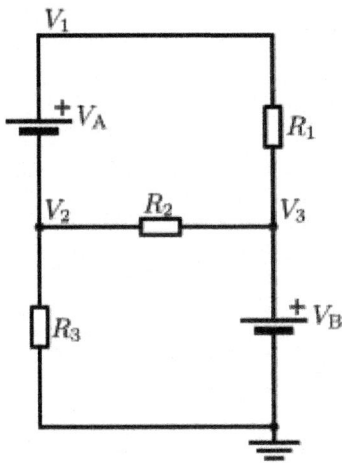

Fig. : In this circuit, V_A is between two unknown voltages, and is therefore a supernode.

By substituting V_1 to the first equation and solving in respect to V_2, we get :

$$V_2 = \frac{(R_1 + R_2) R_3 V_B - R_2 R_3 V_A}{(R_1 + R_2) R_3 + R_1 R_2}$$

MESH ANALYSIS

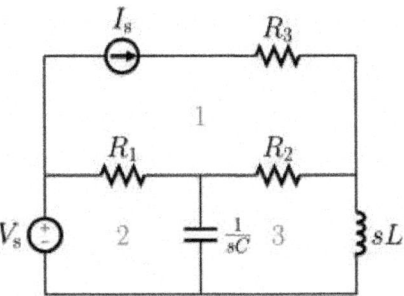

Fig. : Essential meshes of the planar circuit labeled 1, 2, and 3. R_1, R_2, R_3, 1/sc, and Ls represent the impedance of the resistors, capacitor, and inductor values in the s-domain. V_s and i_s are the values of the voltage source and current source, respectively.

Mesh analysis (or the **mesh current method**) is a method that is used to solve planar circuits for the currents (and indirectly the voltages) at any place in the circuit. Planar circuits are circuits that can be drawn on a plane surface with no wires crossing each other. A more general technique, called **loop analysis** (with the corresponding network variables called **loop currents**) can be applied to any circuit, planar or not. Mesh analysis and loop analysis both make use of Kirch-hoff's voltage law to arrive at a set of equations guaranteed to be solvable if the circuit has a solution. Mesh analysis is usually easier to use when the circuit is planar, compared to loop analysis.

Mesh Currents and Essential Meshes

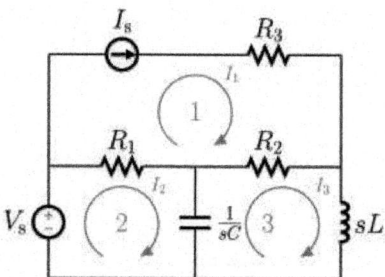

Fig. : Circuit with mesh currents labelled as i_1, i_2, and i_3. The arrows show the direction of the mesh current.

Mesh analysis works by arbitrarily assigning mesh currents in the essential meshes (also referred to as independent meshes). An essential mesh is a loop in the circuit that does not contain any other loop.

A mesh current is a current that loops around the essential mesh and the equations are set solved in terms of them. A mesh current may not correspond to any physically flowing current, but the physical currents are easily found from them. It is usual practice to have all the mesh currents loop in the same direction. This helps prevent errors when writing out the equations. The convention is to have all the mesh currents looping in a clockwise direction.

Solving for mesh currents instead of directly applying Kirchhoff's current law and Kirchhoff's voltage law can greatly reduce the amount of calculation required. This is because there are fewer mesh currents than there are physical branch currents.

Setting Up the Equations

Each mesh produces one equation. These equations are the sum of the voltage drops in a complete loop of the mesh current. For problems more general than those including current and voltage sources, the voltage drops will be the impedance of the electronic component multiplied by the mesh current in that loop.

If a voltage source is present within the mesh loop, the voltage at the source is either added or subtracted depending on if it is a voltage drop or a voltage rise in the direction of the mesh current. For a current source that is not contained between two meshes, the mesh current will take the positive or negative value of the current source depending on if the mesh current is in the same or opposite direction of the current source. The following is the same circuit from above with the equations needed to solve for all the currents in the circuit.

$$\begin{cases} \text{Mesh } 1: I_1 = I_s \\ \text{Mesh } 2: -V_S + R_1\,(I_2 - I_1) + \dfrac{1}{sc}\,(I_2 - I_3) =) \\ \text{Mesh } 3: \dfrac{1}{sc}\,(I_3 - I_2) + R_2\,(I_3 - I_1) + LsI_3 = 0 \end{cases}$$

Once the equations are found, the system of linear equations can be solved by using any technique to solve linear equations.

Special Cases

There are two special cases in mesh current : currents containing a supermesh and currents containing dependent sources.

Supermesh

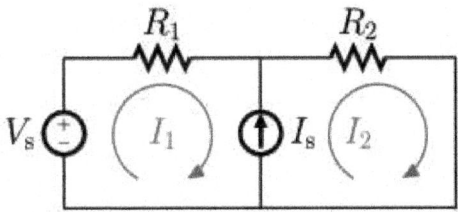

Fig. : Circuit with a supermesh. Supermesh occurs because the current source is in between the essential meshes.

A supermesh occurs when a current source is contained between two essential meshes. The circuit is first treated as if the current source is not there. This leads to one equation that incorporates two mesh currents. Once this equation is formed, an equation is needed that relates the two mesh currents with the current source. This will be an equation where the current source is equal to one of the mesh currents minus the other. The following is a simple example of dealing with a supermesh :

$$\begin{cases} \text{Mesh } 1,2: -V_s + R_1\,I_1 + R_2 I_2 = 0 \\ \text{Current source}: I_s = I_2 - I_1 \end{cases}$$

Dependent Sources

A dependent source is a current source or voltage source that depends on the voltage or current of another element in the circuit. When a dependent source is contained within an essential mesh, the dependent source should be treated like an independent source. After the mesh equation is formed, a dependent source equation is needed. This equation is generally called a constraint equation. This is an equation that relates the dependent source's variable to the voltage or current that the source depends on in the circuit. The following is a simple example of a dependent source.

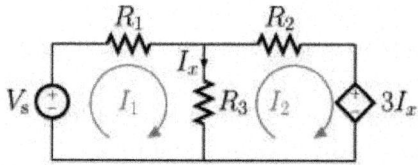

Fig. : Circuit with dependent source. i$_x$ is the current upon which the dependent source depends.

$$\begin{cases} \text{Mesh 1}: -V_s + R_1 I_1 + R_3\,(I_1 - I_2) = 0 \\ \text{Mesh 2}: R_2 I_2 + 3I_x + R_3\,(I_2 - I_1) = 0 \\ \text{Dependednt variable}: I_x = I_1 - I_2 \end{cases}$$

CHOICE OF METHOD

Choice of method is to some extent a matter of taste. If the network is particularly simple or only a specific current or voltage is required then *ad-hoc* application of some simple equivalent circuits may yield the answer without recourse to the more systematic methods.

- Superposition is possibly the most conceptually simple method but rapidly leads to a large number of equations and messy impedance combinations as the network becomes larger.
- Nodal analysis : The number of voltage variables, and hence simultaneous equations to solve, equals the number of nodes minus one. Every voltage source connected to the reference node reduces the number of unknowns (and equations) by one.
- Mesh analysis : The number of current variables, and hence simultaneous equations to solve, equals the number of meshes. Every current source in a mesh reduces the number of unknowns by one. Mesh analysis can only be used with networks which can be drawn as a planar network, that is, with no crossing components.

TRANSFER FUNCTION

A transfer function expresses the relationship between an input and an output of a network. For resistive networks, this will always be a simple real number or an expression which boils down to a real number. Resistive networks are represented by a system of simultaneous algebraic equations. However in the general case of linear networks, the network is represented by a system of simultaneous linear differential equations. In network analysis, rather than use the differential equations directly, it is usual practice to carry out a Laplace transform on them first and then express the result in terms of the Laplace parameter s, which in general is complex. This is described as working in the s-domain. Working with the equations directly would be described as working in the time (or t) domain because the results would be expressed as time varying quantities. The Laplace transform is the mathematical method of transforming between the s-domain and the t-domain.

This approach is standard in control theory and is useful for determining stability of a system, for instance, in an amplifier with feedback.

Two Terminal Component Transfer Functions

For two terminal components the transfer function, or more generally for non-linear elements, the constitutive equation, is the relationship between the current input to the device and the resulting voltage across it. The transfer function, $Z(s)$, will thus have units of impedance – ohms. For the three passive components found in electrical networks, the transfer functions are;

Resistor	$Z(s) = R$
Inductor	$Z(s) = sL$
Capacitor	$Z(s) = \dfrac{1}{sC}$

For a network to which only steady ac signals are applied, s is replaced with $j\omega$ and the more familiar values from ac network theory result.

Resistor	$Z(j\omega) = R$
Inductor	$Z(j\omega) = j\omega L$
Capacitor	$Z(j\omega) = \dfrac{1}{j\omega C}$

Finally, for a network to which only steady dc is applied, s is replaced with zero and dc network theory applies.

Resistor	$Z = R$
Inductor	$Z = 0$
Capacitor	$Z = \infty$

Two-port Network

A **two-port network** (a kind of **four-terminal network** or **quadripole**) is an electrical network (circuit) or device with two *pairs* of terminals to connect to external circuits. Two terminals constitute a **port** if the currents applied to them satisfy the essential requirement known as the **port condition** : the electric current entering one terminal must equal the current emerging from the other terminal on the same port. The ports constitute interfaces where the network connects to other networks, the points where signals are applied or outputs are taken. In a two-port network, often port 1 is considered the input port and port 2 is considered the output port.

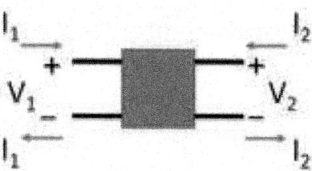

Fig. : Example two-port network with symbol definitions. Notice the port condition is satisfied : the same current flows into each port as leaves that port.

The two-port network model is used in mathematical circuit analysis techniques to isolate portions of larger circuits. A two-port network is regarded as a "black box" with its properties specified by a matrix of numbers. This allows the response of the network to signals applied to the ports to be calculated easily, without solving for all the internal voltages and currents in the network. It also allows similar circuits or devices to be compared easily. For example, transistors are often regarded as two-ports, characterized by their h-parameters which are listed by the manufacturer. Any linear circuit with four terminals can be regarded as a two-port network provided that it does not contain an independent source and satisfies the port conditions.

Examples of circuits analyzed as two-ports are filters, matching networks, transmission lines, transformers, and small-signal models for transistors (such as the hybrid-pi model). The analysis of passive two-port networks is an outgrowth of reciprocity theorems first derived by Lorentz.

In two-port mathematical models, the network is described by a 2 by 2 square matrix of complex numbers. The common models that are used are referred to as *z-parameters*, *y-parameters*, *h-parameters*, *g-parameters*, and *ABCD-parameters*. These are all limited to linear networks since an underlying assumption of their derivation is that any given circuit condition is a linear superposition of various short-circuit and open circuit conditions. They are usually expressed in matrix notation, and they establish relations between the variables

V_1, voltage across port 1

I_1, current into port 1

V_2, voltage across port 2

I_2, current into port 2

which are shown in figure. These current and voltage variables are most useful at low-to-moderate frequencies. At high frequencies (*e.g.*, microwave frequencies), the use of power and energy variables is more appropriate, and the two-port current–voltage approach is replaced by an approach based upon scattering parameters.

General Properties

There are certain properties of two-ports that frequently occur in practical networks and can be used to greatly simplify the analysis. These include :

Reciprocal networks : A network is said to be reciprocal if the voltage appearing at port 2 due to a current applied at port 1 is the same as the voltage appearing at port 1 when the same current is applied to port 2. Exchanging voltage and current results in an equivalent definition of reciprocity. A network that consists entirely of linear passive components (that is, resistors, capacitors and inductors) is always reciprocal. In general, it *will not* be reciprocal if it contains active components such as generators or transistors.

Symmetrical networks : A network is symmetrical if its input impedance is equal to its output impedance. Most often, but not necessarily, symmetrical networks are also physically symmetrical. Sometimes also antimetrical networks are of interest. These are networks where the input and output impedances are the duals of each other.

Lossless network : A lossless network is one which contains no resistors or other dissipative elements.

Impedance Parameters (z-Parameters)

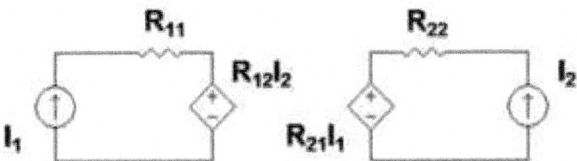

Fig. : z-equivalent two port showing independent variables I_1 and I_2. Although resistors are shown, general impedances can be used instead.

$$\begin{bmatrix} V_1 \\ V_2 \end{bmatrix} = \begin{bmatrix} z_{11} & z_{12} \\ z_{21} & z_{22} \end{bmatrix} \begin{bmatrix} I_1 \\ I_2 \end{bmatrix}$$

where

$$y_{11} \overset{def}{=} \frac{V_1}{I_1}\bigg|_{V_2=0} \qquad y_{12} \overset{def}{=} \frac{V_1}{I_2}\bigg|_{V_1=0}$$

$$y_{21} \overset{def}{=} \frac{V_2}{I_1}\bigg|_{V_2=0} \qquad y_{22} \overset{def}{=} \frac{V_2}{I_2}\bigg|_{V_1=0}$$

Notice that all the z-parameters have dimensions of ohms.

For reciprocal networks $z_{12} = z_{21}$. For symmetrical networks $z_{11} = z_{22}$. For reciprocal lossless networks all the z_{mn} are purely imaginary.

Example : Bipolar Current Mirror with Emitter Degeneration

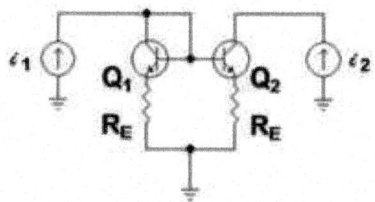

Fig. : Bipolar current mirror : i_1 is the *reference current* and i_2 is the *output current;* lower case symbols indicate these are totalz currents that include the DC components.

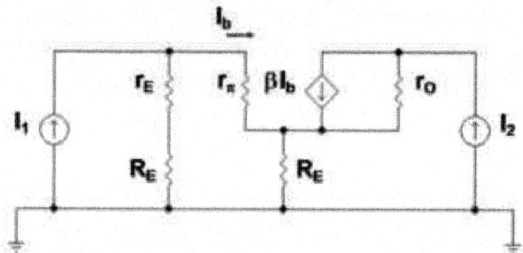

Fig. : Small-signal bipolar current mirror : I_1 is the amplitude of the small-signal *reference current* and I_2 is the amplitude of the small-signal *output current.*

Figure shows a bipolar current mirror with emitter resistors to increase its output resistance. Transistor Q_1 is *diode connected,* which is to say its collector-base voltage is zero. Transistor Q_1 is represented by its emitter resistance $r_E \approx V_T / I_E$ (V_T = thermal voltage, I_E = Q-point emitter current), a simplification made possible because the dependent current source in the hybrid-pi model for Q_1 draws the same current as a resistor $1 / g_m$ connected across r_π. The second transistor Q_2 is represented by its hybrid-pi model. Table 1 below shows the z-parameter expressions that make the z-equivalent circuit of Figure 2 electrically equivalent to the small-signal circuit of Figure.

Table 1	Expression	Approximation
$R_{21} = \left.\dfrac{V_2}{I_1}\right\|_{I_2 = 0}$	$-(\beta r_o = R_E)\dfrac{r_E + R_E}{r_\pi + r_E + 2R_E}$	$-\beta r_o \dfrac{r_E + R_E}{r_\pi + 2R_E}$
$R_{11} + \left.\dfrac{V_1}{I_1}\right\|_{I_2 =)}$	$(r_E + R_E) \| (r_\pi + R_E)$	
$R_{22} + \left.\dfrac{V_2}{I_2}\right\|_{I_1 = 0}$	$\left(1 + \beta \dfrac{R_E}{r_\pi + r_E + 2R_E}\right)r_o + \dfrac{r_\pi + r_E + R_E}{r_\pi + r_E + 2R_E}R_E$	$\left(1 + \beta \dfrac{R_E}{r_\pi + 2R_E}\right)r_o$

$R_{12} + \dfrac{V_1}{I_2}\Big|_{I_1=0}$	$R_E\,\dfrac{r_E + R_E}{r_\pi + r_E + 2R_E}$	$R_E\,\dfrac{r_E + R_E}{r_\pi + 2R_E}$

The negative feedback introduced by resistors R_E can be seen in these parameters. For example, when used as an active load in a differential amplifier, $I_1 \approx -I_2$, making the output impedance of the mirror approximately $R_{22}-R_{21} \approx 2\,\beta$ $r_oR_E/(r_\pi + 2R_E)$ compared to only r_o without feedback (that is with $R_E = 0\,\Omega$). At the same time, the impedance on the reference side of the mirror is approximately R_{11} & nbsp$- R_{12} \approx \dfrac{r_\pi}{r_\pi + 2R_E}\,(r_E + R_E)$, only a moderate value, but still

larger than r_E with no feedback. In the differential amplifier application, a large output resistance increases the difference-mode gain, a good thing, and a small mirror input resistance is desirable to avoid Miller effect.

Admittance Parameters (y-Parameters)

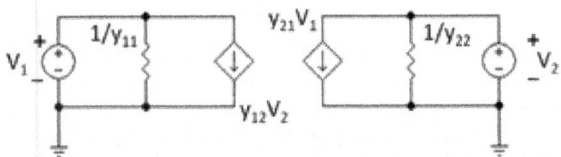

Fig. : Y-equivalent two port showing independent variables V_1 and V_2. Although resistors are shown, general admittances can be used instead.

$$\begin{bmatrix} I_1 \\ I_2 \end{bmatrix} = \begin{bmatrix} y_{11} & y_{12} \\ y_{21} & y_{22} \end{bmatrix}\begin{bmatrix} V_1 \\ V_2 \end{bmatrix}$$

where

$$y_{11} \overset{\text{def}}{=} \frac{V_1}{I_1}\Big|_{V_2=0} \qquad y_{12} \overset{\text{def}}{=} \frac{V_1}{I_2}\Big|_{V_1=0}$$

$$y_{21} \overset{\text{def}}{=} \frac{V_2}{I_1}\Big|_{V_2=0} \qquad y_{22} \overset{\text{def}}{=} \frac{V_2}{I_2}\Big|_{V_1=0}$$

Notice that all the Y-parameters have dimensions of siemens.

For reciprocal networks $y_{12} = y_{21}$. For symmetrical networks $y_{11} = y_{22}$. For reciprocal lossless networks all the y_{mn} are purely imaginary.

Hybrid Parameters (h-Parameters)

$$\begin{bmatrix} V_1 \\ I_2 \end{bmatrix} = \begin{bmatrix} h_{11} & h_{12} \\ h_{21} & h_{22} \end{bmatrix}\begin{bmatrix} I_1 \\ V_2 \end{bmatrix}$$

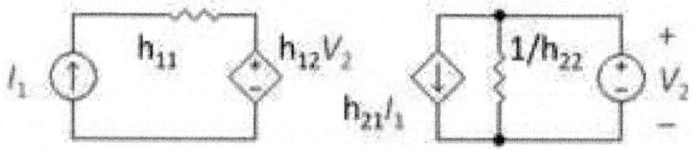

Fig. : H-equivalent two-port showing independent variables I_1 and V_2; h_{22} is reciprocated to make a resistor

where

$$h_{11} \overset{\text{def}}{=} \left. \frac{V_1}{I_1} \right|_{I_2 = 0} \qquad h_{12} \overset{\text{def}}{=} \left. \frac{V_1}{V_2} \right|_{I_1 = 0}$$

$$h_{21} \overset{\text{def}}{=} \left. \frac{I_2}{I_1} \right|_{I_2 = 0} \qquad h_{22} \overset{\text{def}}{=} \left. \frac{I_2}{V_2} \right|_{I_1 = 0}$$

This circuit is often selected when a current amplifier is wanted at the output. The resistors shown in the diagram can be general impedances instead.

Notice that off-diagonal h-parameters are dimensionless, while diagonal members have dimensions the reciprocal of one another.

Example : Common-base Amplifier

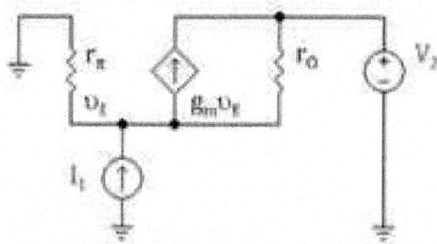

Fig. : Common-base amplifier with AC current source I_1 as signal input and unspecified load supporting voltage V_2 and a dependent current I_2.

Note : Tabulated formulas in Table make the h-equivalent circuit of the transistor from Figure agree with its small-signal low-frequency hybrid-pi model in Figure. Notation : r_π = base resistance of transistor, r_O = output resistance, and g_m = transconductance. The negative sign for h_{21} reflects the convention that I_1, I_2 are positive when directed *into* the two-port. A non-zero value for h_{12} means the output voltage affects the input voltage, that is, this amplifier is **bilateral**. If $h_{12} = 0$, the amplifier is **unilateral**.

Table	Expression	Approximation
$h_{21} = \dfrac{I_2}{I_1}\Big|_{V_2 = 0}$	$\dfrac{\dfrac{\beta}{\beta+1}\,r_o + r_E}{-\dfrac{}{r_o + r_E}}$	$-\dfrac{\beta}{\beta+1}$
$h_{11} = \dfrac{V_1}{I_1}\Big|_{V_2 = 0}$	$r_E \,\|\, r_o$	r_E
$h_{22} = \dfrac{V_1}{V_2}\Big|_{I_1 = 0}$	$\dfrac{1}{(\beta+1)(r_o + r_E)}$	$\dfrac{1}{(\beta+1)r_o}$
$h_{12} = \dfrac{V_1}{V_2}\Big|_{I_1 = 0}$	$\dfrac{r_E}{r_E + r_o}$	$\dfrac{r_E}{r_o} \ll 1$

Inverse Hybrid Parameters (g-Parameters)

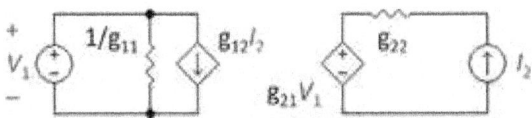

Fig. : G-equivalent two-port showing independent variables V_1 and I_2; g_{11} is reciprocated to make a resistor.

$$\begin{bmatrix} I_1 \\ V_2 \end{bmatrix} = \begin{bmatrix} g_{11} & g_{12} \\ g_{21} & g_{22} \end{bmatrix} \begin{bmatrix} V_1 \\ I_2 \end{bmatrix}$$

where

$$g_{11} \overset{\text{def}}{=} \frac{I_1}{V_1}\Big|_{I_2 = 0} \qquad g_{12} \overset{\text{def}}{=} \frac{I_1}{I_2}\Big|_{V_1 = 0}$$

$$g_{21} \overset{\text{def}}{=} \frac{V_2}{V}\Big|_{I_2 = 0} \qquad g_{22} \overset{\text{def}}{=} \frac{V_2}{I_2}\Big|_{V_1 = 0}$$

Often this circuit is selected when a voltage amplifier is wanted at the output. Notice that off-diagonal g-parameters are dimensionless, while diagonal members have dimensions the reciprocal of one another. The resistors shown in the diagram can be general impedances instead.

Example : Common-base Amplifier

Note : Tabulated formulas in Table make the g-equivalent circuit of the transistor from Figure agree with its small-signal low-frequency hybrid-pi model in

Figure. Notation : r_π = base resistance of transistor, r_O = output resistance, and g_m = transconductance. The negative sign for g_{12} reflects the convention that I_1, I_2 are positive when directed *into* the two-port. A non-zero value for g_{12} means the output current affects the input current, that is, this amplifier is **bilateral**. If $g_{12} = 0$, the amplifier is **unilateral**.

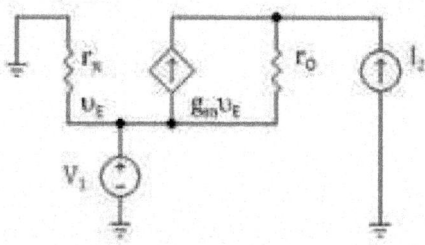

Fig. : Common-base amplifier with AC voltage source V_1 as signal input and unspecified load delivering current I_2 at a dependent voltage V_2.

Table	Expression	Approximation
$g_{21} = \dfrac{V_2}{V_1}\bigg|_{I_2=0}$	$\dfrac{r_o}{r_\pi} + g_m r_o + 1$	$g_m r_o$
$g_{11} = \dfrac{I_1}{V_1}\bigg|_{I_2=0}$	$\dfrac{1}{r_\pi}$	$\dfrac{1}{r_\pi}$
$g_{22} = \dfrac{V_2}{I_1}\bigg|_{V_1=0}$	r_o	r_o
$g_{12} = \dfrac{I_1}{I_2}\bigg|_{V_1=0}$	$-\dfrac{\beta+1}{\beta}$	-1

ABCD-Parameters

The *ABCD*-parameters are known variously as chain, cascade, or transmission line parameters. There are a number of definitions given for *ABCD*-parameters, the most common is,

$$\begin{bmatrix} V_1 \\ I_1 \end{bmatrix} = \begin{bmatrix} A & B \\ C & D \end{bmatrix} \begin{bmatrix} V_2 \\ -I_2 \end{bmatrix}$$

For reciprocal networks $AD - BC = 1$. For symmetrical networks $A = D$. For networks which are reciprocal and lossless, A and D are purely real while B and C are purely imaginary.

This representation is preferred because when the parameters are used to represent a cascade of two-ports, the matrices are written in the same order that a network diagram would be drawn, that is, left to right. However, the examples given below are based on a variant definition;

$$\begin{bmatrix} V_2 \\ I'_2 \end{bmatrix} = \begin{bmatrix} A' & B' \\ C' & D' \end{bmatrix} \begin{bmatrix} V_1 \\ I_1 \end{bmatrix}$$

where

$$A' \stackrel{\text{def}}{=} \frac{V_2}{V_1}\bigg|_{I_1=0} \qquad B' \stackrel{\text{def}}{=} \frac{V_2}{I_1}\bigg|_{V_1=0}$$

$$C' \stackrel{\text{def}}{=} \frac{I_2}{V_1}\bigg|_{I_1=0} \qquad D' \stackrel{\text{def}}{=} \frac{I_2}{I_1}\bigg|_{V_1=0}$$

The negative signs in the definitions of parameters C' and D' arise because I'_2 is defined with the opposite sense to I_2, that is, $I'_2 = -I_2$. The reason for adopting this convention is so that the output current of one cascaded stage is equal to the input current of the next. Consequently, the input voltage/current matrix vector can be directly replaced with the matrix equation of the preceding cascaded stage to form a combined A' B' C' D' matrix.

The terminology of representing the $ABCD$ parameters as a matrix of elements designated a_{11} etc., as adopted by some authors and the inverse A' B' C' D' parameters as a matrix of elements designated b_{11} etc., is used here for both brevity and to avoid confusion with circuit elements.

$$[a] = \begin{bmatrix} a_{11} & a_{12} \\ a_{21} & a_{22} \end{bmatrix} = \begin{bmatrix} A & B \\ C & D \end{bmatrix}$$

$$[b] = \begin{bmatrix} b_{11} & b_{12} \\ b_{21} & b_{22} \end{bmatrix} \begin{bmatrix} A' & B' \\ C' & D' \end{bmatrix}$$

An $ABCD$ matrix has been defined for Telephony four-wire Transmission Systems by P K Webb in British Post Office Research Department Report 630 in 1977.

Table of Transmission Parameters

The table below lists $ABCD$ and inverse $ABCD$ parameters for some simple network elements.

Ele-ment	[a] matrix	[b] matrix	Remarks
Series resistor	$\begin{bmatrix} 1 & R \\ 0 & 1 \end{bmatrix}$	$\begin{bmatrix} 1 & -R \\ 0 & 1 \end{bmatrix}$	R = resistance
Shunt resistor	$\begin{bmatrix} 1 & 0 \\ \dfrac{1}{R} & 1 \end{bmatrix}$	$\begin{bmatrix} 1 & 0 \\ -\dfrac{1}{R} & 1 \end{bmatrix}$	R = resistance
Series conductor	$\begin{bmatrix} 1 & \dfrac{1}{G} \\ 0 & 1 \end{bmatrix}$	$\begin{bmatrix} 1 & -\dfrac{1}{G} \\ 0 & 1 \end{bmatrix}$	G = conductance
Shunt conductor	$\begin{bmatrix} 1 & 0 \\ G & 1 \end{bmatrix}$	$\begin{bmatrix} 1 & 0 \\ -G & 1 \end{bmatrix}$	G = conductance
Series inductor	$\begin{bmatrix} 1 & sL \\ 0 & 1 \end{bmatrix}$	$\begin{bmatrix} 1 & -sL \\ 0 & 1 \end{bmatrix}$	L = inductance s = complex angular frequency
Shunt capacitor	$\begin{bmatrix} 1 & 0 \\ sC & 1 \end{bmatrix}$	$\begin{bmatrix} 1 & 0 \\ -sC & 1 \end{bmatrix}$	C = capacitance s = complex angular frequency
Transmission Line	$\begin{bmatrix} \cosh(\gamma l) & -Z_0 \sinh(\gamma l) \\ -\dfrac{1}{Z_0}\sinh(\gamma l) & \cosh(\gamma l) \end{bmatrix}$	$\begin{bmatrix} \cosh(\gamma l) & Z_0 \sinh(\gamma l) \\ \dfrac{1}{Z_0}\sinh(\gamma l) & \cosh(\gamma l) \end{bmatrix}$	$Z0$ = Characteristic impedance γ = Propagation constant l = Length of transmission line (m)

Scattering Parameters (S-Parameters)

The previous parameters are all defined in terms of voltages and currents at ports. S-parameters are different, and are defined in terms of incident and reflected waves at ports. S-parameters are used primarily at UHF and microwave frequencies where it becomes difficult to measure voltages and currents directly. On the other hand, incident and reflected power are easy to measure using directional couplers. The definition is,

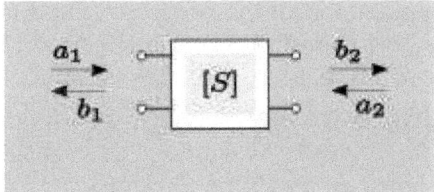

Fig. : Terminology of waves used in S-parameter definition.

$$\begin{bmatrix} b_1 \\ b_2 \end{bmatrix} = \begin{bmatrix} S_{11} & S_{12} \\ S_{21} & S_{22} \end{bmatrix} \begin{bmatrix} a_1 \\ a_2 \end{bmatrix}$$

where the a_k are the incident waves and the b_k are the reflected waves at port k. It is conventional to define the a_k and b_k in terms of the square root of power. Consequently, there is a relationship with the wave voltages.

For reciprocal networks $S_{12} = S_{21}$. For symmetrical networks $S_{11} = S_{22}$. For antimetrical networks $S_{11} = -S_{22}$. For lossless reciprocal networks $|S_{11}| = |S_{22}|$ and $|S_{11}|^2 + |S_{12}|^2 = 1$.

Scattering Transfer Parameters (T-Parameters)

Scattering transfer parameters, like scattering parameters, are defined in terms of incident and reflected waves. The difference is that T-parameters relate the waves at port 1 to the waves at port 2 whereas S-parameters relate the reflected waves to the incident waves. In this respect T-parameters fill the same role as $ABCD$ parameters and allow the T-parameters of cascaded networks to be calculated by matrix multiplication of the component networks. T-parameters, like $ABCD$ parameters, can also be called transmission parameters. The definition is,

$$\begin{bmatrix} a_1 \\ b_1 \end{bmatrix} = \begin{bmatrix} T_{11} & T_{12} \\ T_{21} & T_{22} \end{bmatrix} \begin{bmatrix} b_2 \\ a_2 \end{bmatrix}$$

T-parameters are not so easy to measure directly unlike S-parameters. However, S-parameters are easily converted to T-parameters.

Combinations of Two-port Networks

When two or more two-port networks are connected, the two-port parameters of the combined network can be found by performing matrix algebra on the matrices of parameters for the component two-ports. The matrix operation can be made particularly simple with an appropriate choice of two-port parameters to match the form of connection of the two-ports. For instance, the z-parameters are best for series connected ports.

The combination rules need to be applied with care. Some connections (when dissimilar potentials are joined) result in the port condition being invalidated and the combination rule will no longer apply. This difficulty can be overcome by placing 1 : 1 ideal transformers on the outputs of the problem two-ports. This does not change the parameters of the two-ports, but does ensure that they will continue to meet the port condition when interconnected. An example of this problem is shown for series-series connections in figures 11 and 12 below.

Series-series Connection

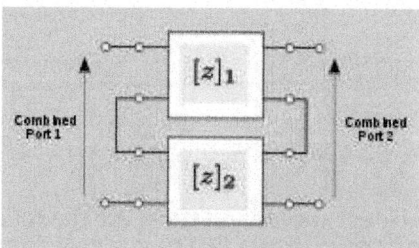

Fig. : Two two-port networks with input ports connected in series and output ports connected in series.

When two-ports are connected in a series-series configuration as shown in figure, the best choice of two-port parameter is the z-parameters. The z-parameters of the combined network are found by matrix addition of the two individual z-parameter matrices.

$$[z] = [z]_1 + [z]_2$$

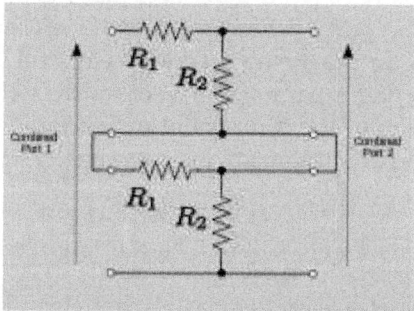

Fig. : Example of an improper connection of two-ports. R_1 of the lower two-port has been by-passed by a short circuit.

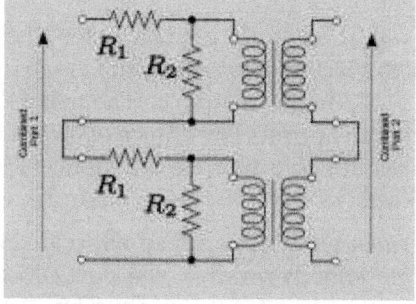

Fig. : Use of ideal transformers to restore the port condition to interconnected networks.

As mentioned above, there are some networks which will not yield directly to this analysis. A simple example is a two-port consisting of a L-network of resistors R_1 and R_2. The z-parameters for this network are;

$$[z]_1 = \begin{bmatrix} R_1 + R_2 & R_2 \\ R_2 & R_2 \end{bmatrix}$$

Figure shows two identical such networks connected in series-series. The total z-parameters predicted by matrix addition are;

$$[z] = [z]_1 + [z]_2 = 2[z]_1 = \begin{bmatrix} 2R_1 + 2R_2 & 2R_2 \\ 2R_2 & 2R_2 \end{bmatrix}$$

However, direct analysis of the combined circuit shows that,

$$[z] = \begin{bmatrix} R_1 + 2R_2 & 2R_2 \\ 2R_2 & 2R_2 \end{bmatrix}$$

The discrepancy is explained by observing that R_1 of the lower two-port has been by-passed by the short-circuit between two terminals of the output ports. This results in no current flowing through one terminal in each of the input ports of the two individual networks. Consequently, the port condition is broken for both the input ports of the original networks since current is still able to flow into the other terminal. This problem can be resolved by inserting an ideal transformer in the output port of at least one of the two-port networks. While this is a common text-book approach to presenting the theory of two-ports, the practicality of using transformers is a matter to be decided for each individual design.

Parallel-parallel Connection

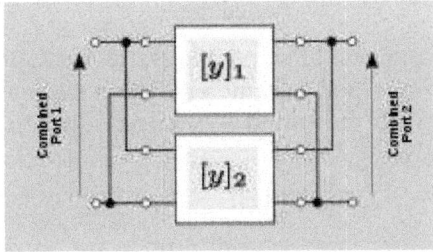

Fig. : Two two-port networks with input ports connected in parallel and output ports connected in parallel.

When two-ports are connected in a parallel-parallel configuration as shown in figure, the best choice of two-port parameter is the y-parameters. The y-parameters of the combined network are found by matrix addition of the two individual y-parameter matrices.

$$[y] = [y]_1 + [y]_2$$

Series-parallel Connection

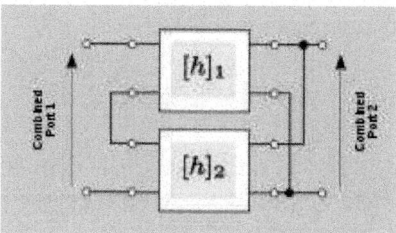

Fig. : Two two-port networks with input ports connected in series and output ports connected in parallel.

When two-ports are connected in a series-parallel configuration as shown in figure, the best choice of two-port parameter is the h-parameters. The h-parameters of the combined network are found by matrix addition of the two individual h-parameter matrices.

$$[h] = [h]_1 + [h]_2$$

Parallel-series Connection

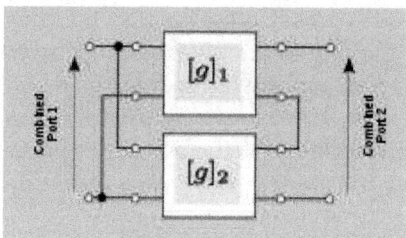

Fig. : Two two-port networks with input ports connected in parallel and output ports connected in series.

When two-ports are connected in a parallel-series configuration as shown in figure, the best choice of two-port parameter is the g-parameters. The g-parameters of the combined network are found by matrix addition of the two individual g-parameter matrices.

$$[g] = [g]_1 + [g]_2$$

Cascade Connection

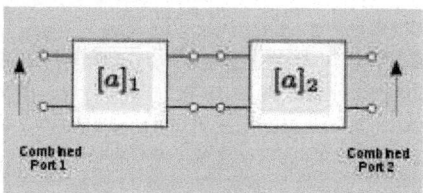

Fig. : Two two-port networks with the first's output port connected to the second's input port.

When two-ports are connected with the output port of the first connected to the input port of the second (a cascade connection) as shown in figure, the best choice of two-port parameter is the $ABCD$-parameters. The a-parameters of the combined network are found by matrix multiplication of the two individual a-parameter matrices.

$$[a] = [a]_1 \cdot [a]_2$$

A chain of n two-ports may be combined by matrix multiplication of the n matrices. To combine a cascade of b-parameter matrices, they are again multiplied, but the multiplication must be carried out in reverse order, so that;

$$[b] = [b]_2 \cdot [b]_1$$

Example : Suppose we have a two-port network consisting of a series resistor R followed by a shunt capacitor C. We can model the entire network as a cascade of two simpler networks :

$$[b]_1 = \begin{bmatrix} 1 & -R \\ 0 & 1 \end{bmatrix}$$

$$[b]_2 = \begin{bmatrix} 1 & 0 \\ -sC & 1 \end{bmatrix}$$

The transmission matrix for the entire network $[b]$ is simply the matrix multiplication of the transmission matrices for the two network elements :

$$[b] = [b]_2 \cdot [b]_1$$

$$= \begin{bmatrix} 1 & 0 \\ -sC & 1 \end{bmatrix}\begin{bmatrix} 1 & -R \\ 0 & 1 \end{bmatrix}$$

$$= \begin{bmatrix} 1 & -R \\ -sC & 1 + sCR \end{bmatrix}$$

Thus :

$$\begin{bmatrix} V_2 \\ I'_2 \end{bmatrix} = \begin{bmatrix} 1 & -R \\ -sC & 1 + sCR \end{bmatrix}\begin{bmatrix} V_1 \\ I_1 \end{bmatrix}$$

Interrelation of Parameters

	[z]	[y]	[h]	[g]	[a]	[b]
[z]	$\begin{bmatrix} z_{11} & z_{12} \\ z_{21} & z_{22} \end{bmatrix}$	$\begin{bmatrix} \frac{y_{22}}{\Delta[y]} & \frac{-y_{12}}{\Delta[y]} \\ \frac{-y_{21}}{\Delta[y]} & \frac{y_{11}}{\Delta[y]} \end{bmatrix}$	$\begin{bmatrix} \frac{\Delta[h]}{h_{22}} & \frac{h_{12}}{h_{22}} \\ \frac{-h_{21}}{h_{22}} & \frac{1}{h_{22}} \end{bmatrix}$	$\begin{bmatrix} \frac{1}{g_{11}} & \frac{-g_{12}}{g_{11}} \\ \frac{g_{21}}{g_{11}} & \frac{\Delta[g]}{g_{11}} \end{bmatrix}$	$\begin{bmatrix} \frac{a_{11}}{a_{21}} & \frac{\Delta[a]}{a_{21}} \\ \frac{1}{a_{21}} & \frac{a_{22}}{a_{21}} \end{bmatrix}$	$\begin{bmatrix} \frac{-b_{22}}{b_{21}} & \frac{-1}{b_{21}} \\ \frac{-\Delta[b]}{b_{21}} & \frac{-b_{11}}{b_{21}} \end{bmatrix}$
[y]	$\begin{bmatrix} \frac{z_{22}}{\Delta[z]} & \frac{-z_{12}}{\Delta[z]} \\ \frac{-z_{21}}{\Delta[z]} & \frac{z_{11}}{\Delta[z]} \end{bmatrix}$	$\begin{bmatrix} y_{11} & y_{12} \\ y_{21} & y_{22} \end{bmatrix}$	$\begin{bmatrix} \frac{1}{h_{11}} & \frac{-h_{12}}{h_{11}} \\ \frac{h_{21}}{h_{11}} & \frac{\Delta[h]}{h_{11}} \end{bmatrix}$	$\begin{bmatrix} \frac{\Delta[g]}{g_{22}} & \frac{g_{12}}{g_{22}} \\ \frac{-g_{21}}{g_{22}} & \frac{1}{g_{22}} \end{bmatrix}$	$\begin{bmatrix} \frac{a_{22}}{a_{12}} & \frac{-\Delta[a]}{a_{12}} \\ \frac{-1}{a_{12}} & \frac{a_{11}}{a_{12}} \end{bmatrix}$	$\begin{bmatrix} \frac{-b_{11}}{b_{12}} & \frac{1}{b_{12}} \\ \frac{\Delta[b]}{b_{12}} & \frac{-b_{22}}{b_{12}} \end{bmatrix}$
[h]	$\begin{bmatrix} \frac{\Delta[z]}{z_{22}} & \frac{z_{12}}{z_{22}} \\ \frac{-z_{21}}{z_{22}} & \frac{1}{z_{22}} \end{bmatrix}$	$\begin{bmatrix} \frac{1}{y_{11}} & \frac{-y_{12}}{Y_{11}} \\ \frac{y_{21}}{y_{11}} & \frac{\Delta[y]}{y_{11}} \end{bmatrix}$	$\begin{bmatrix} h_{11} & h_{12} \\ h_{21} & h_{22} \end{bmatrix}$	$\begin{bmatrix} \frac{g_{22}}{\Delta[g]} & \frac{-g_{12}}{\Delta[g]} \\ \frac{-g_{21}}{\Delta[g]} & \frac{g_{11}}{\Delta[g]} \end{bmatrix}$	$\begin{bmatrix} \frac{a_{12}}{a_{22}} & \frac{\Delta[a]}{a_{22}} \\ \frac{-1}{a_{22}} & \frac{a_{21}}{a_{22}} \end{bmatrix}$	$\begin{bmatrix} \frac{-b_{12}}{b_{11}} & \frac{1}{b_{11}} \\ \frac{-\Delta[b]}{b_{11}} & \frac{-b_{21}}{b_{11}} \end{bmatrix}$
[g]	$\begin{bmatrix} \frac{1}{z_{11}} & \frac{-z_{12}}{z_{11}} \\ \frac{z_{21}}{z_{11}} & \frac{\Delta[z]}{z_{11}} \end{bmatrix}$	$\begin{bmatrix} \frac{\Delta[y]}{y_{22}} & \frac{y_{12}}{y_{22}} \\ \frac{-y_{21}}{y_{22}} & \frac{1}{y_{22}} \end{bmatrix}$	$\begin{bmatrix} \frac{h_{22}}{\Delta[h]} & \frac{-h_{12}}{\Delta[h]} \\ \frac{-h_{21}}{\Delta[h]} & \frac{h_{11}}{\Delta[h]} \end{bmatrix}$	$\begin{bmatrix} g_{11} & g_{12} \\ g_{21} & g_{22} \end{bmatrix}$	$\begin{bmatrix} \frac{a_{21}}{a_{11}} & \frac{-\Delta[a]}{a_{11}} \\ \frac{1}{a_{11}} & \frac{a_{12}}{a_{11}} \end{bmatrix}$	$\begin{bmatrix} \frac{-b_{21}}{b_{22}} & \frac{-1}{b_{22}} \\ \frac{\Delta[b]}{b_{22}} & \frac{-b_{12}}{b_{22}} \end{bmatrix}$
[a]	$\begin{bmatrix} \frac{z_{11}}{z_{21}} & \frac{\Delta[z]}{z_{21}} \\ \frac{1}{z_{21}} & \frac{z_{22}}{z_{21}} \end{bmatrix}$	$\begin{bmatrix} \frac{-y_{22}}{y_{21}} & \frac{-1}{y_{21}} \\ \frac{-\Delta[y]}{y_{21}} & \frac{-y_{11}}{y_{21}} \end{bmatrix}$	$\begin{bmatrix} \frac{-\Delta[h]}{h_{21}} & \frac{-h_{11}}{h_{21}} \\ \frac{-h_{22}}{h_{21}} & \frac{-1}{h_{21}} \end{bmatrix}$	$\begin{bmatrix} \frac{1}{g_{21}} & \frac{g_{22}}{g_{21}} \\ \frac{g_{11}}{g_{21}} & \frac{\Delta[g]}{g_{21}} \end{bmatrix}$	$\begin{bmatrix} a_{11} & a_{12} \\ a_{21} & a_{22} \end{bmatrix}$	$\begin{bmatrix} \frac{b_{22}}{\Delta[b]} & \frac{-b_{12}}{\Delta[b]} \\ \frac{-b_{21}}{\Delta[b]} & \frac{b_{11}}{\Delta[b]} \end{bmatrix}$
[b]	$\begin{bmatrix} \frac{z_{22}}{z_{12}} & \frac{-\Delta[z]}{z_{12}} \\ \frac{-1}{z_{12}} & \frac{z_{11}}{z_{12}} \end{bmatrix}$	$\begin{bmatrix} \frac{-y_{11}}{y_{12}} & \frac{1}{y_{12}} \\ \frac{\Delta[y]}{y_{12}} & \frac{-y_{22}}{y_{12}} \end{bmatrix}$	$\begin{bmatrix} \frac{1}{h_{12}} & \frac{-h_{11}}{h_{12}} \\ \frac{-h_{22}}{h_{12}} & \frac{\Delta[h]}{h_{12}} \end{bmatrix}$	$\begin{bmatrix} \frac{-\Delta[g]}{g_{12}} & \frac{g_{22}}{g_{12}} \\ \frac{g_{11}}{g_{12}} & \frac{-1}{g_{12}} \end{bmatrix}$	$\begin{bmatrix} \frac{a_{22}}{\Delta[a]} & \frac{-a_{12}}{\Delta[a]} \\ \frac{-a_{21}}{\Delta[a]} & \frac{a_{11}}{\Delta[a]} \end{bmatrix}$	$\begin{bmatrix} b_{11} & b_{12} \\ b_{21} & b_{22} \end{bmatrix}$

Where $\Delta[x]$ is the determinant of $[x]$.

Certain pairs of matrices have a particularly simple relationship. The admittance parameters are the matrix inverse of the impedance parameters, the inverse hybrid parameters are the matrix inverse of the hybrid parameters, and the [b] form of the ABCD-parameters is the matrix inverse of the [a] form. That is,

$$[y] = [z]^{-1}$$

$$[g] = [h]^{-1}$$

$$[b] = [a]^{-1}$$

Networks with More Than Two Ports

While two port networks are very common (*e.g.*, amplifiers and filters), other electrical networks such as directional couplers and circulators have more than 2 ports. The following representations are also applicable to networks with an arbitrary number of ports :

- Admittance (*y*) parameters
- Impedance (*z*) parameters
- Scattering (*S*) parameters.

For example three-port impedance parameters result in the following relationship :

$$
\begin{bmatrix} V_1 \\ V_2 \\ V_3 \end{bmatrix} = \begin{bmatrix} Z_{11} & Z_{12} & Z_{13} \\ Z_{21} & Z_{22} & Z_{23} \\ Z_{31} & Z_{32} & Z_{33} \end{bmatrix} \begin{bmatrix} I_1 \\ I_2 \\ I_3 \end{bmatrix}
$$

However the following representations are necessarily limited to two-port devices :

- Hybrid (h) parameters
- Inverse hybrid (g) parameters
- Transmission (*ABCD*) parameters
- Scattering transfer (*T*) parameters.

Collapsing a Two-port to a One Port

A two-port network has four variables with two of them being independent. If one of the ports is terminated by a load with no independent sources, then the load enforces a relationship between the voltage and current of that port. A degree of freedom is lost. The circuit now has only one independent parameter. The two-port becomes a one-port impedance to the remaining independent variable.

For example, consider impedance parameters

$$
\begin{bmatrix} V_1 \\ V_2 \end{bmatrix} = \begin{bmatrix} z_{11} & z_{12} \\ z_{21} & z_{22} \end{bmatrix} \begin{bmatrix} I_1 \\ I_2 \end{bmatrix}
$$

Connecting a load, Z_L onto port 2 effectively adds the constraint

$$
V_2 = - Z_L I_2
$$

The negative sign is because the positive direction for I2 is directed into the two-port instead of into the load. The augmented equations become

$$
V_1 = Z_{11} I_1 + Z_{12} I_2
$$

$$
- Z_L I_2 = Z_{21} I_1 + Z_{22} I_2
$$

The second equation can be easily solved for I_2 as a function of I_1 and that expression can replace I_2 in the first equation leaving V_1 (and V_2 and I_2) as functions of I_1

$$I_2 = -\frac{Z_{21}}{Z_L + Z_{22}} I_1$$

$$V_1 = Z_{11}I_1 - \frac{Z_{12} Z_{21}}{Z_L + Z_{22}} I_1$$

$$= \left(Z_{11} - \frac{Z_{12} Z_{21}}{Z_L + Z_{22}} \right) I_1 = Z_{in} I_1$$

So, in effect, I_1 sees an input impedance Z_{in} and the two-port's effect on the input circuit has been effectively collapsed down to a one-port; *i.e.*, a simple two terminal impedance.

Image Impedance

Image impedance is a concept used in electronic network design and analysis and most especially in filter design. The term image impedance applies to the impedance seen looking into the ports of a network. Usually a two-port network is implied but the concept is capable of being extended to networks with more than two ports. The definition of image impedance for a two-port network is the impedance, Z_{i1}, seen looking into port 1 when port 2 is terminated with the image impedance, Z_{i2}, for port 2. In general, the image impedances of ports 1 and 2 will not be equal unless the network is symmetrical (or anti-symmetrical) with respect to the ports.

Derivation

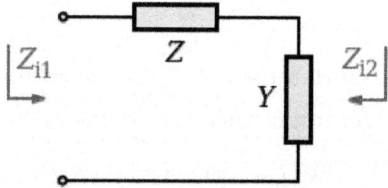

Fig. : Simple 'L' network with series impedance Z and shunt admittance Y. Image impedances Z_{i1} and Z_{i2} are shown.

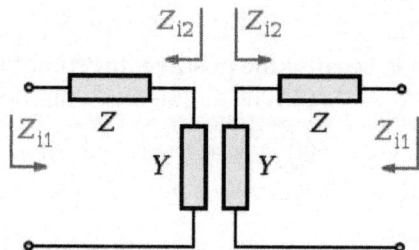

Fig. : Showing how a T section is made from two cascaded L half sections. Z_{i2} is facing Z_{i2} to provide matching impedances.

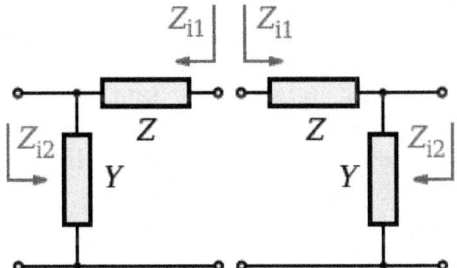

Fig. : Showing how a Π section is made from two cascaded L half sections. Z_{i1} is facing Z_{i1} to provide matching impedances.

The difficulty here is that in order to find Z_{i1} it is first necessary to terminate port 2 with Z_{i2}. However, Z_{i2} is also an unknown at this stage. The problem is solved by terminating port 2 with an identical network : port 2 of the second network is connected to port 2 of the first network and port 1 of the second network is terminated with Z_{i1}. The second network is terminating the first network in Z_{i2} as required. Mathematically, this is equivalent to eliminating one variable from a set of simultaneous equations. The network can now be solved for Z_{i1}. Writing out the expression for input impedance gives;

$$Z_{i1} = Z + \cfrac{1}{2Y + \cfrac{1}{Z + Z_{il}}}$$

and solving for Z_{i1},

$$Z^2{}_{i1} = Z^2 + \frac{Z}{Y}$$

Z_{i2} is found by a similar process, but it is simpler to work in terms of the reciprocal, that is image admittance Y_{i2},

$$Y^2{}_{i2} = Y^2 + \frac{Y}{Z}$$

Also, it can be seen from these expressions that the two image impedances are related to each other by;

$$\frac{Z_{i1}}{Z_{i2}} = \frac{Z}{Y}$$

Usage in Filter Design

When used in filter design, the 'L' network analysed above is usually referred to as a half section. Two half sections in cascade will make either a T section or a Π section depending on which port of the L section comes first. This leads to the terminology of Z_{iT} to mean the Z_{i1} in the above analysis and $Z_{i\Pi}$ to mean Z_{i2}.

Relation to Characteristic Impedance

Image impedance is a similar concept to the characteristic impedance used in the analysis of transmission lines. In fact, in the limiting case of a chain of cascaded networks where the size of each single network is approaching an infinitesimally small element, the mathematical limit of the image impedance expression is the characteristic impedance of the chain. That is,

$$Z^2_i \rightarrow \frac{Z}{Y}$$

The connection between the two can further be seen by noting an alternative, but equivalent, definition of image impedance. In this definition, the image impedance of a network is the input impedance of an infinitely long chain of cascaded identical networks (with the ports arranged so that like impedance faces like). This is directly analogous to the definition of characteristic impedance as the input impedance of an infinitely long line.

Conversely, it is possible to analyse a transmission line with lumped components, such as one utilising loading coils, in terms of an image impedance filter.

Transfer Function

The transfer function of the half section, like the image impedance, is calculated for a network terminated in its image impedances (or equivalently, for a single section in an infinitely long chain of identical sections) and is given by,

$$A(i\omega) = \sqrt{\frac{Z_{I2}}{Z_{I1}}}e^{-\gamma}$$

where γ is called the transmission function, propagation function or transmission parameter and is given by,

$$\gamma = \sinh^{-1} \sqrt{ZY}$$

The $\sqrt{\frac{Z_{I2}}{Z_{I1}}}$ term represents the voltage ratio that would be observed if the maximum available power was transferred from the source to the load. It would be possible to absorb this term into the definition of γ, and in some treatments this approach is taken. In the case of a network with symmetrical image impedances, such as a chain of an even number of identical L sections, the expression reduces to,

$$A(i\omega) = e^{-\gamma}$$

In general, γ is a complex number such that,

$$\gamma = \alpha + i\beta$$

The real part of γ, represents an attenuation parameter, a in nepers and the imaginary part represents a phase change parameter, β in radians. The transmission parameters for a chain of n half sections, provided that like impedance always faces like, is given by;

$$\gamma_n = n\gamma$$

As with the image impedance, the transmission parameters approach those of a transmission line as the filter section become infinitesimally small so that,

$$\gamma \to \sqrt{ZY}$$

with a, β, γ, Z, and Y all now being measured per metre instead of per half section.

Relationship to Two-port Network Parameters

ABCD Parameters

For a reciprocal network ($AD-BC=1$), the image impedances can be expressed in terms of ABCD parameters as,

$$Z_{l1} = \sqrt{\frac{AB}{CD}}$$

$$Z_{l2} = \sqrt{\frac{DB}{CA}}.$$

The image propagation term, γ may be expressed as,

$$\gamma = \cosh^{-1} \sqrt{AD}.$$

Note that the image propagation term for a transmission line segment is equivalent to the propagation constant of the transmission line times the length.

NON-LINEAR NETWORKS

Most electronic designs are, in reality, non-linear. There is very little that does not include some semi-conductor devices. These are invariably non-linear, the transfer function of an ideal semi-conductor pn junction is given by the very non-linear relationship;

$$i = I_0 \left(e^{\frac{v}{V_T}} - 1 \right)$$

where;

- i and v are the instantaneous current and voltage.
- I_o is an arbitrary parameter called the reverse leakage current whose value depends on the construction of the device.
- V_T is a parameter proportional to temperature called the thermal voltage and equal to about 25mV at room temperature.

There are many other ways that non-linearity can appear in a network. All methods utilising linear superposition will fail when non-linear components are present. There are several options for dealing with non-linearity depending on the type of circuit and the information the analyst wishes to obtain.

Constitutive Equations

The diode equation above is an example of an element constitutive equation of the general form,

$$f(v, i) = 0$$

This can be thought of as a non-linear resistor. The corresponding constitutive equations for non-linear inductors and capacitors are respectively;

$$f(v, \varphi) = 0$$

$$f(v, q) = 0$$

where f is any arbitrary function, φ is the stored magnetic flux and q is the stored charge.

Existence, Uniqueness and Stability

An important consideration in non-linear analysis is the question of uniqueness. For a network composed of linear components there will always be one, and only one, unique solution for a given set of boundary conditions. This is not always the case in non-linear circuits. For instance, a linear resistor with a fixed voltage applied to it has only one solution for the current through it. On the other hand, the non-linear tunnel diode has up to three solutions for the current for a given voltage. That is, a particular solution for the current through the diode is not unique, there may be others, equally valid. In some cases there may not be a solution at all : the question of existence of solutions must be considered.

Another important consideration is the question of stability. A particular solution may exist, but it may not be stable, rapidly departing from that point at the slightest stimulation. It can be shown that a network that is absolutely stable for all conditions must have one, and only one, solution for each set of conditions.

Methods

Boolean Analysis of Switching Networks

A switching device is one where the non-linearity is utilised to produce two opposite states. CMOS devices in digital circuits, for instance, have their output connected to either the positive or the negative supply rail and are never found at anything in between except during a transient period when the device is actually switching. Here the non-linearity is designed to be extreme, and the analyst can actually take advantage of that fact. These kinds of networks can be analysed using Boolean algebra by assigning the two states ("on"/"off", "positive"/"negative" or whatever states are being used) to the boolean constants "0" and "1".

The transients are ignored in this analysis, along with any slight discrepancy between the actual state of the device and the nominal state assigned to a boolean value. For instance, boolean "1" may be assigned to the state of +5V. The output of the device may actually be +4.5V but the analyst still considers this to be boolean "1". Device manufacturers will usually specify a range of values in their data sheets that are to be considered undefined (*i.e.* the result will be unpredictable).

The transients are not entirely uninteresting to the analyst. The maximum rate of switching is determined by the speed of transition from one state to the other. Happily for the analyst, for many devices most of the transition occurs in the linear portion of the devices transfer function and linear analysis can be applied to obtain at least an approximate answer.

It is mathematically possible to derive boolean algebras which have more than two states. There is not too much use found for these in electronics, although three-state devices are passingly common.

Separation of Bias and Signal Analyses

This technique is used where the operation of the circuit is to be essentially linear, but the devices used to implement it are non-linear. A transistor amplifier is an example of this kind of network. The essence of this technique is to separate the analysis into two parts. Firstly, the dc biases are analysed using some non-linear method. This establishes the quiescent operating point of the circuit. Secondly, the small signal characteristics of the circuit are analysed using linear network analysis.

Graphical Method of dc Analysis

In a great many circuit designs, the dc bias is fed to a non-linear component via a resistor (or possibly a network of resistors). Since resistors are linear components, it is particularly easy to determine the quiescent operating point of the non-linear device from a graph of its transfer function. The method is as follows : from linear network analysis the output transfer function (that is output voltage against output current) is calculated for the network of resistor(s) and the generator driving them. This will be a straight line and can readily be superimposed on the transfer function plot of the non-linear device. The point where the lines cross is the quiescent operating point.

Perhaps the easiest practical method is to calculate the (linear) network open circuit voltage and short circuit current and plot these on the transfer function of the non-linear device. The straight line joining these two point is the transfer function of the network.

In reality, the designer of the circuit would proceed in the reverse direction to that described. Starting from a plot provided in the manufacturers data sheet for the non-linear device, the designer would choose the desired operating point and then calculate the linear component values required to achieve it.

It is still possible to use this method if the device being biased has its bias fed through another device which is itself non-linear – a diode for instance. In this case however, the plot of the network transfer function onto the device being biased would no longer be a straight line and is consequently more tedious to do.

Small Signal Equivalent Circuit

This method can be used where the deviation of the input and output signals in a network stay within a substantially linear portion of the non-linear devices transfer function, or else are so small that the curve of the transfer function can be considered linear. Under a set of these specific conditions, the non-linear device can be represented by an equivalent linear network. It must be remembered that this equivalent circuit is entirely notional and only valid for the small signal deviations. It is entirely inapplicable to the dc biasing of the device.

For a simple two-terminal device, the small signal equivalent circuit may be no more than two components. A resistance equal to the slope of the v/i curve at the operating point (called the dynamic resistance), and tangent to the curve. A generator, because this tangent will not, in general, pass through the origin. With more terminals, more complicated equivalent circuits are required.

A popular form of specifying the small signal equivalent circuit amongst transistor manufacturers is to use the two-port network parameters known as [h] parameters. These are a matrix of four parameters as with the [z] parameters but in the case of the [h] parameters they are a hybrid mixture of impedances, admittances, current gains and voltage gains. In this model the three terminal transistor is considered to be a two port network, one of its terminals being common to both ports. The [h] parameters are quite different depending on which terminal is chosen as the common one. The most important parameter for transistors is usually the forward current gain, h_{21}, in the common emitter configuration. This is designated h_{fe} on data sheets.

The small signal equivalent circuit in terms of two-port parameters leads to the concept of dependent generators. That is, the value of a voltage or current generator depends linearly on a voltage or current elsewhere in the circuit. For instance, the [z] parameter model leads to dependent voltage generators as shown in this diagram;

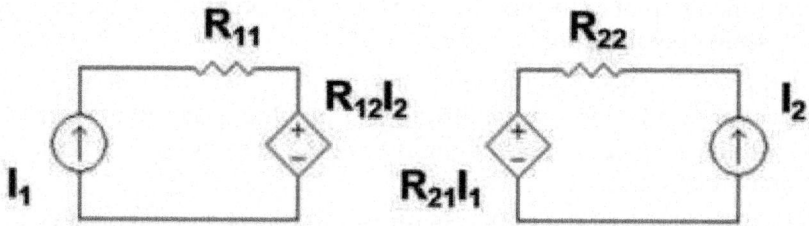

Fig. : [z] parameter equivalent circuit showing dependent voltage generators.

There will always be dependent generators in a two-port parameter equivalent circuit. This applies to the [h] parameters as well as to the [z] and any other kind. These dependencies must be preserved when developing the equations in a larger linear network analysis.

Piecewise Linear Method

In this method, the transfer function of the non-linear device is broken up into regions. Each of these regions is approximated by a straight line. Thus, the transfer function will be linear up to a particular point where there will be a discontinuity. Past this point the transfer function will again be linear but with a different slope.

A well known application of this method is the approximation of the transfer function of a pn junction diode. The actual transfer function of an ideal diode has been given at the top of this (non-linear) section. However, this formula is rarely used in network analysis, a piecewise approximation being used instead. It can be seen that the diode current rapidly diminishes to $-I_o$ as the voltage falls. This current, for most purposes, is so small it can be ignored. With increasing voltage, the current increases exponentially. The diode is modelled as an open circuit up to the knee of the exponential curve, then past this point as a resistor equal to the bulk resistance of the semi-conducting material.

The commonly accepted values for the transition point voltage are 0.7V for silicon devices and 0.3V for germanium devices. An even simpler model of the diode, sometimes used in switching applications, is short circuit for forward voltages and open circuit for reverse voltages.

The model of a forward biased pn junction having an approximately constant 0.7V is also a much used approximation for transistor base-emitter junction voltage in amplifier design.

The piecewise method is similar to the small signal method in that linear network analysis techniques can only be applied if the signal stays within certain bounds. If the signal crosses a discontinuity point then the model is no longer valid for linear analysis purposes. The model does have the advantage over small signal however, in that it is equally applicable to signal and dc bias. These can therefore both be analysed in the same operations and will be linearly superimposable.

Time-varying Components

In linear analysis, the components of the network are assumed to be unchanging, but in some circuits this does not apply, such as sweep oscillators, voltage controlled amplifiers, and variable equalisers. In many circumstances the change in component value is periodic. A non-linear component excited with a periodic signal, for instance, can be represented as periodically varying *linear* component. Sidney Darlington disclosed a method of analysing such periodic time varying circuits. He developed canonical circuit forms which are analogous to the canonical forms of Ronald Foster and Wilhelm Cauer used for analysing linear circuits.

INTRODUCTION TO NETWORK THEOREMS

Anyone who's studied geometry should be familiar with the concept of a *theorem*: a relatively simple rule used to solve a problem, derived from a more intensive analysis using fundamental rules of mathematics. At least hypothetically, any problem in math can be solved just by using the simple rules of arithmetic (in fact, this is how modern digital computers carry out the most complex mathematical calculations : by repeating many cycles of additions and subtractions !), but human beings aren't as consistent or as fast as a digital computer. We need "shortcut" methods in order to avoid procedural errors.

In electric network analysis, the fundamental rules are Ohm's Law and Kirchhoff's Laws. While these humble laws may be applied to analyze just about any circuit configuration (even if we have to resort to complex algebra to handle multiple unknowns), there are some "shortcut" methods of analysis to make the math easier for the average human.

As with any theorem of geometry or algebra, these network theorems are derived from fundamental rules.

Millman's Theorem

In Millman's Theorem, the circuit is re-drawn as a parallel network of branches, each branch containing a resistor or series battery/resistor combination. Millman's Theorem is applicable only to those circuits which can be re-drawn accordingly. Here again is our example circuit used for the last two analysis methods :

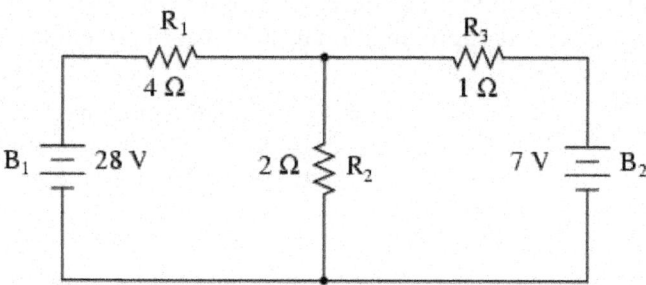

And here is that same circuit, re-drawn for the sake of applying Millman's Theorem :

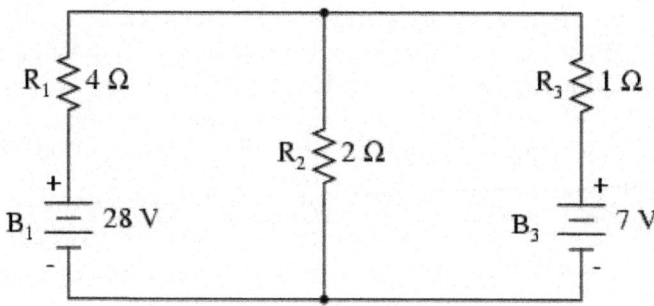

By considering the supply voltage within each branch and the resistance within each branch, Millman's Theorem will tell us the voltage across all branches.

Please note that I've labelled the battery in the rightmost branch as "B_3" to clearly denote it as being in the third branch, even though there is no "B_2" in the circuit!

Millman's Theorem is nothing more than a long equation, applied to any circuit drawn as a set of parallel-connected branches, each branch with its own voltage source and series resistance :

Millman's Theorem Equation

$$\frac{\dfrac{E_{B1}}{R_1} + \dfrac{E_{B2}}{R_2} + \dfrac{E_{B3}}{R_3}}{\dfrac{1}{R_1} + \dfrac{1}{R_2} + \dfrac{1}{R_3}} = \text{Voltage across all branches}$$

Substituting actual voltage and resistance figures from our example circuit for the variable terms of this equation, we get the following expression :

$$\frac{\dfrac{28\,V}{4\,\Omega} + \dfrac{0\,V}{2\,\Omega} + \dfrac{7\,V}{1\,\Omega}}{\dfrac{1}{4\,\Omega} + \dfrac{1}{2\,\Omega} + \dfrac{1}{1\,\Omega}} = 8\,V$$

The final answer of 8 volts is the voltage seen across all parallel branches, like this :

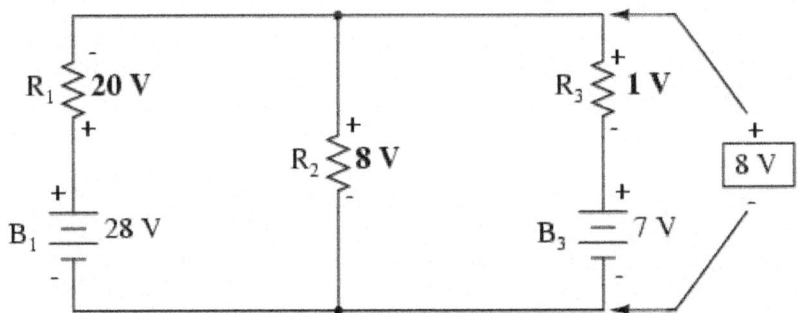

The polarity of all voltages in Millman's Theorem are referenced to the same point. In the example circuit above, I used the bottom wire of the parallel circuit as my reference point, and so the voltages within each branch (28 for the R_1 branch, 0 for the R_2 branch, and 7 for the R_3 branch) were inserted into the equation as positive numbers. Likewise, when the answer came out to 8 volts (positive), this meant that the top wire of the circuit was positive with respect to the bottom wire (the original point of reference). If both batteries had been connected backwards (negative ends up and positive ends down), the voltage for branch 1 would have been entered into the equation as a -28 volts, the voltage for branch 3 as -7 volts, and the resulting answer of -8 volts would have told us that the top wire was negative with respect to the bottom wire (our initial point of reference).

To solve for resistor voltage drops, the Millman voltage (across the parallel network) must be compared against the voltage source within each branch, using the principle of voltages adding in series to determine the magnitude and polarity of voltage across each resistor :

$$E_{R1} = 8\ V - 28\ V = -20\ V\ \text{(negative in top)}$$
$$E_{R2} = 8\ V - 0\ V = 8\ V\ \text{(positive on top)}$$
$$E_{R3} = 8\ V - 7\ V = 1\ V\ \text{(positive on top)}$$

To solve for branch currents, each resistor voltage drop can be divided by its respective resistance (I=E/R) :

$$I_{R1} = \frac{20\ V}{4\ \Omega} = 5\ A$$

$$I_{R2} = \frac{8\ V}{2\ \Omega} = 4\ A$$

$$I_{R3} = \frac{1\ V}{1\ \Omega} = 1\ A$$

The direction of current through each resistor is determined by the polarity across each resistor, *not* by the polarity across each battery, as current can be forced backwards through a battery, as is the case with B_3 in the example circuit. This is important to keep in mind, since Millman's Theorem doesn't provide as direct an indication of "wrong" current direction as does the Branch Current or Mesh Current methods. You must pay close attention to the polarities of resistor voltage drops as given by Kirchhoff's Voltage Law, determining direction of currents from that.

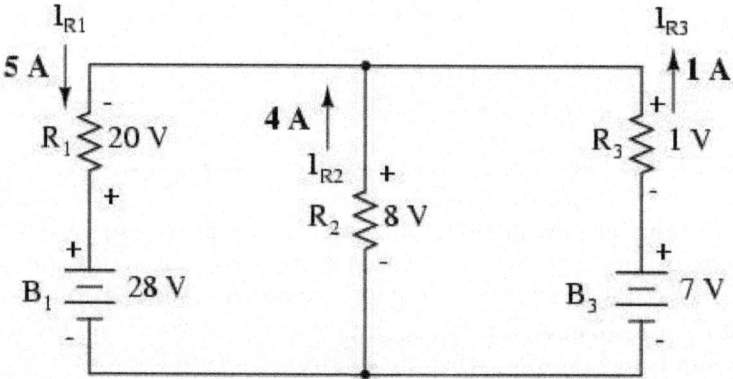

Millman's Theorem is very convenient for determining the voltage across a set of parallel branches, where there are enough voltage sources present to preclude solution via regular series-parallel reduction method. It also is easy in the sense that it doesn't require the use of simultaneous equations. However, it is limited in that it only applied to circuits which can be re-drawn to fit this form. It cannot be used, for example, to solve an unbalanced bridge circuit. And, even in cases where Millman's Theorem can be applied, the solution of individual resistor voltage drops can be a bit daunting to some, the Millman's Theorem equation only providing a single figure for branch voltage.

As you will see, each network analysis method has its own advantages and disadvantages. Each method is a tool, and there is no tool that is perfect for all jobs. The skilled technician, however, carries these methods in his or her mind like a mechanic carries a set of tools in his or her tool box. The more tools you have equipped yourself with, the better prepared you will be for any eventuality.

-Y and Y- Conversions

In many circuit applications, we encounter components connected together in one of two ways to form a three-terminal network : the "Delta," or Δ (also known as the "Pi," or π) configuration, and the "Y" (also known as the "T") configuration.

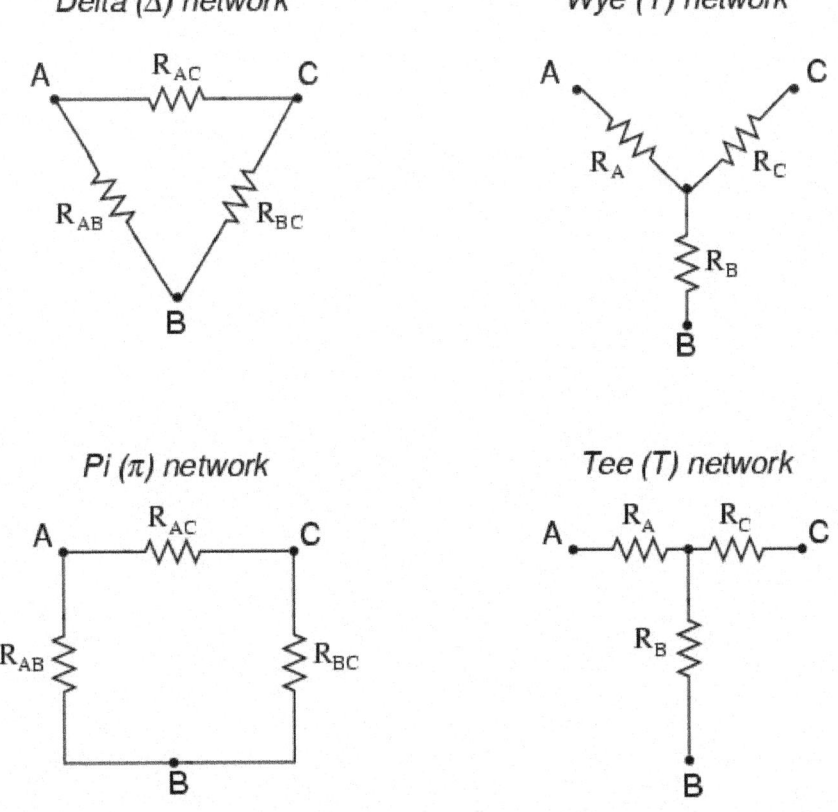

It is possible to calculate the proper values of resistors necessary to form one kind of network (Δ or Y) that behaves identically to the other kind, as analyzed from the terminal connections alone. That is, if we had two separate resistor networks, one Δ and one Y, each with its resistors hidden from view, with nothing but the three terminals (A, B, and C) exposed for testing, the resistors could be sized for the two networks so that there would be no way to electrically determine one network apart from the other. In other words, equivalent Δ and Y networks behave identically.

There are several equations used to convert one network to the other :

To convert a Delta (Δ) to a Wye (Y)	*To convert a Wye (Y) to a Delta (Δ)*
$$R_A = \dfrac{R_{AB}\,R_{AC}}{R_{AB} + R_{AC} + R_{BC}}$$	$$R_{AB} = \dfrac{R_A R_B + R_A R_C + R_B R_C}{R_C}$$
$$R_B = \dfrac{R_{AB}\,R_{BC}}{R_{AB} + R_{AC} + R_{BC}}$$	$$R_{BC} = \dfrac{R_A R_B + R_A R_C + R_B R_C}{R_A}$$
$$R_C = \dfrac{R_{AC}\,R_{BC}}{R_{AB} + R_{AC} + R_{BC}}$$	$$R_{AC} = \dfrac{R_A R_B + R_A R_C + R_B R_C}{R_B}$$

Δ and Y networks are seen frequently in 3-phase AC power systems (a topic covered in volume II of this book series), but even then they're usually balanced networks (all resistors equal in value) and conversion from one to the other need not involve such complex calculations. When would the average technician ever need to use these equations?

A prime application for Δ-Y conversion is in the solution of unbalanced bridge circuits :

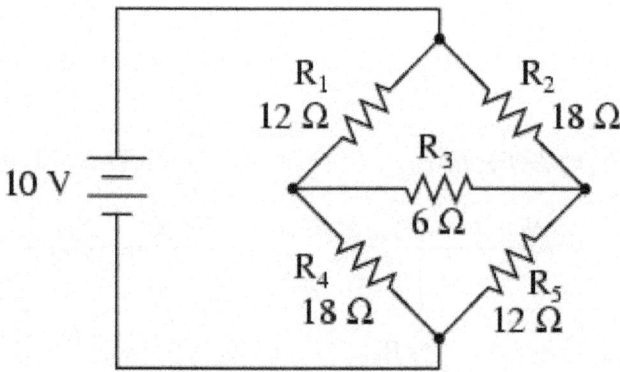

Solution of this circuit with Branch Current or Mesh Current analysis is fairly involved, and neither the Millman nor Superposition Theorems are of any help, since there's only one source of power. We could use Thevenin's or Norton's Theorem, treating R_3 as our load, but what fun would that be?

If we were to treat resistors R_1, R_2, and R_3 as being connected in a Δ configuration (R_{ab}, R_{ac}, and R_{bc}, respectively) and generate an equivalent Y network to replace them, we could turn this bridge circuit into a (simpler) series/parallel combination circuit :

Selecting Delta (Δ) network to convert:

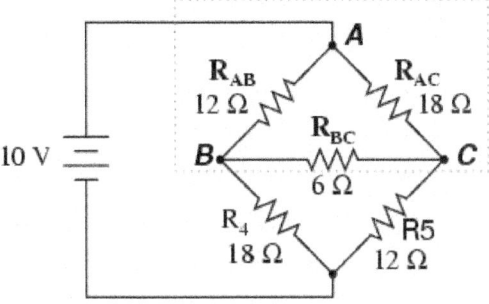

After the Δ-Y conversion . . .

Δ converted to a Y

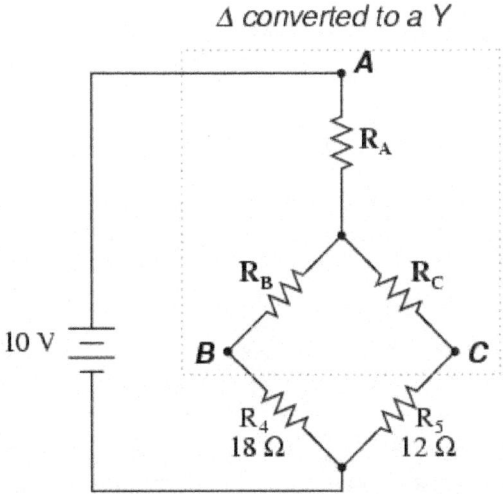

If we perform our calculations correctly, the voltages between points A, B, and C will be the same in the converted circuit as in the original circuit, and we can transfer those values back to the original bridge configuration.

$$R_A = \frac{(12\,\Omega)(18\,\Omega)}{(12\,\Omega)+(18\,\Omega)+(6\,\Omega)} = \frac{216}{36} = 6\,\Omega$$

$$R_B = \frac{(12\,\Omega)(6\,\Omega)}{(12\,\Omega)+(18\,\Omega)+(6\,\Omega)} = \frac{72}{36} = 2\,\Omega$$

$$R_C = \frac{(18\,\Omega)(6\,\Omega)}{(12\,\Omega)+(18\,\Omega)+(6\,\Omega)} = \frac{72}{36} = 3\,\Omega$$

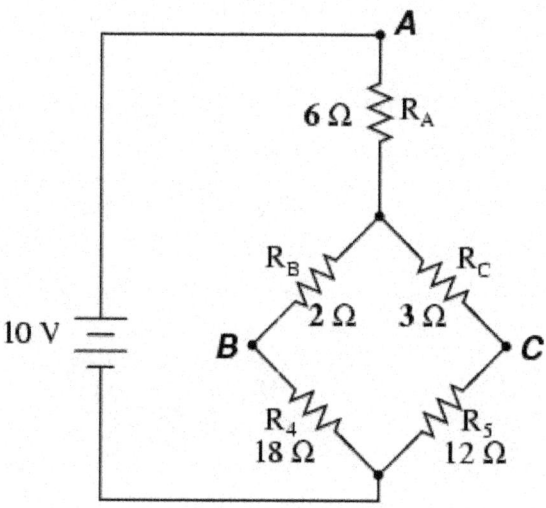

Resistors R_4 and R_5, of course, remain the same at 18 Ω and 12 Ω, respectively. Analyzing the circuit now as a series/parallel combination, we arrive at the following figures :

	R_A	R_B	R_C	R_4	R_5	
E	4.118	588.24m	1.176	5.294	4.706	Volts
I	686.27m	294.12m	392.16m	294.12m	392.16m	Amps
R	6	2	3	18	12	Ohms

	$R_B + R_4$	$R_C + R_5$	$R_B + R_4$ // $R_C + R_5$	Total	
E	5.882	5.882	5.882	10	Volts
I	294.12m	392.16m	686.27m	686.27m	Amps
R	20	15	8.571	14.571	Ohms

We must use the voltage drops figures from the table above to determine the voltages between points A, B, and C, seeing how the add up (or subtract, as is the case with voltage between points B and C) :

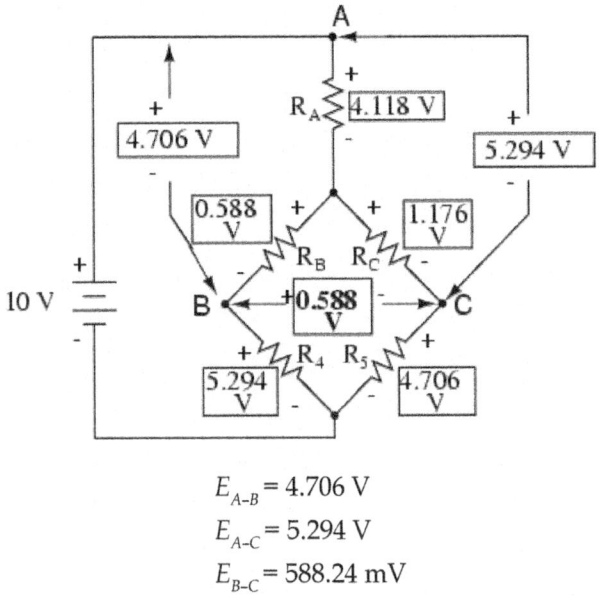

$$E_{A-B} = 4.706 \text{ V}$$
$$E_{A-C} = 5.294 \text{ V}$$
$$E_{B-C} = 588.24 \text{ mV}$$

Now that we know these voltages, we can transfer them to the same points A, B, and C in the original bridge circuit :

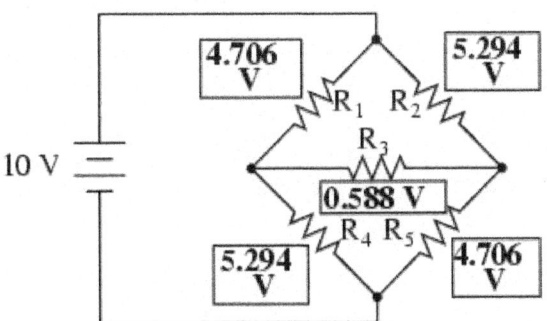

Voltage drops across R_4 and R_5, of course, are exactly the same as they were in the converted circuit.

At this point, we could take these voltages and determine resistor currents through the repeated use of Ohm's Law (I=E/R) :

$$I_{R1} = \frac{4.706 \text{ V}}{12 \text{ } \Omega} = 392.16 \text{ mA}$$

$$I_{R2} = \frac{5.294 \text{ V}}{18 \text{ } \Omega} = 294.12 \text{ mA}$$

$$I_{R3} = \frac{588.24 \text{ mV}}{6 \text{ } \Omega} = 98.04 \text{ mA}$$

$$I_{R4} = \frac{5.294 \text{ V}}{18 \, \Omega} = 294.12 \text{ mA}$$

$$I_{R5} = \frac{4.706 \text{ V}}{12 \, \Omega} = 392.16 \text{ mA}$$

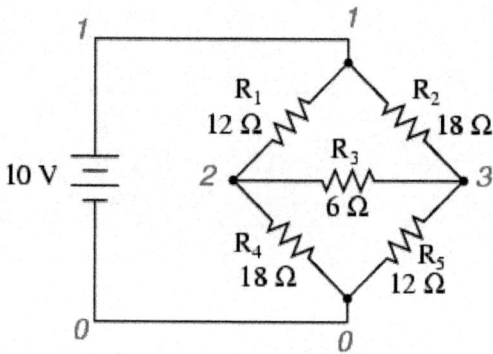

```
unbalanced bridge circuit
v1 1 0
r1 1 2 12
r2 1 3 18
r3 2 3 6
r4 2 0 18
r5 3 0 12
.dc v1 10 10 1
.print dc v(1,2) v(1,3) v(2,3) v(2,0) v(3,0)
.end
```

```
v1          v(1,2)      v(1,3)      v(2,3)      v(2)        v(3)
1.000E+01   4.706E+00   5.294E+00   5.882E-01   5.294E+00   4.706E+00
```

The voltage figures, as read from left to right, represent voltage drops across the five respective resistors, R_1 through R_5. I could have shown currents as well, but since that would have required insertion of "dummy" voltage sources in the SPICE netlist, and since we're primarily interested in validating the Δ-Y conversion equations and not Ohm's Law, this will suffice.

Chapter 3

MAGNETIC CIRCUITS

A **magnetic circuit** is made up of one or more closed loop paths containing a magnetic flux. The flux is usually generated by permanent magnets or electromagnets and confined to the path by magnetic cores-consisting of ferromagnetic materials like iron, although there may be air gaps or other materials in the path. Magnetic circuits are employed to efficiently channel magnetic fields in many devices such as electric motors, generators, transformers, relays, lifting electromagnets, SQUIDs, galvanometers, and magnetic recording heads.

The concept of a "magnetic circuit" exploits a one-to-one correspondence between the equations of the magnetic field in an unsaturated ferromagnetic material to that of an electrical circuit. Using this concept the magnetic fields of complex devices such as transformers can be quickly solved using the methods and techniques developed for electrical circuits.

Some examples of magnetic circuits are :

- horseshoe magnet with iron keeper (low-reluctance circuit)
- horseshoe magnet with no keeper (high-reluctance circuit)
- electric motor (variable-reluctance circuit).

MAGNETOMOTIVE FORCE

In physics, the **magnetomotive force** is a quantity appearing in the equation for the magnetic flux in a magnetic circuit, sometimes known as Hopkinson's law :

$$F = \Phi R,$$

where Φ is the magnetic flux and R is the reluctance of the circuit. It can be seen that the magnetomotive force plays a role in this equation analogous to the voltage V in Ohm's law : $V = IR$.

Magnetomotive force is analogous to electromotive force, emf (= difference in electric potential, or voltage, between the terminals of a source of electricity, *e.g.*, a battery from which no current is being drawn) since it is the cause of magnetic flux in a magnetic circuit; *i.e.,*

1. $\mathcal{F} = NI$

where N is the number of turns in the coil and

     I is the electric current through the circuit

2. $\mathcal{F} = \Phi \mathcal{R}$

where Φ is the magnetic flux and

     $\mathcal{R}$ is the reluctance

3. $\mathcal{F} = Hl$

     where H is the magnetizing force (the strength of the magnetizing field) and l is the mean length of a solenoid or the circumference of a toroid.

MAGNETIC FLUX

In physics, specifically electromagnetism, the magnetic flux (often denoted Φ or ΦB) through a surface is the surface integral of the normal component of the magnetic field B passing through that surface. The SI unit of magnetic flux is the weber (Wb) (in derived units : volt-seconds), and the CGS unit is the maxwell. Magnetic flux is usually measured with a fluxmeter, which contains measuring coils and electronics, that evaluates the change of voltage in the measuring coils to calculate the magnetic flux.

     magnetic interaction is described in terms of a vector field, where each point in space (and time) is associated with a vector that determines what force a moving charge would experience at that point. Since a vector field is quite difficult to visualize at first, in elementary physics one may instead visualize this field with field lines. The magnetic flux through some surface, in this simplified picture, is proportional to the number of field lines passing through that surface (in some contexts, the flux may be defined to be precisely the number of field lines passing through that surface; although technically misleading, this distinction is not important). Note that the magnetic flux is the *net* number of field lines passing through that surface; that is, the number passing through in one direction minus the number passing through in the other direction. In more advanced physics, the field line analogy is dropped and the magnetic flux is properly defined as the surface integral of the normal component of the magnetic field passing through a surface. If the magnetic field is constant, the magnetic flux passing through a surface of vector area S is

$$\Phi_B = \mathbf{B} . \mathbf{S} = BS \cos \theta,$$

where **B** is the magnitude of the magnetic field (the magnetic flux density) having the unit of Wb/m^2 (tesla), **S** is the area of the surface, and θ is the angle between the magnetic field lines and the normal (perpendicular) to **S**. For a varying magnetic field, we first consider the magnetic flux through an infinitesimal area element dS, where we may consider the field to be constant :

$$d\Phi_B = \mathbf{B} \cdot d\mathbf{S}.$$

A generic surface, S, can then be broken into infinitesimal elements and the total magnetic flux through the surface is then the surface integral

$$\Phi_B = \iint_S \mathbf{B} \cdot d\mathbf{S}.$$

From the definition of the magnetic vector potential **A** and the fundamental theorem of the curl the magnetic flux may also be defined as :

$$\Phi_B = \oint_{\partial S} \mathbf{A} \cdot d\ell,$$

where the line integral is taken over the boundary of the surface S, which is denoted ∂S.

Magnetic Flux Through a Closed Surface

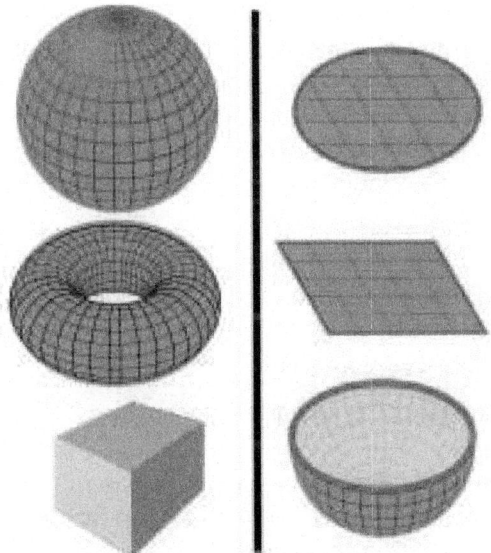

Fig.: Some examples of closed surfaces (left) and open surfaces (right). Left : Surface of a sphere, surface of a torus, surface of a cube. Right : Disk surface, square surface, surface of a hemisphere. (The surface is blue, the boundary is red.)

Gauss's law for magnetism, which is one of the four Maxwell's equations, states that the total magnetic flux through a closed surface is equal to zero. (A "closed surface" is a surface that completely encloses a volume(s) with no holes.) This law is a consequence of the empirical observation that magnetic monopoles have never been found.

In other words, Gauss's law for magnetism is the statement :

$$\Phi_B = \oiint_S \mathbf{B} \cdot d\mathbf{S} = 0$$

for any closed surface S.

Magnetic Flux Through an Open Surface

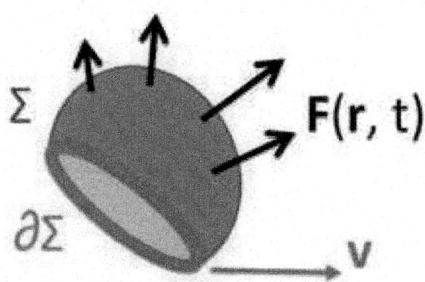

Fig. : For an open surface Σ, the electromotive force along the surface boundary, $\partial\Sigma$, is a combination of the boundary's motion, with velocity v, through a magnetic field B (illustrated by the generic F field in the diagram) and the induced electric field caused by the changing magnetic field.

While the magnetic flux through a closed surface is always zero, the magnetic flux through an open surface need not be zero and is an important quantity in electromagnetism. For example, a change in the magnetic flux passing through a loop of conductive wire will cause an electromotive force, and therefore an electric current, in the loop. The relationship is given by Faraday's law :

$$\varepsilon = \oint_{\partial\Sigma}(\mathbf{E} + v \times \mathbf{B})\cdot d\ell = -\frac{d\Phi_B}{dt},$$

where

    ε is the electromotive force (EMF),

    Φ_B is the magnetic flux through the open surface Σ,

    $\partial\Sigma$ is the boundary of the open surface Σ; note that the surface, in general, may be in motion and deforming, and so is generally a function of time. The electromotive force is induced along this boundary.

    $d\ell$ is an infinitesimal vector element of the contour $\partial\Sigma$,

    v is the velocity of the boundary $\partial\Sigma$,

    E is the electric field,

    B is the magnetic field.

    The two equations for the EMF are, firstly, the work per unit charge done against the Lorentz force in moving a test charge around the (possibly moving) surface boundary $\partial\Sigma$ and, secondly, as the change of magnetic flux through the open surface Σ. This equation is the principle behind an electrical generator.

Comparison with Electric Flux

By way of contrast, Gauss's law for electric fields, another of Maxwell's equations, is

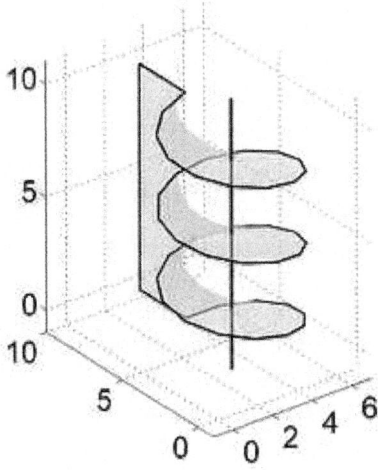

Fig. : Area defined by an electric coil with three turns.

$$\Phi_E = \iint_S \mathbf{E} \cdot d\mathbf{S} = \frac{Q}{\epsilon_0}$$

where :

   E is the electric field,

   S is any closed surface,

   Q is the total electric charge inside the surface S,

   ε_0 is the electric constant (a universal constant, also called the "permittivity of free space").

   Note that the flux of **E** through a closed surface is *not* always zero; this indicates the presence of "electric monopoles", that is, free positive or negative charges.

HOPKINSON'S LAW : THE MAGNETIC ANALOGY TO OHM'S LAW

In electronic circuits, Ohm's law is an empirical relation between the EMF ε applied across an element and the current I it generates through that element. It is written as :

$$\varepsilon = IR$$

where R is the electrical resistance of that material. **Hopkinson's law** is a counterpart to Ohm's law used in magnetic circuits. The law is named after the British electrical engineer, John Hopkinson. It states that

$$F = \phi R_m$$

where F is the magnetomotive force (MMF) across a magnetic element, ϕ is the magnetic flux through the magnetic element, and R_m is the magnetic reluctance of that element. (It shall be shown later that this relationship is due to the empirical relationship between the *H*-field and the magnetic field *B*, $B=\mu H$, where μ is the permeability of the material.) Like Ohm's law, Hopkinson's law can be interpreted either as an empirical equation that works for some materials, or it may serve as a definition of reluctance.

MAGNETIC RELUCTANCE

Magnetic reluctance, or **magnetic resistance**, is a concept used in the analysis of magnetic circuits. It is analogous to resistance in an electrical circuit, but rather than dissipating electric energy it stores magnetic energy. In likeness to the way an electric field causes an electric current to follow the path of least resistance, a magnetic field causes magnetic flux to follow the path of least magnetic reluctance. It is a scalar, extensive quantity, akin to electrical resistance. The units for magnetic reluctance are inverse Henries, H^{-1}.

History

The term was coined in May 1888 by Oliver Heaviside. The notion of "magnetic resistance" was first mentioned by James Joule and the term "magnetomotive force" (MMF) was first named by Bosanquet. The idea for a magnetic flux law, similar to Ohm's law for closed electric circuits, is attributed to H. Rowland.

Reluctance is usually represented by a cursive capital R.

Definition

In a DC field, the reluctance is the ratio of the "magnetomotive force" (MMF) in a magnetic circuit to the magnetic flux in this circuit. In a pulsating DC or AC field, the reluctance is the ratio of the amplitude of the "magnetomotive force" (MMF) in a magnetic circuit to the amplitude of the magnetic flux in this circuit.

The definition can be expressed as follows :

$$R = \frac{F}{\Phi}$$

where :

R("R") is the reluctance in ampere-turns per weber (a unit that is equivalent to turns per henry). "Turns" refers to the winding number of an electrical conductor comprising an inductor.

f("F") is the magnetomotive force (MMF) in ampere-turns

Φ ("Phi") is the magnetic flux in webers.

It is sometimes known as Hopkinson's law and is analogous to Ohm's Law with resistance replaced by reluctance, voltage by MMF and current by magnetic flux.

Magnetic flux always forms a closed loop, as described by Maxwell's equations, but the path of the loop depends on the reluctance of the surrounding materials. It is concentrated around the path of least reluctance. Air and vacuum have high reluctance, while easily magnetized materials such as soft iron have low reluctance. The concentration of flux in low-reluctance materials forms strong temporary poles and causes mechanical forces that tend to move the materials towards regions of higher flux so it is always an attractive force (pull).

The reluctance of a uniform magnetic circuit can be calculated as :

$$R = \frac{l}{\mu_0 \, \mu_r A}$$

or

$$R = \frac{l}{\mu A}$$

where :

l is the length of the circuit in metres

μ_0 is the permeability of vacuum, equal to $4\pi \times 10^{-7}$ henry per metre

μ_r is the relative magnetic permeability of the material (dimensionless)

μ is the permeability of the material ($\mu = \mu_0 \, \mu_r$)

A is the cross-sectional area of the circuit in square metres

The inverse of reluctance is called *permeance*.

$$P = \frac{1}{R}$$

Its SI derived unit is the henry (the same as the unit of inductance, although the two concepts are distinct).

Applications

- Constant air gaps can be created in the core of certain transformers to reduce the effects of saturation. This increases the reluctance of the magnetic circuit, and enables it to store more energy before core saturation. This effect is also used in the flyback transformer.
- Variable air gaps can be created in the cores by a movable keeper to create a Flux Switch that alters the amount of magnetic flux in a magnetic circuit without varying the constant magnetomotive force in that circuit.
- Variation of reluctance is the principle behind the reluctance motor (or the variable reluctance generator) and the Alexanderson alternator. Another way of saying this is that the *reluctance forces* strive for a maximally aligned magnetic circuit and a minimal air gap distance.
- Multimedia loudspeakers are typically shielded magnetically, in order to reduce magnetic interference caused to televisions and other CRTs. The

speaker magnet is covered with a material such as soft iron to minimize the stray magnetic field.

Reluctance can also be applied to :

- Reluctance motors
- Variable reluctance (magnetic) pickups

Microscopic Origins of Reluctance

The reluctance of a magnetically uniform magnetic circuit element can be calculated as :

$$R = \frac{l}{\mu A}$$

where

l is the length of the element in metres

$\mu = \mu_r \mu_0$ is the permeability of the material (μ_r is the relative permeability of the material (dimensionless), and μ_0 is the permeability of free space)

A is the cross-sectional area of the circuit in square metres

This is similar to the equation for electrical resistance in materials, with permeability being analogous to conductivity; the reciprocal of the permeability is known as magnetic reluctivity and is analogous to resistivity. Longer, thinner geometries with low permeabilities lead to higher reluctance. Low reluctance, like low resistance in electric circuits, is generally preferred.

Summary of Analogy Between Magnetic Circuits and Electrical Circuits

The following table summarizes the mathematical analogy between electrical circuit theory and magnetic circuit theory. This is mathematical analogy and not a physical one. Objects in the same row have the same mathematical role; the physics of the two theories are very different. For example, current is the flow of electrical charge, while magnetic flux is **not** the flow of any quantity.

Analogy between 'magnetic circuits' and electrical circuits					
Magnetic			Electric		
Name	Symbol	Units	Name	Symbol	Units
Magnetomotive force (MMF)	$f = \int \mathbf{H} \cdot d\mathbf{l}$	ampere-turn	Electromotive force (EMF)	$\varepsilon = \int \mathbf{E} \cdot d\mathbf{l}$	volt
Magnetic field	H	ampere/meter	Electric field	E	volt/meter = newton/coulomb

Magnetic flux	ϕ	weber	Electric current	I	ampere
Hopkinson's law or Rowland's law	$F = \phi R_m$	ampere-turn	Ohm's law	$\varepsilon = IR$	
Reluctance	R_m	1/henry	Electrical resistance	R	ohm
Permeance	$p = \dfrac{1}{R_m}$	henry	Electric conductance	$G = 1/R$	1/ohm = mho = siemen
Relation between B and H	$B = \mu H$		Microscopic Ohm's law	$J = \sigma E$	
Magnetic flux density B	B	tesla	Current density	J	ampere/square meter
Permeability	μ	henry/meter	Electrical conductivity	σ	siemen/meter

Limitations of the Analogy

When using the analogy between magnetic circuits and electric circuits, the limitations of this analogy must be kept in mind. Electric and magnetic circuits are only superficially similar because of the similarity between Hopkinson's law and Ohm's law. Magnetic circuits have significant differences, which must be taken into account in their construction :

- Electric currents represent the flow of particles (electrons) and carry power, which is dissipated as heat in resistances. Magnetic fields don't represent the "flow" of anything, and no power is dissipated in reluctances.

- The current in typical electric circuits is confined to the circuit, with very little "leakage". In typical magnetic circuits not all of the magnetic field is confined to the magnetic circuit; there is significant "leakage flux" in the space outside the magnetic cores, which must be taken into account but is difficult to calculate.

- Most importantly, magnetic circuits are non-linear; the reluctance in a magnetic circuit is not constant, as resistance is, but varies depending on the magnetic field. At high magnetic fluxes the ferromagnetic materials used for the cores of magnetic circuits saturate, limiting the magnetic flux, so above this level the reluctance increases rapidly. The reluctance also increases at low fluxes. In addition, ferromagnetic materials suffer from hysteresis so the flux in them depends not just on the instantaneous MMF but also on the history of MMF. After the source of the magnetic flux is turned off, remanent magnetism is left in ferromagnetic circuits, creating a flux with no MMF.

Circuit Laws

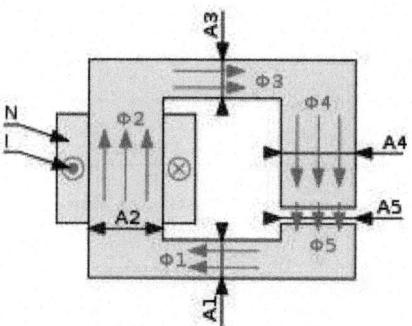

Fig. : Magnetic circuit.

Magnetic circuits obey other laws that are similar to electrical circuit laws. For example, the total reluctance R_T of reluctances R_1, R_2, ... in series is :

$$R_T = R_1 + R_2 + ...$$

This also follows from Ampère's law and is analogous to Kirchhoff's voltage law for adding resistances in series. Also, the sum of magnetic fluxes Φ_1, Φ_2,...into any node is always zero :

$$\Phi_1 + \Phi_2 + \cdots = 0$$

This follows from Gauss's law and is analogous to Kirchhoff's current law for analyzing electrical circuits.

Together, the three laws above form a complete system for analysing magnetic circuits, in a manner similar to electric circuits. Comparing the two types of circuits shows that :

- The equivalent to resistance R is the *reluctance R_m*
- The equivalent to current I is the *magnetic flux Φ*
- The equivalent to voltage V is the *magnetomotive Force F.*

Magnetic circuits can be solved for the flux in each branch by application of the magnetic equivalent of Kirchhoff's Voltage Law (KVL) for pure source/resistance circuits. Specifically, whereas KVL states that the voltage excitation applied to a loop is equal to the sum of the voltage drops (resistance times current) around the loop, the magnetic analogue states that the magnetomotive force (achieved from ampere-turn excitation) is equal to the sum of MMF drops (product of flux and reluctance) across the rest of the loop. (If there are multiple loops, the current in each branch can be solved through a matrix equation – much as a matrix solution for mesh circuit branch currents is obtained in loop analysis – after which the individual branch currents are obtained by adding and/or subtracting the constituent loop currents as indicated by the adopted sign convention and loop orientations.) Per Ampère's law, the excitation is the product of the current and the number of complete loops made and is measured in ampere-turns. Stated more generally :

$$F = N\,I = \oint \vec{H} \cdot \vec{dl}$$

(Note that, per Stokes's theorem, the closed line integral of H ·dl around a contour is equal to the open surface integral of curl H ·dA across the surface bounded by the closed contour. Since, from Maxwell's equations, curl H = J, the closed line integral of H ·dl evaluates to the total current passing through the surface. This is equal to the excitation, NI, which also measures current passing through the surface, thereby verifying that the net current flow through a surface is zero ampere-turns in a closed system that conserves energy.)

More complex magnetic systems, where the flux is not confined to a simple loop, must be analysed from first principles by using Maxwell's equations.

History

The term reluctance was coined in May 1888 by Oliver Heaviside. The notion of "magnetic resistance" was first mentioned by James Joule and the term "magneto-motive force" (MMF) was first named by Bosanquet. The idea for a magnetic flux law, similar to Ohm's law for closed electric circuits, is attributed to H. Rowland.

Applications

- Air gaps can be created in the cores of certain transformers to reduce the effects of saturation. This increases the reluctance of the magnetic circuit, and enables it to store more energy before core saturation. This effect is also used in the flyback transformer.
- Variation of reluctance is the principle behind the reluctance motor (or the variable reluctance generator) and the Alexanderson alternator.
- Multimedia loudspeakers are typically shielded magnetically, in order to reduce magnetic interference caused to televisions and other CRTs. The speaker magnet is covered with a material such as soft iron to minimize the stray magnetic field.

Reluctance can also be applied to variable reluctance (magnetic) pickups.

MAGNETIC CAPACITIVITY

Magnetic capacitivity (SI Unit : H) is a component used in the gyrator-capacitor model of magnetic systems.

This element, denoted as C_M, is an extensive property and is defined as :

$$C_M = \mu_r \mu_0 \frac{S}{l}$$

Where : $\mu_r \mu_0 = \mu$ is the magnetic permeability, S is the element cross-section, and l is the element length.

For phasor analysis, the magnetic permeability[1] and the magnetic capacitivity are complex values[1, 2].

Magnetic capacitivity is also equal to magnetic flux divided by the difference of magnetic potential across the element.

$$C_M = \frac{\Phi}{\phi_{M1} - \phi_{M2}}$$

Where :

$\phi_{M1} - \phi_{M2}$ is the difference of the magnetic potentials.

The notion of magnetic capacitivity is employed in the gyrator-capacitor model in a way analogous to capacitance in electrical circuits.

MAGNETIC CAPACITANCE

Magnetic capacitance (capacitive magnetic reactance) (SI Unit : $-\Omega^{-1}$) is a magnetic "reactance" which prevents magnetic "current" in oscillating magnetic circuits from rising. This is associated with high reluctance.

For harmonic regimes it is equal to :

$$x_C = \frac{1}{\omega C_M}$$

Where :

$\quad C_M$ is the magnetic capacitivity (SI Unit : $-s\ \Omega^{-1}$)

$\quad \omega$ is the angular frequency of the magnetic circuit

In complex form it is written as an imaginary number :

$$-j\,x_C = -j\,\frac{1}{\omega C_M} = \frac{1}{j\omega C_M}$$

The electrical potential energy sustained by Magnetic capacitivity varies with the frequency of oscillations in magnetic fields. The average power in a given period is equal to zero. The magnetic capacitivity is a reactive part of the magnetic circuit [1, 2].

MAGNETIC COMPLEX RELUCTANCE

Magnetic complex reluctance (SI Unit : H^{-1}) is a measurement of a passive magnetic circuit (or element within that circuit) dependent on sinusoidal magnetomotive force (SI Unit : $At\cdot Wb^{-1}$) and sinusoidal magnetic flux (SI Unit : $T\cdot m^2$), and this is determined by deriving the ratio of their complex *effective* amplitudes.[Ref. 1-3]

$$Z_\mu = \frac{\dot{N}}{\dot{\Phi}}\,\frac{\dot{N}_m}{\dot{\Phi}_m} = z_\mu e^{j\phi}$$

As seen above, magnetic complex reluctance is a phasor represented as *uppercase Z mu* where :

$\dot{N}$ and $\dot{N}_m$ represent the magnetomotive force (complex effective amplitude)

$\dot{\Phi}$ and $\dot{\Phi}_m$ represent the magnetic flux (complex effective amplitude)

z_μ, *lowercase z mu*, is the real part of magnetic complex reluctance

The "lossless" magnetic reluctance, *lowercase z mu*, is equal to the absolute value (modulus) of the magnetic complex reluctance. The argument distinguishing the "lossy" magnetic complex reluctance from the "lossless" magnetic reluctance is equal to the natural number *e* raised to a power equal to :

$$j\phi = j(\beta - \alpha)$$

Where :

* j is the imaginary number
* β is the phase of the magnetomotive force
* α is the phase of the magnetic flux
* ϕ is the phase difference.

The "lossy" magnetic complex reluctance represents a magnetic circuit element's resistance to not only magnetic flux but also to *changes* in magnetic flux. When applied to harmonic regimes, this formality is similar to Ohm's Law in ideal AC circuits. In magnetic circuits, magnetic complex reluctance is equal to :

$$Z_\mu = \frac{1}{\dot{\mu}\mu_0} \frac{l}{S}$$

Where :

* l is the length of the circuit element
* S is the cross-section of the circuit element
* $\dot{\mu}\mu_0$ is the complex magnetic permeability

MAGNETIC CORE

A **magnetic core** is a piece of magnetic material with a high permeability used to confine and guide magnetic fields in electrical, electromechanical and magnetic devices such as electromagnets, transformers, electric motors, generators, inductors, magnetic recording heads, and magnetic assemblies. It is made of ferromagnetic metal such as iron, or ferrimagnetic compounds such as ferrites. The high permeability, relative to the surrounding air, causes the magnetic field lines to be concentrated in the core material. The magnetic field is often created by a coil of wire around the core that carries a current. The presence of the core can increase the magnetic field of a coil by a factor of several thousand over what it would be without the core.

The use of a magnetic core can enormously concentrate the strength and increase the effect of magnetic fields produced by electric currents and permanent magnets. The properties of a device will depend crucially on the following factors :

- The geometry of the magnetic core.
- The amount of air gap in the magnetic circuit.
- The properties of the core material (especially permeability and hysteresis).
- The operating temperature of the core.
- Whether the core is laminated to reduce eddy currents.

In many applications it is undesirable for the core to retain magnetization when the applied field is removed. This property, called *hysteresis* can cause energy losses in applications such as transformers. Therefore 'soft' magnetic materials with low hysteresis, such as silicon steel, rather than the 'hard' magnetic materials used for permanent magnets, are usually used in cores.

Commonly Used Structures

Air Core

A coil not containing a magnetic core is called an air core coil. This includes coils wound on a plastic or ceramic form in addition to those made of stiff wire that are self-supporting and have air inside them. Air core coils generally have a much lower inductance than similarly sized ferromagnetic core coils, but are used in radio frequency circuits to prevent energy losses called core losses that occur in magnetic cores. The absence of normal core losses permits a higher Q factor, so air core coils are used in high frequency resonant circuits, such as up to a few megahertz. However, losses such as proximity effect and dielectric losses are still present.

Straight Cylindrical Rod

Most commonly made of ferrite or a similar material, and used in radios especially for tuning an inductor. The rod sits in the middle of the coil, and small adjustments of the rod's position will fine tune the inductance. Often the rod is threaded to allow adjustment with a screwdriver. In radio circuits, a blob of wax or resin is used once the inductor has been tuned to prevent the core from moving.

The presence of the high permeability core increases the inductance but the field must still spread into the air at the ends of the rod. The path through the air ensures that the inductor remains linear. In this type of inductor radiation occurs at the end of the rod and electromagnetic interference may be a problem in some circumstances.

Single "I" Core

Like a cylindrical rod but square, rarely used on its own. This type of core is most likely to be found in car ignition coils.

"C" or "U" Core

U and C-shaped cores are used with I or another C or U core to make a square closed core, the simplest closed core shape. Windings may be put on one or both legs of the core.

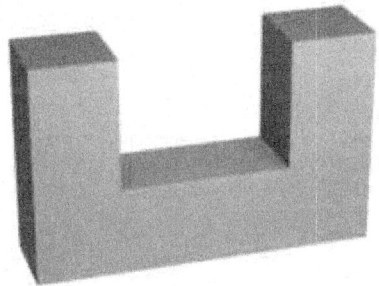

Fig. : A U-shaped core, with sharp corners.

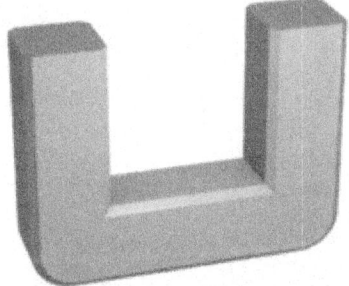

Fig. : The C-shaped core, with rounded corners.

"E" Core

E-shaped core are more symmetric solutions to form a closed magnetic system. Most of the time, the electric circuit is wound around the center leg, whose section area is twice that of each individual outer leg.

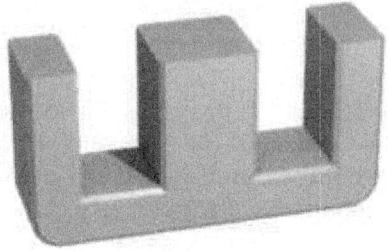

Fig. : Classical E core.

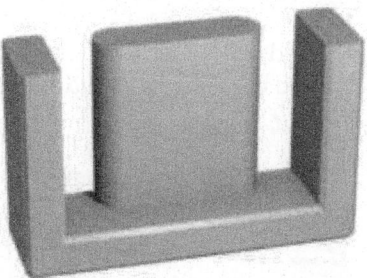

Fig. : The *EFD'* core allows for construction of inductors or transformers with a lower profile.

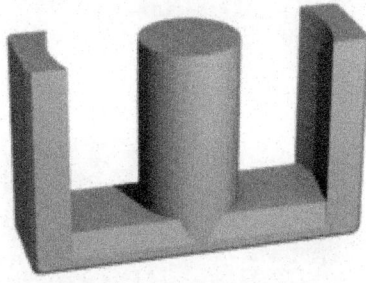

Fig. : The *ER* core has a cylindrical central leg.

Fig. : The *EP* core is halfway between a *E* and a *pot* core.

"E" and "I" Core

Sheets of suitable iron stamped out in shapes like the (sans-serif) letters "E" and "I", are stacked with the "I" against the open end of the "E" to form a 3-legged structure. Coils can be wound around any leg, but usually the center leg is used. This type of core is frequently used for power transformers, autotransformers, and inductors.

Fig. : Construction of an inductor using two *ER* cores, a plastic bobbin and two clips. The bobbin has pins to be soldered to a printed circuit board.

Fig. : Exploded view of the previous figure showing the structure.

Pair of "E" Cores

Again used for iron cores. Similar to using an "E" and "I" together, a pair of "E" cores will accommodate a larger coil former and can produce a larger inductor or transformer. If an air gap is required, the centre leg of the "E" is shortened so that the air gap sits in the middle of the coil to minimise fringing and reduce electromagnetic interference.

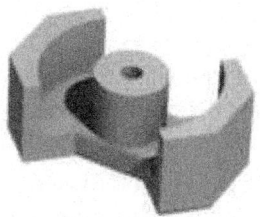

Fig. : a pot core of 'RM' type.

Pot Core

Usually ferrite or similar. This is used for inductors and transformers. The shape of a pot core is round with an internal hollow that almost completely encloses the coil. Usually a pot core is made in two halves which fit together around a coil former (bobbin). This design of core has a shielding effect, preventing radiation and reducing electromagnetic interference.

Fig. : A toroidal core.

Toroidal Core

This design is based on a toroid (the same shape as a doughnut). The coil is wound through the hole in the torus and around the outside. An ideal coil is distributed evenly all around the circumference of the torus. The symmetry of this geometry creates a magnetic field of circular loops inside the core, and the lack of sharp bends will constrain virtually all of the field to the core material. This not only makes a highly efficient transformer, but also reduces the electromagnetic interference radiated by the coil.

It is popular for applications where the desirable features are : high specific power per mass and volume, low mains hum, and minimal electromagnetic interference. One such application is the power supply for a hi-fi audio amplifier. The main drawback that limits their use for general purpose applications, is the inherent difficulty of winding wire through the center of a torus.

Unlike a split core (a core made of two elements, like a pair of E cores), specialized machinery is required for automated winding of a toroidal core. Toroids have less audible noise, such as mains hum, because the magnetic forces do not exert bending moment on the core. The core is only in compression or tension, and the circular shape is more stable mechanically.

Ring or Bead

The ring is essentially identical in shape and performance to the toroid, except that inductors commonly pass only through the center of the core, without wrapping around the core multiple times.

The ring core may also be composed of two separate C-shaped hemispheres secured together within a plastic shell, permitting it to be placed on finished cables with large connectors already installed, that would prevent threading the cable through the small inner diameter of a solid ring.

Planar Core

A planar core consists of two flat pieces of magnetic material, one above and one below the coil. It is typically used with a flat coil that is part of a printed circuit board. This design is excellent for mass production and allows a high power, small volume transformer to be constructed for low cost. It is not as ideal as either a **pot core** or **toroidal core** but costs less to produce.

Fig. : A planar 'E' core.

Fig. : A planar inductor.

Fig. : Exploded view that shows the spiral track made directly on the printed circuit board.

AL Value

The AL value of a core configuration is frequently specified by manufacturers. The relationship between inductance and A_L number in the linear portion of the magnetisation curve is defined to be :

$$L = n^2 A_L$$

where n is the number of turns, L is the inductance (*e.g.* in nH) and A_L is expressed in inductance per turn squared (*e.g.* in nH/n^2).

Core Loss

When the core is subjected to a *changing* magnetic field, as it is in devices that use AC current such as transformers, inductors, and AC motors and alternators, some of the power that would ideally be transferred through the device is lost in the core, dissipated as heat and sometimes noise. This is due primarily to two processes :

- **Hysteresis** - When the magnetic field through the core changes, the magnetization of the core material changes by expansion and contraction of the tiny magnetic domains it is composed of, due to movement of the domain walls. This process causes losses, because the domain walls get "snagged" on defects in the crystal structure and then "snap" past them, dissipating energy as heat. This is called hysteresis loss. It can be seen in the graph of the *B* field versus the *H* field for the material, which has the form of a closed loop. The amount of energy lost in the material in one cycle of the applied field is proportional to the area inside the hysteresis loop. Since the energy lost in each cycle is constant, hysteresis power losses increase proportionally with frequency.

- **Eddy currents** - If the core is electrically conductive, the changing magnetic field induces circulating loops of current in it, called eddy currents, due to electromagnetic induction. The loops flow perpendicular to the magnetic field axis. The energy of the currents is dissipated as heat in the resistance of the core material. The power loss is proportional to the area of the loops and inversely proportional to the resistivity of the core material. Eddy current losses can be reduced by making the core out of thin laminations which have an insulating coating, or alternately, making the core of a nonconductive magnetic material, like ferrite.

Magnetic Core Material

Having no magnetically active core material (an "air core") provides very low inductance in most situations, so a wide range of high-permeability materials are used to concentrate the field. Most high-permeability material are ferromagnetic or ferrimagnetic.

Soft Iron

"Soft" (annealed) iron is used in magnetic assemblies, electromagnets and in some electric motors; and it can create a concentrated field that is as much as 50,000 times more intense than an air core.

Iron is desirable to make magnetic cores, as it can withstand high levels of magnetic field without saturating (up to 2.16 teslas at ambient temperature.)

It is also used because, unlike "hard" iron, it does not remain magnetised when the field is removed, which is often important in applications where the magnetic field is required to be repeatedly switched.

Unfortunately, due to the electrical conductivity of the metal, at AC frequencies a bulk block or rod of soft iron can often suffer from large eddy currents circulating within it that waste energy and cause undesirable heating of the iron.

Laminated Silicon Steel

Because iron is a relatively good conductor, it cannot be used in bulk form with a rapidly changing field, such as in a transformer, as intense eddy currents would appear due to the magnetic field, resulting in huge losses (this is used in induction heating).

Two techniques are commonly used together to increase the resistivity of iron : lamination and alloying of the iron with silicon.

Lamination

Figure : Typical EI Lamination Pair.

Laminated magnetic cores are made of thin, insulated iron sheets, lying, as much as possible, parallel with the lines of flux. Using this technique, the magnetic core is equivalent to many individual magnetic circuits, each one receiving only a small fraction of the magnetic flux (because their section is a fraction of the whole core section). Because eddy currents flow around lines of flux, the laminations prevent most of the eddy currents from flowing at all, restricting any flow to much smaller, thinner and thus higher resistance regions. From this, it can be seen that the thinner the laminations, the lower the eddy currents.

Silicon Alloying

A small addition of silicon to iron (around 3%) results in a dramatic increase of the resistivity, up to four times higher. Further increase in silicon concentration impairs the steel's mechanical properties, causing difficulties for rolling due to brittleness.

Among the two types of silicon steel, grain-oriented (GO) and grain non-oriented (GNO), GO is most desirable for magnetic cores. It is anisotropic, offering better magnetic properties than GNO in one direction. As the magnetic field in inductor and transformer cores is static (compared to that in electric motors), it is possible to use GO steel in the preferred orientation.

Carbonyl Iron

Powdered cores made of carbonyl iron, a highly pure iron, have high stability of parameters across a wide range of temperatures and magnetic flux levels, with excellent Q factors between 50 kHz and 200 MHz. Carbonyl iron powders are basically constituted of micrometer-size spheres of iron coated in a thin layer of electrical insulation. This is equivalent to a microscopic laminated magnetic circuit, hence reducing the eddy currents, particularly at very high frequencies.

A popular application of carbonyl iron-based magnetic cores is in high-frequency and broadband inductors and transformers.

Iron Powder

Powdered cores made of hydrogen reduced iron have higher permeability but lower Q. They are used mostly for electromagnetic interference filters and low-frequency chokes, mainly in switched-mode power supplies.

Ferrite

Ferrite ceramics are used for high-frequency applications. The ferrite materials can be engineered with a wide range of parameters. As ceramics, they are essentially insulators, which prevents eddy currents, although losses such as hysteresis losses can still occur.

Vitreous Metal

Amorphous metal is a variety of alloys that are non-crystalline or glassy. These are being used to create high efficiency transformers. The materials can be highly responsive to magnetic fields for low hysteresis losses and they can also have lower conductivity to reduce eddy current losses. China is currently making wide spread industrial and power grid usage of these transformers for new installations.

MAGNETIC EFFECTIVE RESISTANCE

Magnetic effective resistance (SI Unit : $-\Omega^{-1}$) is the real component of complex magnetic impedance of a circuit in the gyrator-capacitor model. This causes a magnetic circuit to lose magnetic potential energy.

Active power in a magnetic circuit equals the product of magnetic effective resistance r_M and magnetic current squared I^2_M.

$$P = r_M I^2_M$$

The magnetic effective resistance on a complex plane appears as the side of the resistance triangle for magnetic circuit of an alternating current. The effective magnetic resistance is bounding with the effective magnetic conductance g_M by the expression

$$g_M = \frac{r_M}{z^2_M}$$

where z_M is the full magnetic impedance of a magnetic circuit.

MAGNETIC IMPEDANCE

Magnetic impedance (SI Unit : $-\Omega^{-1}$) is the ratio of a sinusoidal magnetic tension N_m to a sinusoidal magnetic current I_{Mm} in a gyrator-capacitor model. Analogous to electrical impedance, magnetic impedance is likewise a complex variable.

$$z_M = \frac{N}{I_M} = \frac{N_m}{I_{Mn}}$$

Magnetic impedance is also called the *full* magnetic resistance. It is derived from :

$r_M = z_m \cos \phi$, the effective magnetic resistance (real)

$x_M = z_M \sin \phi$, the reactive magnetic resistance (imaginary)

The phase angle ϕ of the magnetic impedance is equal to :

$$\phi = \arctan \frac{x_M}{r_M}$$

MAGNETIC INDUCTANCE

Magnetic inductance (inductive magnetic reactance) (SI Unit : $-\Omega^{-1}$) is a component in the gyrator-capacitor model for magnetic systems.

For phasor analysis the magnetic inductive reactance is :

$$x_L = \omega L_M$$

Where :

L_M is the magnetic inductivity (SI Unit : $-s \cdot \Omega^{-1}$)

ω is the angular frequency of the magnetic circuit

In the complex form it is a positive imaginary number :

$$jx_L = j\omega L_M$$

The magnetic potential energy sustained by magnetic inductivity varies with the frequency of oscillations in electric fields. The average power in a given period is equal to zero. Due to its dependence on frequency, magnetic inductance is mainly observable in magnetic circuits which operate at VHF and/or UHF frequencies.

The notion of magnetic inductivity is employed in analysis and computation of circuit behaviour in the gyrator-capacitor model in a way analogous to inductance in electrical circuits.

MAGNETIC TENSION FORCE

The **magnetic tension force** is a restoring force (SI unit : Pa·m^{-1}) that acts to straighten bent magnetic field lines. It equals :

$$\frac{(\mathbf{B}\cdot\nabla)\,\mathbf{B}}{\mu_0}\;(\text{S.I.}) \qquad\qquad \frac{(\mathbf{B}\cdot\nabla)\mathbf{B}}{4\pi}\;(\text{c.g.s.})$$

It is analogous to rubber bands and their restoring force. The force is directed antiradially. Although magnetic tension is referred to as a force, it is actually a pressure gradient (Pa m^{-1}) which is also a force density (N m^{-3}).

The magnetic pressure is the energy density of the magnetic field and it increases as magnetic field lines convene with each other. In contrast, magnetic tension force is determined by how much the magnetic pressure changes with distance. Magnetic tension forces also rely on vector current densities **J** and their interaction with the magnetic field **B**. Plotting magnetic tension along adjacent field lines can give a picture as to their divergence and convergence with respect to each other as well as current densities **J**.

Use in Plasma Physics

Magnetic tension is particularly important in plasma physics and magnetohydrodynamics, where it controls dynamics of some systems and the shape of magnetized structures. In magnetohydrodynamics, the magnetic tension force can be derived from the momentum equation or plasma physics

$$\rho\left(\frac{\partial}{\partial t}+\mathbf{V}\cdot\nabla\right)\mathbf{V}=\mathbf{J}\times\mathbf{B}-\nabla p$$

using the relation $\mu_0\mathbf{J}=\nabla\times\mathbf{B}$. The first term on the right hand side of the above equation represents electromagnetic forces and the second term represents pressure gradient forces.

Chapter 4

THREE-PHASE ELECTRIC POWER

Three-phase electric power is a common method of alternating-current electric power generation, transmission, and distribution. It is a type of polyphase system and is the most common method used by electrical grids worldwide to transfer power. It is also used to power large motors and other heavy loads. A three-phase system is usually more economical than an equivalent single-phase or two-phase system at the same voltage because it uses less conductor material to transmit electrical power. The three-phase system was independently invented by Galileo Ferraris, Mikhail Dolivo-Dobrovolsky and Nikola Tesla in the late 1880s.

DETAILS

In a three-phase system, three circuit conductors carry three alternating currents (of the same frequency) which reach their instantaneous peak values at one third of a cycle from each other. Taking one current as the reference, the other two currents are delayed in time by one third and two thirds of one cycle of the electric current. This delay between phases has the effect of giving constant power transfer over each cycle of the current and also makes it possible to produce a rotating magnetic field in an electric motor.

The sum of the currents is always zero and each line returns the current from the other two. Thus a three-phase system can operate with only three wires. Three-phase systems may also have a fourth wire, particularly in low-voltage distribution, which is the neutral wire. The neutral allows three separate single-phase supplies to be provided at a constant voltage and is commonly used for supplying groups of domestic properties which are each single-phase loads. The connections are arranged so that as far as possible in each group equal power is drawn from each phase. Further up the supply chain in high-voltage distribution the currents are usually well balanced and it is therefore normal to omit the neutral wire.

Three-phase has properties that make it very desirable in electric power systems :

- The phase currents tend to cancel out one another, summing to zero in the case of a linear balanced load. This makes it possible to reduce the size of the neutral conductor because it carries little to no current; all the phase conductors carry the same current and so can be the same size, for a balanced load.
- Power transfer into a linear balanced load is constant, which helps to reduce generator and motor vibrations.
- Three-phase systems can produce a rotating magnetic field with a specified direction and constant magnitude, which simplifies the design of electric motors.

Most household loads are single-phase. In North American residences, three-phase power might feed a multiple-unit apartment block, but the household loads are connected only as single phase. In lower-density areas, only a single phase might be used for distribution. Some large European appliances may be powered by three-phase power, such as electric stoves and clothes dryers.

Wiring for the three phases is typically identified by colour codes which vary by country. Connection of the phases in the right order is required to ensure the intended direction of rotation of three-phase motors. For example, pumps and fans may not work in reverse. Maintaining the identity of phases is required if there is any possibility two sources can be connected at the same time; a direct interconnection between two different phases is a short-circuit.

GENERATION AND DISTRIBUTION

At the power station, an electrical generator converts mechanical power into a set of three AC electric currents, one from each coil (or winding) of the generator. The windings are arranged such that the currents vary sinusoidally at the same frequency but with the peaks and troughs of their wave forms offset to provide three complementary currents with a phase separation of one-third cycle (120° or $2\pi/3$ radians). The generator frequency is typically 50 or 60 Hz, varying by country.

Further information : Mains power systems

At the power station, transformers change the voltage from generators to a level suitable for transmission minimizing losses.

After further voltage conversions in the transmission network, the voltage is finally transformed to the standard utilization before power is supplied to customers.

Most automotive alternators generate three phase AC and rectify it to DC with a diode bridge.

TRANSFORMER CONNECTIONS

A "delta" connected transformer winding is connected between phases of a three-phase system. A "wye" ("star") transformer connects each winding from a phase wire to a common neutral point.

In an "open delta" or "V" system, only two sets of transformers are used. A closed delta system can operate as an open delta if one of the transformers has failed or needs to be removed. In open delta, each transformer must carry current for its respective phases as well as current for the third phase, therefore capacity is reduced to 87%. With one of three transformers missing and the remaining two at 87% efficiency, the capacity is 58% ((2/3) × 87%).

Where a delta-fed system must be grounded for protection from surge voltages, a grounding transformer (usually a zigzag transformer) may be connected to allow ground fault currents to return from any phase to ground. Another variation is a "corner grounded" delta system, which is a closed delta that is grounded at one of the junctions of transformers.

THREE-WIRE AND FOUR-WIRE CIRCUITS

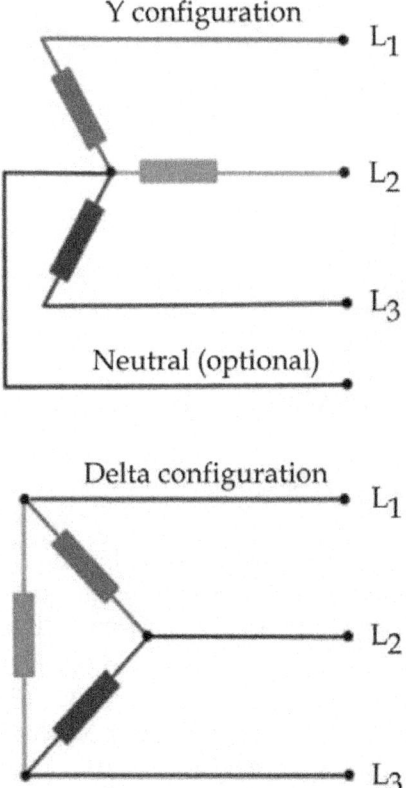

Fig. : Wye (Y) and Delta (Δ) circuits.

There are two basic three-phase configurations : delta and wye (star). Either type can be wired for three or four wires. The fourth wire, if present, is provided as a neutral. The '3-wire' and '4-wire' designations do not count the ground wire used on many transmission lines which is solely for fault protection and does not carry current under non-fault conditions.

A four-wire system with symmetrical voltages between phase and neutral is obtained when the neutral is connected to the "common star point" of an all supply windings. All three phases will have the same magnitude of voltages to the neutral in such a system. Other non-symmetrical systems have been used. In a high-leg delta system, one winding of a delta transformer feeding the system is center-tapped and connected to neutral. This setup produces three voltages. If the voltage between center tap and the two adjacent phases is 100%, the voltage across any two phases is 200% and neutral to "high leg" is ≈ 173%.

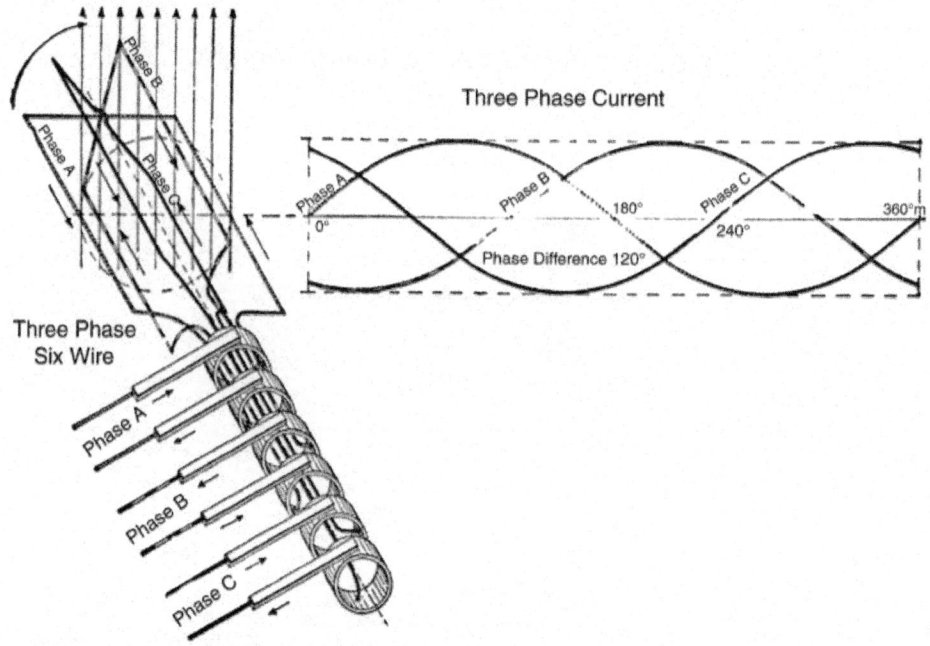

Left : Elementary six-wire three-phase alternator, with each phase using a separate pair of transmission wires. Right : Elementary three-wire three-phase alternator, showing how the phases can share only three wires.

The four-wire wye system is used when ground referenced voltages or the flexibility of more voltage selections are required. Faults on one phase to ground will cause a protection event (fuse or breaker open) locally and not involve other phases or other connected equipment. An example of application is a local distribution in Europe, where each customer is fed a phase and a neutral. When a set of customers sharing the neutral draw unequal currents, the common neutral wire carries a current as a result of the imbalance. Electrical engineers try to design the system so the loads are balanced as much as possible. By distributing a large number of houses over all three phases, on average a nearly balanced load is seen at the point of supply.

In a *three-phase, four-wire, delta* (high-leg delta) system, the neutral is a center tap in one of the delta phase supply windings. This can also be supplied by two single-phase transformers in a *V* formation (open delta).

BALANCED CIRCUITS

In the perfectly balanced case all three lines share equivalent loads. Examining the circuits we can derive relationships between line voltage and current, and load voltage and current for wye and delta connected loads.

In a balanced system each line will produce equal voltage magnitudes at phase angles equally spaced from each other. With V_1 as our reference and V_3 lagging V_2 lagging V_1, using angle notation, we have :

$$V_1 = V_{LN} \angle 0°$$
$$V_1 = V_{LN} \angle - 120°$$
$$V_1 = V_{LN} \angle + 120°$$

These Voltages feed into either a wye or delta connected load.

Wye

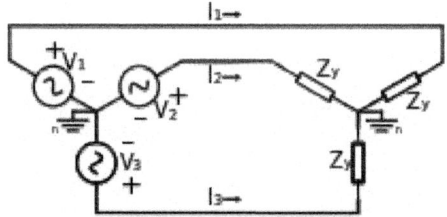

Fig. : Three phase AC generator connected as a wye source to a wye connected load.

$$I_1 = \frac{V_1}{|Z_{total}|} \angle (-\theta)$$

$$I_2 = \frac{V_2}{|Z_{total}|} \angle (-120° - \theta)$$

$$I_3 = \frac{V_3}{|Z_{total}|} \angle (120° - \theta)$$

where Z_{total} is the sum of line and load impedances ($Z_{total} = Z_{LN} + Z_Y$), and θ is the phase of the total impedance (Z_{total}).

The phase angle difference between voltage and current of each phase is not necessarily 0 and is dependent on the type of load impedance, Z_y. Inductive and capacitive loads will cause current to either lag or lead the voltage. However, the relative phase angle between each pair of lines (1 to 2, 2 to 3, and 3 to 1) will still be –120 degrees.

By performing Kirchhoff's Current Law (KCL) on the neutral node, the three phase currents sum up to the total current in the neutral line. In the balanced case :

$$I_1 + I_2 + I_3 = I_n = 0$$

DELTA

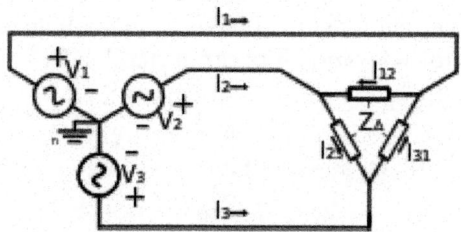

Fig. : Three phase AC generator connected as a wye source to a delta connected load.

$$V_{12} = V_1 - V_2 = (V_{LN} \angle °) - (V_{LN} \angle -120°)$$

$$= \sqrt{3}\, V_{LN} \angle 30° = \sqrt{3}V_1 \angle (phase_{V_1} + 30°)$$

$$V_{23} = V_2 - V_3 = (V_{LN} \angle -120°) - (V_{LN} \angle 120°)$$

$$= \sqrt{3}V_{LN} \angle -90° = \sqrt{3}V_2 \angle (phase_{V_2} + 30°)$$

$$V_{31} = V_3 - V_1 = (V_{LN} \angle 120°) - (V_{LN} \angle 0°)$$

$$= \sqrt{3}V_{LN} \angle 150° = \sqrt{3}V_3 \angle (phase_{V_3} + 30°)$$

Further :

$$I_{12} = \frac{V_{12}}{|Z_\Delta|} \angle (30° - \theta)$$

$$I_{23} = \frac{V_{23}}{|Z_\Delta|} \angle (-90° - \theta)$$

$$I_{31} = \frac{V_{31}}{|Z_\Delta|} \angle (150° - \theta)$$

where θ is the phase of delta impedance (Z_Δ).

Relative angles are preserved, so I_{31} lags I_{23} lags I_{12} by 120 degrees. Calculating line currents by using KCL at each delta node gives :

$$I_1 = I_{12} - I_{31} = I_{12} - I_{12} \angle 120°$$

$$= \sqrt{3}I_{12} \angle (phase_{I_{12}} - 30°) + \sqrt{3}I_{12} \angle (-\theta)$$

And similarly for each other line :

$$I_2 = \sqrt{3}I_{23} \angle (phase\ I_{23} - 30°) = \sqrt{3}I_{23} \angle (-120° - \theta)$$

$$I_3 = \sqrt{3}I_{31} \angle (phase\ I_{31} - 30°) = \sqrt{3}I_{31} \angle (120° - \theta)$$

again, θ is the phase of delta impedance (Z_Δ).

SINGLE-PHASE LOADS

Single-phase loads may be connected across any two phases, or a load can be connected from phase to neutral. Distributing single-phase loads among the phases of a three-phase system balances the load and makes most economical use of conductors and transformers.

In a symmetrical three-phase four-wire, wye system, the three phase conductors have the same voltage to the system neutral. The voltage between line conductors is $\sqrt{3}$ times the phase conductor to neutral voltage.

$$V_{L-L} = \sqrt{3}\,V_{L-N}$$

The currents returning from the customers' premises to the supply transformer all share the neutral wire. If the loads are evenly distributed on all three phases, the sum of the returning currents in the neutral wire is approximately zero. Any unbalanced phase loading on the secondary side of the transformer will use the transformer capacity inefficiently.

If the supply neutral is broken, phase-to-neutral voltage is no longer maintained. Phases with higher relative loading will experience reduced voltage and phases with lower relative loading will experience elevated voltage, up to the phase-to-phase voltage.

A high-leg delta provides phase-to-neutral relationship of $V_{L-L} = 2\,V_{L-N}$, however, L-N load is imposed on one phase. A transformer manufacturer's page suggests that L-N loading to not exceed 5% of transformer capacity.

$\sqrt{3}$ is ≈ 1.73, so if V_{L-N} was defined as 100%, V_{L-L} would be $\approx 100\% \times 1.73 = 173\%$. If V_{L-L} was set as 100%, then $V_{L-N} \approx 57.7\%$

UNBALANCED LOADS

When the currents on the three live wires of a three-phase system are not equal or are not at an exact 120° phase angle, the power-loss is greater than for a perfectly balanced system. The method of symmetrical components is used to analyze unbalanced systems.

NON-LINEAR LOADS

With linear loads, the neutral only carries the current due to imbalance between the phases. Devices that utilize rectifier-capacitor front-end such as switch-mode power supplies, computers, office equipment and such produce third order harmonics that are in-phase on all the supply phases. Consequently, such harmonic currents add in the neutral which can cause the neutral current to exceed the phase current.

THREE-PHASE LOADS

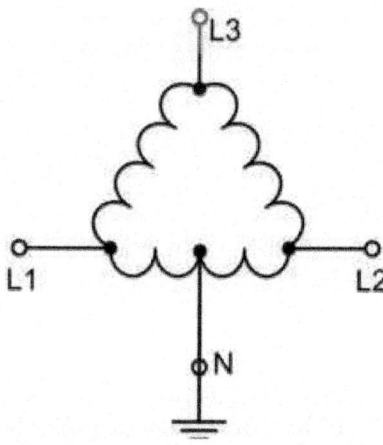

Fig. : A transformer for a high-leg delta system; 200 V 3-phase motors would be connected to L1, L2 and L3. 200 V Single-phase load would be connected L1 and L2. Single phase 100 V load between either L1 or L2 and neutral (N). L3 (wild or high leg) will be 173.2 V to neutral.

An important class of three-phase load is the electric motor. A three-phase induction motor has a simple design, inherently high starting torque and high efficiency. Such motors are applied in industry for many applications. A three-phase motor is more compact and less costly than a single-phase motor of the same voltage class and rating and single-phase AC motors above 10 HP (7.5 kW) are uncommon. Three-phase motors also vibrate less and hence last longer than single-phase motors of the same power used under the same conditions.

Line frequency flicker in light can be reduced by evenly spreading three phases across line frequency operated light sources so that illuminated area is provided light from all three phases. The effect of line frequency flicker is detrimental to super slow motion cameras used in sports event broadcasting. Three phase lighting has been applied successfully at the 2008 Beijing Olympics to provide consistent light level for each frame for SSM cameras. Resistance heating loads such as electric boilers or space heating may be connected to three-phase systems. Electric lighting may also be similarly connected.

Rectifiers may use a three-phase source to produce a six-pulse DC output. The output of such rectifiers is much smoother than rectified single phase and, unlike single-phase, does not drop to zero between pulses. Such rectifiers may be used for battery charging, electrolysis processes such as aluminium production or for operation of DC motors. "Zig-zag" transformers may make the equivalent of six-phase full-wave rectification, twelve pulses per cycle, and this method is occasionally employed to reduce the cost of the filtering components, while improving the quality of the resulting DC.

One example of a three-phase load is the electric arc furnace used in steelmaking and in refining of ores.

In Germany, a 1965 publication shows some "full size" stoves are designed for a three-phase feed. However, the individual heating units may be connected between phase and neutral to allow for connection by three individual circuits on the same single-phase supply.

PHASE CONVERTERS

Phase converters are used when three-phase equipment needs to be operated on a single-phase power source. They are used when three-phase power is not available or cost is not justifiable. Such converters may also allow the frequency to be varied (resynthesis) allowing speed control. Some railway locomotives use a single-phase source to drive three-phase motors fed through an electronic drive.

Mechanical

One method to generate three-phase power from a single-phase source is the rotary phase converter, essentially a three-phase motor with special starting arrangements and power factor correction that produces balanced three-phase voltages. When properly designed, these rotary converters can allow satisfactory operation of a three-phase motor on a single-phase source. In such a device, the energy storage is performed by the inertia (flywheel effect) of the rotating components. An external flywheel is sometimes found on one or both ends of the shaft.

A three-phase generator can be driven by a single-phase motor. This motor-generator combination can provide a frequency changer function as well as phase conversion, but requires two machines with all their expense and losses. The motor-generator method can also form an uninterruptable power supply when used in conjunction with a large flywheel and a standby generator set.

Non-mechanical

A second method that was popular in the 1940s and 1950s was the *transformer method*. At that time, capacitors were more expensive than transformers, so an autotransformer was used to apply more power through fewer capacitors. Separated it from another common method, the static converter, as both methods have no moving parts, which separates them from the rotary converters.

Another method often attempted is with a device referred to as a static phase converter. This method of running three-phase equipment is commonly attempted with motor loads though it only supplies 2/3 power and can cause the motor loads to run hot and in some cases overheat. This method does not work when sensitive circuitry is involved such as CNC devices or in induction and rectifier-type loads.

Variable-frequency drives (also known as solid-state inverters) are used to provide precise speed and torque control of three-phase motors. Some models can be powered by a single-phase supply. VFDs work by converting the supply voltage to DC and then converting the DC to a suitable three-phase source for the motor.

Digital phase converters are designed for fixed-frequency operation from a single-phase source. Similar to a variable-frequency drive, they use a microprocessor to control solid-state power switching components to maintain balanced three-phase voltages.

ALTERNATIVES TO THREE-PHASE

- Split-phase electric power is used when three-phase power is not available and allows double the normal utilization voltage to be supplied for high-power loads.

- Two-phase electric power, like three-phase, gives constant power transfer to a linear load. For loads that connect each phase to neutral, assuming the load is the same power draw, the two-wire system has a neutral current that is greater than neutral current in a three-phase system. Also motors are not entirely linear, which means that despite the theory, motors running on three-phase tend to run smoother than those on two-phase. The generators in the Adams Power Plant at Niagara Falls that were installed in 1895 were the largest generators in the world at the time and were two-phase machines. True two-phase power distribution is obsolete for "new work" applications, but still exists for "old work" applications, perhaps most particularly in Buffalo and Niagara Falls, NY, Toronto and Niagara Falls, Ontario, Philadelphia and Reading, PA, and Camden, NJ. "New work" three-phase installations may be supplied by old two-phase feeders, and "old work" two-phase installations may be supplied by new three-phase feeders using a Scott-T transformer, invented by Charles F. Scott. Special-purpose systems may use a two-phase system for frequency control.

- *Monocyclic power* was a name for an asymmetrical modified two-phase power system used by General Electric around 1897, championed by Charles Proteus Steinmetz and Elihu Thomson. This system was devised to avoid patent infringement. In this system, a generator was wound with a full-voltage single-phase winding intended for lighting loads and with a small fraction (usually ¼ of the line voltage) winding that produced a voltage in quadrature with the main windings. The intention was to use this "power wire" additional winding to provide starting torque for induction motors, with the main winding providing power for lighting loads. After the expiration of the Westinghouse patents on symmetrical two-phase and three-phase power distribution systems, the monocyclic system fell out of use; it was difficult to analyze and did not last long enough for satisfactory energy metering to be developed.

- High-phase-order systems for power transmission have been built and tested. Such transmission lines typically would use six phases or twelve phases.

High-phase-order transmission lines allow transfer of slightly less than pro-portionately higher power through a given volume without the expense of a high-voltage direct current (HVDC) converter at each end of the line. However, they require correspondingly more pieces of equipment.

THREE-PHASE AC RAILWAY ELECTRIFICATION

Three-phase AC railway electrification was used in Italy, Switzerland and the United States in the early twentieth century. Italy was the major user, from 1901 until 1976, although lines through two tunnels also used the system; the Simplon Tunnel in Switzerland from 1906 to 1930, and the Cascade Tunnel of the Great Northern Railway in the United States from 1909 to 1939. The first line was in Switzerland, from Burgdorf to Thun (40 km or 25 mi), since 1899.

Advantages

The system provides regenerative braking with the power fed back to the system, so is particularly suitable for mountain railways (provided the grid or another locomotive on the line can accept the power). The locomotives use three-phase induction motors. Lacking brushes and commutators, they require less mainte-nance. The early Italian and Swiss systems used a low frequency ($16\frac{2}{3}$ Hz), and a relatively low voltage (3,000 or 3,600 volts) compared with later AC systems.

Disadvantages

The overhead wiring, generally having two separate overhead lines and the rail for the third phase, was more complicated, and the low-frequency used required a separate generation or conversion and distribution system. Train speed was re-stricted to one to four speeds, with two or four speeds obtained by pole-changing or cascade operation or both.

Current Use

The system is only used today for four rack (mountain) railways, where the overhead wiring is less complicated and restrictions on the speeds available less important. The four systems are as follows :

- The Corcovado Rack Railway in Rio de Janeiro Brazil.
- The Gornergratbahn in Switzerland.
- The Jungfraubahn in Switzerland.
- The Petit train de la Rhune in France, still using the original locomotives of 1912.

They are nowadays industrial rather than low frequency (50 Hz, or 60 Hz (Brazil)), using between 725 and 3000 volts.

Convertor Systems

This category does not cover railways with a single-phase (or DC) supply which is converted to three-phase on the locomotive or power car, *e.g.* most railway equipment from the 1990s and earlier using solid-state converters. The Kando system of the 1930s developed by Kálmán Kandó and used in Hungary and Italy used rotating converters on the locomotive to convert the single-phase supply to three phases, as did the Phase-splitting system on the Norfolk and Western Railroad in the USA.

Locomotives

Usually the locomotives had one, two or four motors on the body chassis (not on the bogies), and did not require gearing. The induction motors are designed to run at a particular synchronous speed, and when they run above the synchronous speed downhill, power is fed back to the system. Pole changing and cascade (con-cantation) working was used to allow two or four different speeds, and resistances (often liquid rheostats) were required for starting. In Italy freight locomotives used plain cascade with two speeds, 25 and 50 km/h (16 and 31 mph); while express locomotives used cascade combined with pole-changing giving four speeds, 37, 50, 75 and 100 km/h (23, 31, 46 and 62 mph). With the use of 3,000 or 3,600 volts at $16\frac{2}{3}$ (16.7) Hz, the supply could be fed directly to the motor without an onboard transformer.

Generally the motor(s) fed a single axle with other wheels linked by connecting rods, as the induction motor is sensitive to speed variations and with non-linked motors on several axles the motors on worn wheels would do little or even no work as they would rotate faster. This motor charactership led to a mishap in the Cascade Tunnel to a GN east-bound freight train with four electric locos, two on the head and two pushing. The two pushers suddenly lost power and the train gradually slowed to a stop. But the lead unit engineer was unaware that his train had stopped, and held the controller on the power position until the usual time to transit the tunnel had elapsed. Not seeing daylight, he finally shut down the locomotive, and found that the wheels of his stationary loco had ground through two-thirds of the rail web.

Overhead Wiring

Generally two separate overhead wires are used, with the rail for the third phase, though occasionally three overhead wires are used. At junctions, crossovers and crossings the two lines must be kept apart, with a continuous supply to the locomotive, which must have two live conductors wherever it stops. Hence two collectors per overhead phase are used, but the possibility of bridging a dead section and causing a short circuit from the front collector of one phase to the back collector of the other phase must be avoided. The resistance of the rails used for the third phase or return is higher for AC than for DC because of the "skin" effect, but lower for the low frequency used than for industrial frequency. Losses are also increased, though not in the same proportion, as the impedance is largely reactive.

The locomotive needs to pick up power from two (or three) overhead conductors. Early locomotives on the Italian State Railways used a wide bow collector which covered both wires but later locomotives used two pantographs side by side. In the United States, a pair of trolley poles were used. They worked well with a maximum speed of 15 miles per hour (24 km/h). The dual conductor pantograph system is used on four mountain railways that continue to use three-phase power (Corcovado Rack Railway in Rio de Janeiro, Brazil, Jungfraubahn and Gornergratbahn in Switzerland and the Petit train de la Rhune in France).

Chapter 5

TRANSFORMER

A **transformer** is an electrical device that transfers energy between two circuits through electromagnetic induction. A transformer may be used as a safe and efficient voltage converter to change the AC voltage at its input to a higher or lower voltage at its output. Other uses include current conversion, isolation with or without changing voltage and impedance conversion.

A transformer most commonly consists of two windings of wire that are wound around a common core to provide tight electromagnetic coupling between the windings. The core material is often a laminated iron core. The coil that receives the electrical input energy is referred to as the primary winding, while the output coil is called the secondary winding.

An alternating electric current flowing through the primary winding (coil) of a transformer generates a varying electromagnetic field in its surroundings which causes a varying magnetic flux in the core of the transformer. The varying electromagnetic field in the vicinity of the secondary winding induces an electromotive force in the secondary winding, which appears a voltage across the output terminals. If a load impedance is connected across the secondary winding, a current flows through the secondary winding drawing power from the primary winding and its power source.

A transformer cannot operate with direct current; although, when it is connected to a DC source, a transformer typically produces a short output pulse as the current rises.

INVENTION

The invention of transformers during the late 1800s allowed for longer-distance, cheaper, and more energy efficient transmission, distribution, and utilization of electrical energy. In the early days of commercial electric power, the main energy source was direct current (DC), which operates at low-voltage high-current. According to Joule's Law, energy losses are directly proportional to the square of current. This law revealed that even a tiny decrease in current or rise in voltage can cause a substantial lowering in energy losses and costs. Thus, the historical pursuit for a high-voltage low-current electricity transmission system took shape. Although high voltage transmission systems offered many benefits, the future fate of high-voltage alternating current still remained unclear for several reasons : high-voltage sources had a much higher risk of causing severe electrical injuries; many essential appliances could only function at low voltage. Regarded as one of the most influential electrical innovations of all time, the introduction of transformers had successfully reduced the safety concerns associated with alternating current and had the ability to lower voltage to a value that was required by most essential appliances.

APPLICATIONS

Transformers perform voltage conversion; isolation protection; and impedance matching. In terms of voltage conversion, transformers can step-up voltage/step-down current from generators to high-voltage transmission lines, and step-down voltage/step-up current to local distribution circuits or industrial customers. The step-up transformer is used to increase the secondary voltage relative to the primary voltage, whereas the step-down transformer is used to decrease the secondary voltage relative to the primary voltage. Transformers range in size from thumbnail-sized used in microphones to units weighing hundreds of tons interconnecting the power grid. A broad range of transformer designs are used in electronic and electric power applications, including miniature, audio, isolation, high-frequency, power conversion transformers, etc.

BASIC PRINCIPLES

The functioning of a transformer is based on two principles of the laws of electromagnetic induction : An electric current through a conductor, such as a wire, produces a magnetic field surrounding the wire, and a changing magnetic field in the vicinity of a wire induces a voltage across the ends of that wire.

The magnetic field excited in the primary coil gives rise to self-induction as well as mutual induction between coils. This self-induction counters the excited field to such a degree that the resulting current through the primary winding is very small when no load draws power from the secondary winding.

The physical principles of the inductive behaviour of the transformer are most readily understood and formalized when making some assumptions to construct a simple model which is called the *ideal transformer*. This model differs from *real transformers* by assuming that the transformer is perfectly constructed and by neglecting that electrical or magnetic losses occur in the materials used to construct the device.

Ideal Transformer

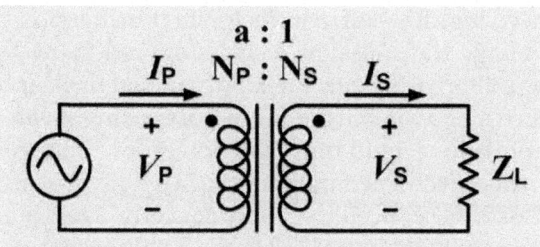

Fig. : Ideal transformer with a source and a load. N_P and N_S are the number of turns in the primary and secondary windings respectively.

The assumptions to characterize the ideal transformer are :

- The windings of the transformer have no resistance. Thus, there is no copper loss in the winding, and hence no voltage drop.
- Flux is confined within the magnetic core. Therefore, it is the same flux that links the input and output windings.
- Permeability of the core is infinitely high which implies that net mmf (amp-turns) must be zero (otherwise there would be infinite flux) hence $I_P N_P - I_S N_S = 0$.
- The transformer core does not suffer magnetic hysteresis or eddy currents, which cause inductive loss.

If the secondary winding of an ideal transformer has no load, no current flows in the primary winding.

The circuit diagram (right) shows the conventions used for an ideal, *i.e.* loss-less and perfectly-coupled transformer having primary and secondary windings with N_P and N_S turns, respectively.

The ideal transformer induces secondary voltage V_S as a proportion of the primary voltage V_P and respective winding turns as given by the equation

$$\frac{V_P}{V_S} = \frac{N_P}{N_S} = a,$$

where,

        a is the winding *turns ratio*, the value of these ratios being respectively higher and lower than unity for step-down and step-up transformers,

        V_P designates source impressed voltage,

        V_S designates output voltage, and,

According to this formalism, when the number of turns in the primary coil is greater than the number of turns in the secondary coil, the secondary voltage is smaller than the primary voltage. On the other hand, when the number of turns in the primary coil is less than the number of turns in the secondary, the secondary voltage is greater than the primary voltage.

Any load impedance Z_L connected to the ideal transformer's secondary winding allows energy to flow without loss from primary to secondary circuits. The resulting input and output apparent power are equal as given by the equation

$$I_P V_P = I_S V_S.$$

Combining the two equations yields the following ideal transformer identity

$$\frac{V_P}{V_S} = \frac{I_s}{I_P} = a.$$

This formula is a reasonable approximation for the typical commercial transformer, with voltage ratio and winding turns ratio both being inversely proportional to the corresponding current ratio.

The load impedance Z_L and secondary voltage V_S determine the secondary current I_S as follows

$$I_S = \frac{V_S}{Z_L}.$$

The apparent impedance Z_L' of this secondary circuit load *referred* to the primary winding circuit is governed by a squared turns ratio multiplication factor relationship derived as follows

$$Z'_L = \frac{V_P}{I_P} = \frac{aV_S}{I_s/a} = a^2 \frac{V_S}{I_S} = a^2 Z_L.$$

For an ideal transformer, the power supplied to the primary and the power dissipated by the load are equal. If $Z_L = R_L$ where R_L is a pure resistance then the power is given by :

$$P = \frac{V_S^2}{R_L} = \frac{V_P^2}{a^2 R_L}$$

The primary current is given by the following equation :

$$I_P = \frac{V_P}{a^2 Z_L}$$

Induction Law

A varying electrical current passing through the primary coil creates a varying magnetic field around the coil which induces a voltage in the secondary winding. The primary and secondary windings are wrapped around a core of very high magnetic permeability, usually iron, so that most of the magnetic flux passes through both the primary and secondary coils. The current through a load connected to the secondary winding and the voltage across it are in the directions indicated in the figure.

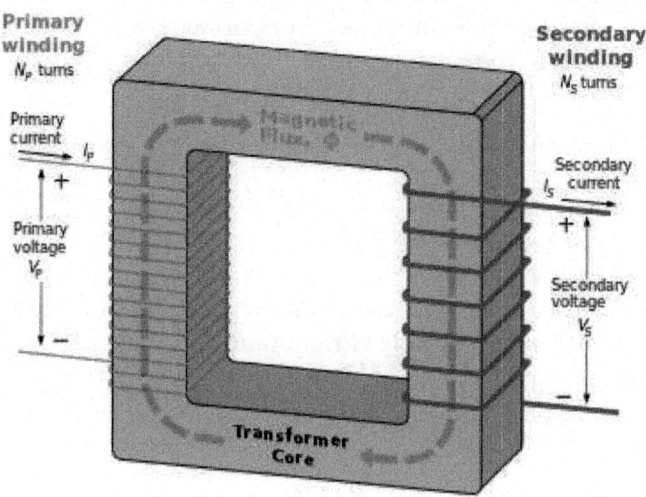

Fig. : Ideal transformer and induction law.

The voltage induced across the secondary coil may be calculated from Faraday's law of induction, which states that :

$$V_S = N_S \frac{d\Phi}{dt}.$$

where V_s is the instantaneous voltage, N_s is the number of turns in the secondary coil, and $d\Phi/dt$ is the derivative of the magnetic flux Φ through one turn of the coil. If the turns of the coil are oriented perpendicularly to the magnetic field lines, the flux is the product of the magnetic flux density B and the area A through which it cuts. The area is constant, being equal to the cross-sectional area of the transformer core, whereas the magnetic field varies with time according to the excitation of the primary. Since the same magnetic flux passes through both the primary and secondary coils in an ideal transformer, the instantaneous voltage across the primary winding equals

$$V_P = N_P \frac{d\Phi}{dt}.$$

Taking the ratio of the above two equations gives the same voltage ratio and turns ratio relationship shown above, that is,

$$\frac{V_P}{V_S} = \frac{N_P}{N_S} = a.$$

The changing magnetic field induces an emf across each winding. The primary emf, acting as it does in opposition to the primary voltage, is sometimes termed the counter emf. This is in accordance with Lenz's law, which states that induction of emf always opposes development of any such change in magnetic field.

As still lossless and perfectly-coupled, the transformer still behaves as described above in the ideal transformer.

Polarity

The relationships of the instantaneous polarity at each of the terminals of the windings of a transformer depend on the direction the windings are wound around the core. Identically wound windings produce the same polarity of voltage at the corresponding terminals. This relationship is usually denoted by the dot convention in transformer circuit diagrams, nameplates, and on terminal markings, which marks the terminals having an in-phase relationship.

Real Transformer

The ideal transformer model neglects the following basic linear aspects in real transformers.

Core losses, collectively called magnetizing current losses, consist of :

- Hysteresis losses due to non-linear application of the voltage applied in the transformer core, and
- Eddy current losses due to joule heating in the core that are proportional to the square of the transformer's applied voltage.

Whereas windings in the ideal model have no impedance, the windings in a real transformer have finite non-zero impedances in the form of :

- Joule losses due to resistance in the primary and secondary windings
- Leakage flux that escapes from the core and passes through one winding only resulting in primary and secondary reactive impedance.

If a voltage is applied across the primary terminals of a real transformer while the secondary winding is open without load, the real transformer must be viewed as a simple inductor with an impedance Z :

$$Z_p = j\omega L_p$$
$$I_p = V_p / Z_p .$$

LEAKAGE INDUCTANCE

Leakage inductance derives from the electrical property of an imperfectly-coupled transformer whereby each winding behaves as a self-inductance constant in series with the winding's respective ohmic resistance constant, these four winding constants also interacting with the transformer's mutual inductance constant. The winding self-inductance constant and associated leakage inductance is due to leakage flux not linking with all turns of each imperfectly-coupled winding.

The leakage flux alternately stores and discharges magnetic energy with each electrical cycle acting as an inductor in series with each of the primary and secondary circuits.

Leakage inductance depends on the geometry of the core and the windings. Voltage drop across the leakage reactance results in often undesirable supply regulation with varying transformer load. But it can also be useful for harmonic isolation (attenuating higher frequencies) of some loads.

Leakage inductance applies to any imperfectly-coupled magnetic circuit device including especially motors.

Leakage Inductance and Coupling Coefficient

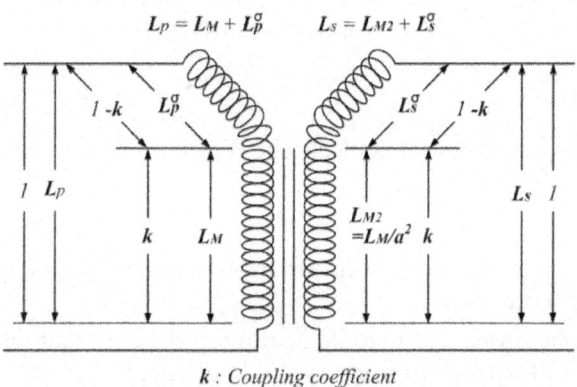

k : Coupling coefficient

Fig. : L_p^σ and L_s^σ are primary and secondary leakage inductances.

The magnetic circuit's flux that does not interlink both windings is the leakage flux corresponding to primary leakage inductance L_p^σ and secondary leakage inductance L_s^σ. These leakage inductances are defined in terms of transformer winding open-circuit inductances as well as the transformer's coupling coefficient k, the primary open-circuit self-inductance being given by

$$L_{oc}^{pri} = L_P = L_P^\sigma + L_M$$

where

$$L_P^\sigma = L_P \cdot (1 - k)$$
$$L_M = L_P \cdot k$$

and

L_{oc}^{pri} = Primary inductance

L_P = Primary self-inductance

L_P^σ = Primary leakage inductance

L_M = Magnetizing inductance referred to the primary.

It therefore follows that the transformer secondary open-circuit self, magnetizing and leakage inductances are given by

$$L_{oc}^{sec} = L_S = L_S \cdot (1 - k) + L_S \cdot k$$

where

L_{oc}^{sec} = Secondary leakage inductance L_S^σ + Magnetizing inductance L_M / a^2

L_S = Secondary self-inductance

L_S^σ = Secondary leakage inductance = $L_S \cdot (1 - k)$

$\dfrac{L_M}{a^2}$ = Magnetizing inductance referred to the secondary = $L_S \cdot k$

a = Winding turns ratio.

The electric validity of the above transformer diagram depends strictly on open circuit conditions for the respective winding inductances considered, more generalized circuit conditions being as developed in the next two sections.

Leakage Factor and Inductance

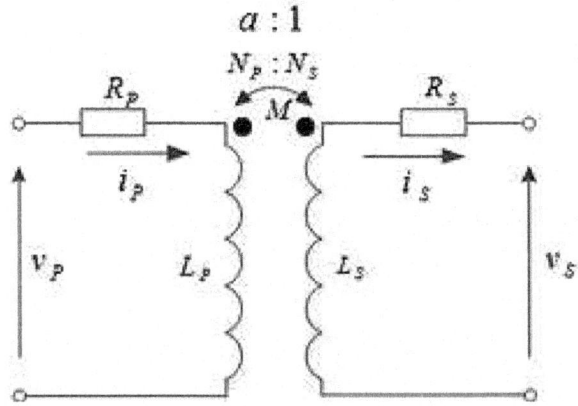

Fig. : Real transformer circuit diagram.

A real linear two-winding transformer can be represented by two mutual inductance coupled circuit loops linking the transformer's five impedance constants as shown in the diagram at right, where,

- M is mutual inductance
- L_P & L_S are primary and secondary winding self-inductances
- R_P & R_S are primary and secondary winding resistances
- Constants M, L_P, L_S, R_P & R_S are measurable at the transformer's terminals
- Coupling coefficient k is given as

   $k = M / \sqrt{L_P L_S}$, with $0 < k < 1$

- Winding turns ratio a is in practice given as

   $a = N_P / N_S = v_P / c_s = i_s / i_p = \sqrt{L_P / L_S}$.

The two circuit loops can be expressed by the following voltage and flux linkage equations,

$$v_p = R_p i_p + \frac{d\Psi_p}{dt}$$

$$v_s = - R_s i_s - \frac{d\Psi_s}{dt}$$

$$\Psi_p = L_p i_p - M i_s$$

$$\Psi_s = L_s i_s - M i_p ,$$

where :

- ψ is flux linkage
- $d\psi/dt$ is derivative of flux linkage with respect to time.

These equations can be developed to show that, neglecting associated winding resistances, the ratio of a winding circuit's inductances and currents with the other winding short circuited and at no-load is as follows,

$$\sigma = 1 - \frac{M^2}{L_p L_s} = 1 - k^2 = \frac{L_{SC}}{L_{oc}} = \frac{L_{sc}^{sec}}{L_S} = \frac{L_{sc}^{pri}}{L_p} = \frac{i_{oc}}{i_{sc}} ,$$

where,

- σ is the leakage factor or Heyland factor
- i_{oc} & i_{sc} are no-load and short circuit currents
- L_{oc} & L_{sc} are no-load and short circuit inductances.

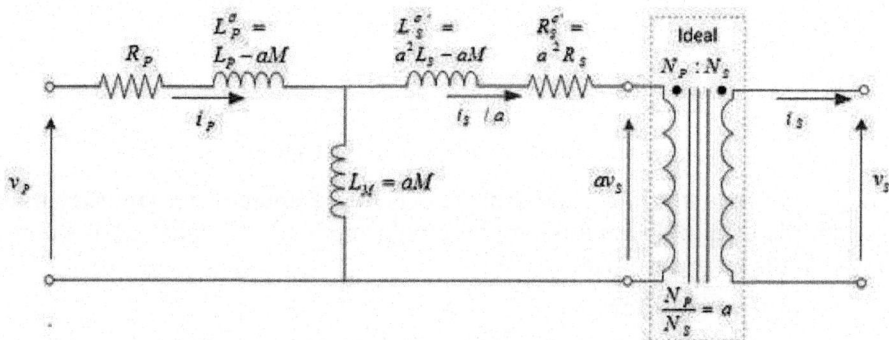

Fig. : Real transformer equivalent circuit.

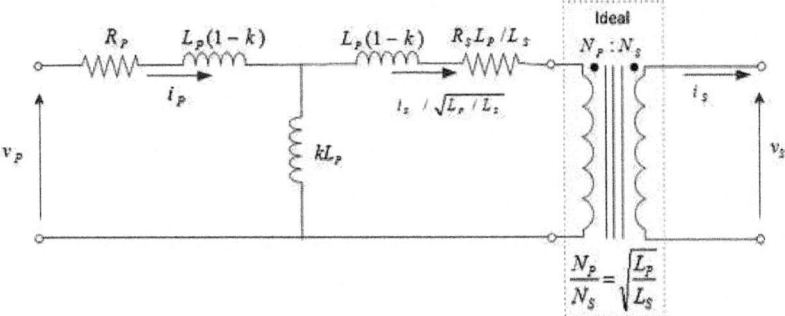

Fig. : Real transformer equivalent circuit in terms of coupling coefficient k.

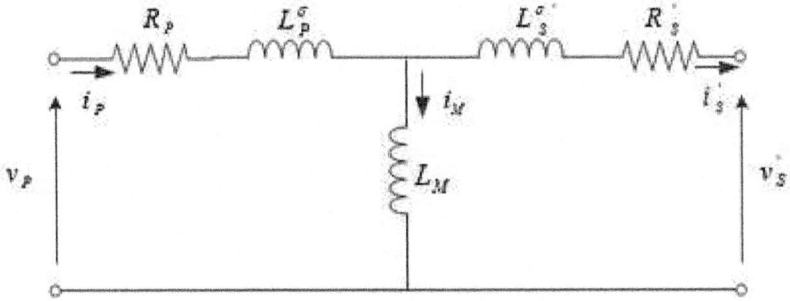

Fig. : Simplified real transformer equivalent circuit.

The transformer can thus be further defined in terms of the three inductance constants as follows,

$$L_M = a_M$$
$$L^{\sigma}_P = L_S - aM,$$
$$L^{\sigma}_S = L_S - aM,$$

where,

- L_M is magnetizing inductance, corresponding to magnetizing reactance X_M
- L_p^{σ} & L_s^{σ} are primary & secondary leakage inductances, corresponding to primary & secondary leakage reactances X_p^{σ} & X_s^{σ}.

The transformer can be expressed more conveniently as the first shown equivalent circuit with secondary constants referred (*i.e.*, with prime superscript notation) to the primary,

$$L^{\sigma\prime}_s = a^2 L_2 - aM$$
$$R'_s = a^2 R_s$$
$$V'_s = aV_s$$
$$I'_s = I_s / a.$$

Since

$$k = M / \sqrt{L_P L_S}$$

and

$$a = \sqrt{L_P / L_S} ,$$

we have

$$aM = \sqrt{L_P / L_S} * k * \sqrt{L_P L_S} = kL_P ,$$

which allows expression as second shown equivalent circuit with winding leakage and magnetizing inductance constants as follows,

$$L^{\sigma}_P = L^{\sigma}_S = L_P * (1 - k)$$
$$L_M = kL_P .$$

Expanded Leakage Factor

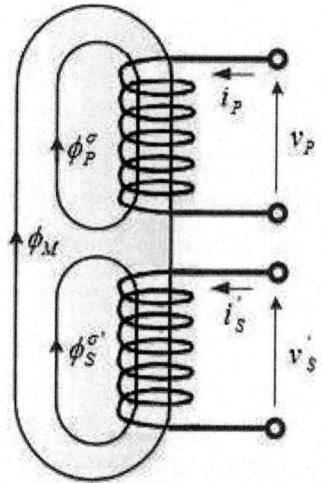

Fig. : Magnetizing and leakage flux in a magnetic circuit.

The real transformer can be simplified as shown in third shown equivalent circuit, with secondary constants referred to the primary and without ideal transformer isolation, where,

$$i_M = i_P - i_S'$$

i_M is magnetizing current excited by flux Φ_M that links both primary and secondary windings.

Referring to the flux diagram at right, the winding-specific leakage factor equations can be defined as follows,

$$\sigma_P = \Phi_P^{\sigma}/\Phi_M = L_P^{\sigma}/L_M$$
$$\sigma_S = \Phi_S^{\sigma'}/\Phi_M = L_S^{\sigma'}/L_M$$

$$\Phi_P = \Phi_M + \Phi_P{}^\sigma = (1 + \sigma_P)\Phi_M$$

$$\Phi_S{}' = \Phi_M + \Phi_S{}^{\sigma'} = (1 + \sigma_S)\Phi_M$$

$$L_P = L_M + L_P{}^\sigma = (1 + \sigma_P)L_M$$

$$L_S{}' = L_M + L_S{}^{\sigma'} = (1 + \sigma_S)L_{M'}$$

where :

- σ_P is primary leakage factor
- σ_S is secondary leakage factor
- Φ is magnetic flux.

The leakage factor σ can thus be expanded in terms of the interrelationship of above winding-specific inductance and leakage factor equations as follows :

$$\sigma = 1 - \frac{M^2}{L_P L_S} = 1 - \frac{a^2 M^2}{L_P a^2 L_S} = 1 - \frac{L_M^2}{L_P L'_S} = 1 - \frac{1}{\dfrac{L_P}{L_M} \cdot \dfrac{L'_S}{L_M}} + 1 - \frac{1}{(1 + \sigma_P)(1 + \sigma_S)} \, .$$

Leakage Inductance in Practice

Leakage inductance can be an undesirable property, as it causes the voltage to change with loading. In many cases it is useful. Leakage inductance has the useful effect of limiting the current flows in a transformer (and load) without itself dissipating power (excepting the usual non-ideal transformer losses). Transformers are generally designed to have a specific value of leakage inductance such that the leakage reactance created by this inductance is a specific value at the desired frequency of operation.

Commercial transformers are usually designed with a short-circuit leakage reactance impedance of between 3% and 10%. If the load is resistive and the leakage reactance is small (<10%) the output voltage will not drop by more than 0.5% at full load, ignoring other resistances and losses.

High leakage reactance transformers are used for some negative resistance applications, such as neon signs, where a voltage amplification (transformer action) is required as well as current limiting. In this case the leakage reactance is usually 100% of full load impedance, so even if the transformer is shorted out it will not be damaged. Without the leakage inductance, the negative resistance characteristic of these gas discharge lamps would cause them to conduct excessive current and be destroyed.

Transformers with variable leakage inductance are used to control the current in arc welding sets. In these cases, the leakage inductance limits the current flow to the desired magnitude.

EQUIVALENT CIRCUIT

Referring to the diagram, a practical transformer's physical behaviour may be represented by an equivalent circuit model, which can incorporate an ideal transformer.

Winding joule losses and leakage reactances are represented by the following series loop impedances of the model :

- Primary winding : R_P, X_P
- Secondary winding : R_S, X_S.

In normal course of circuit equivalence transformation, R_S and X_S are in practice usually referred to the primary side by multiplying these impedances by the turns ratio squared, $(N_P/N_S)^2 = a^2$.

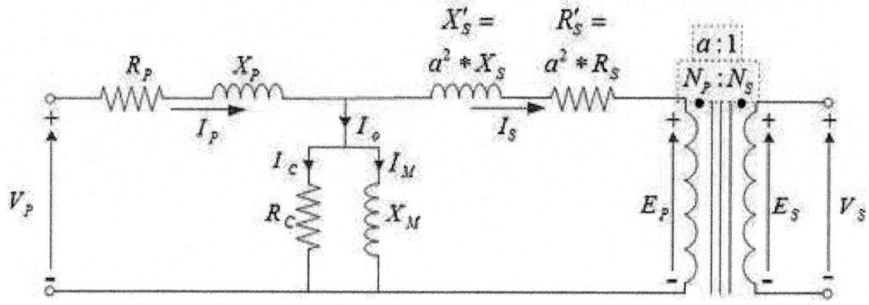

Fig. : Real transformer equivalent circuit.

Core loss and reactance is represented by the following shunt leg impedances of the model :

- Core or iron losses : R_C
- Magnetizing reactance : X_M.

R_C and X_M are collectively termed the *magnetizing branch* of the model.

Core losses are caused mostly by hysteresis and eddy current effects in the core and are proportional to the square of the core flux for operation at a given frequency. The finite permeability core requires a magnetizing current I_M to maintain mutual flux in the core. Magnetizing current is in phase with the flux, the relationship between the two being non-linear due to saturation effects. However, all impedances of the equivalent circuit shown are by definition linear and such non-linearity effects are not typically reflected in transformer equivalent circuits. With sinusoidal supply, core flux lags the induced emf by 90°. With open-circuited secondary winding, magnetizing branch current I_0 equals transformer no-load current.

The resulting model, though sometimes termed 'exact' equivalent circuit based on linearity assumptions, retains a number of approximations. Analysis may be simplified by assuming that magnetizing branch impedance is relatively high and relocating the branch to the left of the primary impedances. This introduces error but allows combination of primary and referred secondary resistances and reactances by simple summation as two series impedances.

Transformer equivalent circuit impedance and transformer ratio parameters can be derived from the following tests : open-circuit test, short-circuit test, winding resistance test, and transformer ratio test.

Basic Transformer Parameters and Construction

Effect of Frequency

Transformer Universal emf Equation

If the flux in the core is purely sinusoidal, the relationship for either winding between its **rms voltage** E_{rms} of the winding, and the supply frequency f, number of turns N, core cross-sectional area a in m^2 and peak magnetic flux density B_{peak} in Wb/m^2 or T (tesla) is given by the universal emf equation :

$$E_{rms} = \frac{2\pi f N a B_{peak}}{\sqrt{2}} = 4.44\ f N a B_{peak}$$

If the flux does not contain even harmonics the following equation can be used for **half-cycle average voltage** E_{avg} of any waveshape :

$$E_{avg} = 4 f N a B_{peak}$$

The time-derivative term in Faraday's Law shows that the flux in the core is the integral with respect to time of the applied voltage. Hypothetically an ideal transformer would work with direct-current excitation, with the core flux increasing linearly with time. In practice, the flux rises to the point where magnetic saturation of the core occurs, causing a large increase in the magnetizing current and overheating the transformer. All practical transformers must therefore operate with alternating (or pulsed direct) current.

The emf of a transformer at a given flux density increases with frequency. By operating at higher frequencies, transformers can be physically more compact because a given core is able to transfer more power without reaching saturation and fewer turns are needed to achieve the same impedance. However, properties such as core loss and conductor skin effect also increase with frequency. Aircraft and military equipment employ 400 Hz power supplies which reduce core and winding weight. Conversely, frequencies used for some railway electrification systems were much lower (*e.g.* 16.7 Hz and 25 Hz) than normal utility frequencies (50–60 Hz) for historical reasons concerned mainly with the limitations of early electric traction motors. As such, the transformers used to step-down the high over-head line voltages (*e.g.* 15 kV) were much heavier for the same power rating than those designed only for the higher frequencies.

Operation of a transformer at its designed voltage but at a higher frequency than intended will lead to reduced magnetizing current. At a lower frequency, the magnetizing current will increase. Operation of a transformer at other than its design frequency may require assessment of voltages, losses, and cooling to establish if safe operation is practical. For example, transformers may need to be equipped with 'volts per hertz' over-excitation relays to protect the transformer from overvoltage at higher than rated frequency.

One example of state-of-the-art design is traction transformers used for electric multiple unit and high-speed train service operating across the country border and using different electrical standards, such transformers' being restricted to be positioned below the passenger compartment. The power supply to, and converter equipment being supply by, such traction transformers have to accommodate different input frequencies and voltage (ranging from as high as 50 Hz down to 16.7 Hz and rated up to 25 kV) while being suitable for multiple AC asynchronous motor and DC converters & motors with varying harmonics mitigation filtering requirements.

Large power transformers are vulnerable to insulation failure due to transient voltages with high-frequency components, such as caused in switching or by lightning.

Energy Losses

A theoretical (ideal) transformer does not experience energy losses, *i.e.* it is 100% efficient. The power dissipated by its load would be equal to the power supplied by its primary source. In contrast, a real transformer is typically 95 to 99% efficient, due to several loss mechanisms, including winding resistance, winding capacitance, leakage flux, core losses, and hysteresis loss. Larger transformers are generally more efficient than small units, and those rated for electricity distribution usually perform better than 98%.

Experimental transformers using superconducting windings achieve efficiencies of 99.85%. The increase in efficiency can save considerable energy in a large heavily loaded transformer; the trade-off is in the additional initial and running cost of the superconducting design.

As transformer losses vary with load, it is often useful to express these losses in terms of no-load loss, full-load loss, half-load loss, and so on. Hysteresis and Eddy current losses are constant at all load levels and dominate overwhelmingly without load, while variable winding joule losses dominating increasingly as load increases. The no-load loss can be significant, so that even an idle transformer constitutes a drain on the electrical supply. Designing energy efficient transformers for lower loss requires a larger core, good-quality silicon steel, or even amorphous steel for the core and thicker wire, increasing initial cost. The choice of construction represents a trade-off between initial cost and operating cost.

Transformer losses arise from :

Winding joule losses : Current flowing through winding conductors causes joule heating. As frequency increases, skin effect and proximity effect causes winding resistance and, hence, losses to increase.

Core losses : **Hysteresis losses** : Each time the magnetic field is reversed, a small amount of energy is lost due to hysteresis within the core. According to Steinmetz's formula, the heat energy due to hysteresis is given by

$$W_h \approx \eta \beta_{max}^{1.6} \text{ , and,}$$

hysteresis loss is thus given by

$$P_h \approx W_h f \approx \eta f \beta_{max}^{1.6}$$

where, f is the frequency, η is the hysteresis coefficient and β_{max} is the maximum flux density, the empirical exponent of which varies from about 1.4 to 1 .8 but is often given as 1.6 for iron.

Eddy current losses : Ferromagnetic materials are also good conductors and a core made from such a material also constitutes a single short-circuited turn throughout its entire length. Eddy currents therefore circulate within the core in a plane normal to the flux, and are responsible for resistive heating of the core material. The eddy current loss is a complex function of the square of supply frequency and inverse square of the material thickness. Eddy current losses can be reduced by making the core of a stack of plates electrically insulated from each other, rather than a solid block; all transformers operating at low frequencies use laminated or similar cores.

Magnetostriction related transformer hum : Magnetic flux in a ferromagnetic material, such as the core, causes it to physically expand and contract slightly with each cycle of the magnetic field, an effect known as magnetostriction, the frictional energy of which produces an audible noise known as mains hum or transformer hum. This transformer hum is especially objectionable in transformers supplied at power frequencies and in high-frequency flyback transformers associated with PAL system CRTs.

Stray losses : Leakage inductance is by itself largely lossless, since energy supplied to its magnetic fields is returned to the supply with the next half-cycle. However, any leakage flux that intercepts nearby conductive materials such as the transformer's support structure will give rise to eddy currents and be converted to heat. There are also radiative losses due to the oscillating magnetic field but these are usually small.

Mechanical Vibration and Audible Noise Transmission

In addition to magnetostriction, the alternating magnetic field causes fluctuating forces between the primary and secondary windings. This energy incites vibration transmission in interconnected metalwork, thus amplifying audible transformer hum.

Core form and Shell form Transformers

Closed-core transformers are constructed in 'core form' or 'shell form'. When windings surround the core, the transformer is core form; when windings are surrounded by the core, the transformer is shell form. Shell form design may be more prevalent than core form design for distribution transformer applications due to the relative ease in stacking the core around winding coils. Core form design tends to, as a general rule, be more economical, and therefore more prevalent, than shell form design for high voltage power transformer applications at the lower end of their voltage and power rating ranges (less than or equal to, nominally, 230 kV or 75 MVA). At higher voltage and power ratings, shell form transformers tend to be more prevalent. Shell form design tends to be preferred for extra high voltage and higher MVA applications because, though more labour-intensive to manufacture, shell form transformers are characterized as having inherently better kVA-to-weight ratio, better short-circuit strength characteristics and higher immunity to transit damage.

Construction

Cores

Laminated Steel Cores

Transformers for use at power or audio frequencies typically have cores made of high permeability silicon steel. The steel has a permeability many times that of free space and the core thus serves to greatly reduce the magnetizing current and confine the flux to a path which closely couples the windings. Early transformer developers soon realized that cores constructed from solid iron resulted in prohibitive eddy current losses, and their designs mitigated this effect with cores consisting of bundles of insulated iron wires. Later designs constructed the core by stacking layers of thin steel laminations, a principle that has remained in use. Each lamination is insulated from its neighbors by a thin non-conducting layer of insulation. The universal transformer equation indicates a minimum cross-sectional area for the core to avoid saturation.

The effect of laminations is to confine eddy currents to highly elliptical paths that enclose little flux, and so reduce their magnitude. Thinner laminations reduce losses, but are more laborious and expensive to construct. Thin laminations are generally used on high-frequency transformers, with some of very thin steel laminations able to operate up to 10 kHz.

One common design of laminated core is made from interleaved stacks of E-shaped steel sheets capped with I-shaped pieces, leading to its name of 'E-I transformer'. Such a design tends to exhibit more losses, but is very economical to manufacture. The cut-core or C-core type is made by winding a steel strip around a rectangular form and then bonding the layers together. It is then cut in two, forming two C shapes, and the core assembled by binding the two C halves together with a steel strap. They have the advantage that the flux is always oriented parallel to the metal grains, reducing reluctance.

A steel core's remanence means that it retains a static magnetic field when power is removed. When power is then reapplied, the residual field will cause a high inrush current until the effect of the remaining magnetism is reduced, usually after a few cycles of the applied AC waveform. Overcurrent protection devices such as fuses must be selected to allow this harmless inrush to pass. On transformers connected to long, overhead power transmission lines, induced currents due to geomagnetic disturbances during solar storms can cause saturation of the core and operation of transformer protection devices.

Distribution transformers can achieve low no-load losses by using cores made with low-loss high-permeability silicon steel or amorphous (non-crystalline) metal alloy. The higher initial cost of the core material is offset over the life of the transformer by its lower losses at light load.

Solid Cores

Powdered iron cores are used in circuits such as switch-mode power supplies that operate above mains frequencies and up to a few tens of kilohertz. These materials combine high magnetic permeability with high bulk electrical resistivity. For frequencies extending beyond the VHF band, cores made from non-conductive magnetic ceramic materials called ferrites are common. Some radio-frequency transformers also have movable cores (sometimes called 'slugs') which allow adjustment of the coupling coefficient (and bandwidth) of tuned radio-frequency circuits.

Toroidal Cores

Toroidal transformers are built around a ring-shaped core, which, depending on operating frequency, is made from a long strip of silicon steel or permalloy wound into a coil, powdered iron, or ferrite. A strip construction ensures that the grain boundaries are optimally aligned, improving the transformer's efficiency by reducing the core's reluctance. The closed ring shape eliminates air gaps inherent in the construction of an E-I core. The cross-section of the ring is usually square or rectangular, but more expensive cores with circular cross-sections are also available. The primary and secondary coils are often wound concentrically to cover the entire surface of the core. This minimizes the length of wire needed, and also provides screening to minimize the core's magnetic field from generating electromagnetic interference.

Toroidal transformers are more efficient than the cheaper laminated E-I types for a similar power level. Other advantages compared to E-I types, include smaller size (about half), lower weight (about half), less mechanical hum (making them superior in audio amplifiers), lower exterior magnetic field (about one tenth), low off-load losses (making them more efficient in standby circuits), single-bolt mounting, and greater choice of shapes. The main disadvantages are higher cost and limited power capacity. Because of the lack of a residual gap in the magnetic path, toroidal transformers also tend to exhibit higher inrush current, compared to laminated E-I types.

Ferrite toroidal cores are used at higher frequencies, typically between a few tens of kilohertz to hundreds of megahertz, to reduce losses, physical size, and weight of inductive components. A drawback of toroidal transformer construction is the higher labor cost of winding. This is because it is necessary to pass the entire length of a coil winding through the core aperture each time a single turn is added to the coil. As a consequence, toroidal transformers rated more than a few kVA are uncommon. Small distribution transformers may achieve some of the benefits of a toroidal core by splitting it and forcing it open, then inserting a bobbin containing primary and secondary windings.

Air Cores

A physical core is not an absolute requisite and a functioning transformer can be produced simply by placing the windings near each other, an arrangement termed an 'air-core' transformer. The air which comprises the magnetic circuit is essentially lossless, and so an air-core transformer eliminates loss due to hysteresis in the core material. The leakage inductance is inevitably high, resulting in very poor regulation, and so such designs are unsuitable for use in power distribution. They have however very high bandwidth, and are frequently employed in radio-frequency applications, for which a satisfactory coupling coefficient is maintained by carefully overlapping the primary and secondary windings. They're also used for resonant transformers such as Tesla coils where they can achieve reasonably low loss in spite of the high leakage inductance.

Windings

The conducting material used for the windings depends upon the application, but in all cases the individual turns must be electrically insulated from each other to ensure that the current travels throughout every turn. For small power and signal transformers, in which currents are low and the potential difference between adjacent turns is small, the coils are often wound from enamelled magnet wire, such as Formvar wire. Larger power transformers operating at high voltages may be wound with copper rectangular strip conductors insulated by oil-impregnated paper and blocks of pressboard.

High-frequency transformers operating in the tens to hundreds of kilohertz often have windings made of braided Litz wire to minimize the skin-effect and proximity effect losses. Large power transformers use multiple-stranded conductors as well, since even at low power frequencies non-uniform distribution of current would otherwise exist in high-current windings. Each strand is individually insulated, and the strands are arranged so that at certain points in the winding, or throughout the whole winding, each portion occupies different relative positions in the complete conductor. The transposition equalizes the current flowing in each strand of the conductor, and reduces eddy current losses in the winding itself. The stranded conductor is also more flexible than a solid conductor of similar size, aiding manufacture.

The windings of signal transformers minimize leakage inductance and stray capacitance to improve high-frequency response. Coils are split into sections, and those sections interleaved between the sections of the other winding.

Power-frequency transformers may have taps at intermediate points on the winding, usually on the higher voltage winding side, for voltage adjustment. Taps may be manually reconnected, or a manual or automatic switch may be provided for changing taps. Automatic on-load tap changers are used in electric power transmission or distribution, on equipment such as arc furnace transformers, or for automatic voltage regulators for sensitive loads. Audio-frequency transformers, used for the distribution of audio to public address loudspeakers, have taps to allow adjustment of impedance to each speaker. A center-tapped transformer is often used in the output stage of an audio power amplifier in a push-pull circuit. Modulation transformers in AM transmitters are very similar.

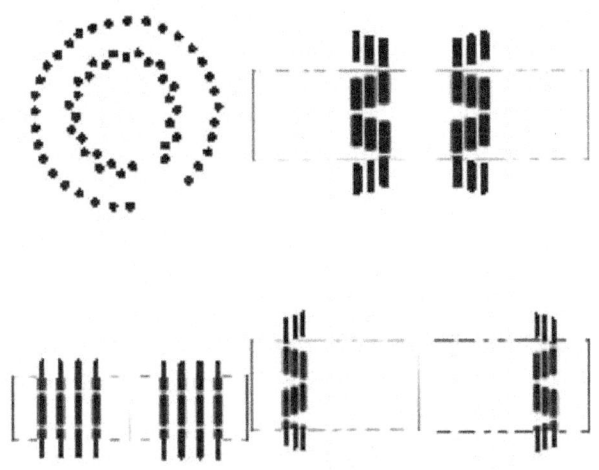

Fig. : Cut view through transformer windings. White : insulator. Green spiral : Grain oriented silicon steel. Black : Primary winding made of oxygen-free copper. Red : Secondary winding. Top left : Toroidal transformer. Right : C-core, but E-core would be similar. The black windings are made of film. Top : Equally low capacitance between all ends of both windings. Since most cores are at least moderately conductive they also need insulation. Bottom : Lowest capacitance for one end of the secondary winding needed for low-power high-voltage transformers. Bottom left : Reduction of leakage inductance would lead to increase of capacitance.

Dry-type transformer winding insulation systems can be either of standard open-wound 'dip-and-bake' construction or of higher quality designs that include vacuum pressure impregnation (VPI), vacuum pressure encapsulation (VPE), and cast coil encapsulation processes. In the VPI process, a combination of heat, vacuum and pressure is used to thoroughly seal, bind, and eliminate entrained air voids in the winding polyester resin insulation coat layer, thus increasing resistance to corona. VPE windings are similar to VPI windings but provide more protection against environmental effects, such as from water, dirt or corrosive ambients, by multiple dips including typically in terms of final epoxy coat.

Cooling

To place the cooling problem in perspective, the accepted rule of thumb is that the life expectancy of insulation in all electric machines including all transformers is halved for about every 7 °C to 10 °C increase in operating temperature, this life expectancy halving rule holding more narrowly when the increase is between about 7 °C to 8 °C in the case of transformer winding cellulose insulation.

Small dry-type and liquid-immersed transformers are often self-cooled by natural convection and radiation heat dissipation. As power ratings increase, transformers are often cooled by forced-air cooling, forced-oil cooling, water-cooling, or combinations of these. Large transformers are filled with transformer oil that both cools and insulates the windings. Transformer oil is a highly refined mineral oil that cools the windings and insulation by circulating within the transformer tank. The mineral oil and paper insulation system has been extensively studied and used for more than 100 years. It is estimated that 50% of power transformers will survive 50 years of use, that the average age of failure of power transformers is about 10 to 15 years, and that about 30% of power transformer failures are due to insulation and overloading failures. Prolonged operation at elevated temperature degrades insulating properties of winding insulation and dielectric coolant, which not only shortens transformer life but can ultimately lead to catastrophic transformer failure. With a great body of empirical study as a guide, transformer oil testing including dissolved gas analysis provides valuable maintenance information. This underlines the need to monitor, model, forecast and manage oil and winding conductor insulation temperature conditions under varying, possibly difficult, power loading conditions.

Building regulations in many jurisdictions require indoor liquid-filled transformers to either use dielectric fluids that are less flammable than oil, or be installed in fire-resistant rooms. Air-cooled dry transformers can be more economical where they eliminate the cost of a fire-resistant transformer room.

The tank of liquid filled transformers often has radiators through which the liquid coolant circulates by natural convection or fins. Some large transformers employ electric fans for forced-air cooling, pumps for forced-liquid cooling, or have heat exchangers for water-cooling. An oil-immersed transformer may be equipped with a Buchholz relay, which, depending on severity of gas accumulation due to internal arcing, is used to either alarm or de-energize the transformer. Oil-immersed transformer installations usually include fire protection measures such as walls, oil containment, and fire-suppression sprinkler systems. Another protection means consists in fast depressurization systems which are activated by the first dynamic pressure peak of the shock wave, avoiding transformer explosion before static pressure increases. Many explosions are reported to have been avoided thanks to this technology.

Polychlorinated biphenyls have properties that once favored their use as a dielectric coolant, though concerns over their environmental persistence led to a widespread ban on their use. Today, non-toxic, stable silicone-based oils, or fluorinated hydrocarbons may be used where the expense of a fire-resistant liquid offsets additional building cost for a transformer vault. PCBs for new equipment was banned in 1981 and in 2000 for use in existing equipment in United Kingdom Legislation enacted in Canada between 1977 and 1985 essentially bans PCB use in transformers manufactured in or imported into the country after 1980, the maximum allowable level of PCB contamination in existing mineral oil transformers being 50 ppm.

Some transformers, instead of being liquid-filled, have their windings enclosed in sealed, pressurized tanks and cooled by nitrogen or sulfur hexafluoride gas.

Experimental power transformers in the 500-to-1,000 kVA range have been built with liquid nitrogen or helium cooled superconducting windings, which eliminates winding losses without affecting core losses.

Insulation Drying

Construction of oil-filled transformers requires that the insulation covering the windings be thoroughly dried of residual moisture before the oil is introduced. Drying is carried out at the factory, and may also be required as a field service. Drying may be done by circulating hot air around the core, or by vapour-phase drying (VPD) where an evapourated solvent transfers heat by condensation on the coil and core.

For small transformers, resistance heating by injection of current into the windings is used. The heating can be controlled very well, and it is energy efficient. The method is called low-frequency heating (LFH) since the current used is at a much lower frequency than that of the power grid, which is normally 50 or 60 Hz. A lower frequency reduces the effect of inductance, so the voltage required can be reduced. The LFH drying method is also used for service of older transformers.

Bushings

Larger transformers are provided with high-voltage insulated bushings made of polymers or porcelain. A large bushing can be a complex structure since it must provide careful control of the electric field gradient without letting the transformer leak oil.

Classification Parameters

Transformers can be classified in many ways, such as the following :

- *Power capacity* : From a fraction of a volt-ampere (VA) to over a thousand MVA.
- *Duty of a transformer* : Continuous, short-time, intermittent, periodic, varying.
- *Frequency range* : Power-frequency, audio-frequency, or radio-frequency.
- *Voltage class* : From a few volts to hundreds of kilovolts.
- *Cooling type* : Dry and liquid-immersed - self-cooled, forced air-cooled; liquid-immersed - forced oil-cooled, water-cooled.
- *Circuit application* : Such as power supply, impedance matching, output voltage and current stabilizer or circuit isolation.
- *Utilization* : Pulse, power, distribution, rectifier, arc furnace, amplifier output, etc.

- *Basic magnetic form* : Core form, shell form.
- *Constant-potential transformer descriptor* : Step-up, step-down, isolation.
- *General winding configuration* : By EIC vector group - various possible two-winding combinations of the phase designations delta, wye or star, and zigzag or interconnected star; other - autotransformer, Scott-T, zigzag grounding transformer winding.
- *Rectifier phase-shift winding configuration* : 2-winding, 6-pulse; 3-winding, 12-pulse; . . . n-winding, [n-1]*6-pulse; polygon; etc.

Applications

Transformers are used to increase voltage before transmitting electrical energy over long distances through wires. Wires have resistance which loses energy through joule heating at a rate corresponding to square of the current. By transforming power to a higher voltage transformers enable economical transmission of power and distribution. Consequently, transformers have shaped the electricity supply industry, permitting generation to be located remotely from points of demand. All but a tiny fraction of the world's electrical power has passed through a series of transformers by the time it reaches the consumer.

Transformers are also used extensively in electronic products to step-down the supply voltage to a level suitable for the low voltage circuits they contain. The transformer also electrically isolates the end user from contact with the supply voltage.

Signal and audio transformers are used to couple stages of amplifiers and to match devices such as microphones and record players to the input of amplifiers. Audio transformers allowed telephone circuits to carry on a two-way conversation over a single pair of wires. A balun transformer converts a signal that is referenced to ground to a signal that has balanced voltages to ground, such as between external cables and internal circuits.

History

Discovery of Induction

Electromagnetic induction, the principle of the operation of the transformer, was discovered independently and almost simultaneously by Joseph Henry and Michael Faraday in 1831. Although Henry's work likely having preceded Faraday's work by a few months, Faraday was the first to publish the results of his experiments and thus receive credit for the discovery. The relationship between emf and magnetic flux is an equation now known as Faraday's law of induction :

$$|\varepsilon| = \left|\frac{d\Phi_B}{dt}\right|$$

where $|\varepsilon|$ is the magnitude of the emf in volts and Φ_B is the magnetic flux through the circuit in webers.

Faraday performed the first experiments on induction between coils of wire, including winding a pair of coils around an iron ring, thus creating the first toroidal closed-core transformer. However he only applied individual pulses of current to his transformer, and never discovered the relation between the turns ratio and emf in the windings.

Induction Coils

The first type of transformer to see wide use was the induction coil, invented by Rev. Nicholas Callan of Maynooth College, Ireland in 1836. He was one of the first researchers to realize the more turns the secondary winding has in relation to the primary winding, the larger the induced secondary emf will be. Induction coils evolved from scientists' and inventors' efforts to get higher voltages from batteries. Since batteries produce direct current (DC) rather than AC, induction coils relied upon vibrating electrical contacts that regularly interrupted the current in the primary to create the flux changes necessary for induction. Between the 1830s and the 1870s, efforts to build better induction coils, mostly by trial and error, slowly revealed the basic principles of transformers.

First Alternating Current Transformers

By the 1870s, efficient generators producing alternating current (AC) were available, and it was found AC could power an induction coil directly, without an interrupter.

In 1876, Russian engineer Pavel Yablochkov invented a lighting system based on a set of induction coils where the primary windings were connected to a source of AC. The secondary windings could be connected to several 'electric candles' (arc lamps) of his own design. The coils Yablochkov employed functioned essentially as transformers.

In 1878, the Ganz factory, Budapest, Hungary, began manufacturing equipment for electric lighting and, by 1883, had installed over fifty systems in Austria-Hungary. Their AC systems used arc and incandescent lamps, generators, and other equipment.

Lucien Gaulard and John Dixon Gibbs first exhibited a device with an open iron core called a 'secondary generator' in London in 1882, then sold the idea to the Westinghouse company in the United States. They also exhibited the invention in Turin, Italy in 1884, where it was adopted for an electric lighting system. However, the efficiency of their open-core bipolar apparatus remained very low.

Early Series Circuit Transformer Distribution

Induction coils with open magnetic circuits are inefficient at transferring power to loads. Until about 1880, the paradigm for AC power transmission from a high voltage supply to a low voltage load was a series circuit. Open-core transformers with a ratio near 1 : 1 were connected with their primaries in series to allow use of a high voltage for transmission while presenting a low voltage to the lamps.

The inherent flaw in this method was that turning off a single lamp (or other electric device) affected the voltage supplied to all others on the same circuit. Many adjustable transformer designs were introduced to compensate for this problematic characteristic of the series circuit, including those employing methods of adjusting the core or bypassing the magnetic flux around part of a coil. Efficient, practical transformer designs did not appear until the 1880s, but within a decade, the transformer would be instrumental in the War of Currents, and in seeing AC distribution systems triumph over their DC counterparts, a position in which they have remained dominant ever since.

Closed-core Transformers and Parallel Power Distribution

In the autumn of 1884, Károly Zipernowsky, Ottó Bláthy and Miksa Déri (ZBD), three engineers associated with the Ganz factory, had determined that open-core devices were impracticable, as they were incapable of reliably regulating voltage. In their joint 1885 patent applications for novel transformers (later called ZBD transformers), they described two designs with closed magnetic circuits where copper windings were either a) wound around iron wire ring core or b) surrounded by iron wire core. The two designs were the first application of the two basic transformer constructions in common use to this day, which can as a class all be termed as either core form or shell form (or alternatively, core type or shell type), as in a) or b), respectively. The Ganz factory had also in the autumn of 1884 made delivery of the world's first five high-efficiency AC transformers, the first of these units having been shipped on September 16, 1884. This first unit had been manufactured to the following specifications : 1,400 W, 40 Hz, 120 :72 V, 11.6 : 19.4 A, ratio 1.67 : 1, one-phase, shell form.

In both designs, the magnetic flux linking the primary and secondary windings travelled almost entirely within the confines of the iron core, with no intentional path through air. The new transformers were 3.4 times more efficient than the open-core bipolar devices of Gaulard and Gibbs. The ZBD patents included two other major interrelated innovations : one concerning the use of parallel connected, instead of series connected, utilization loads, the other concerning the ability to have high turns ratio transformers such that the supply network voltage could be much higher (initially 1,400 to 2,000 V) than the voltage of utilization loads (100 V initially preferred). When employed in parallel connected electric distribution systems, closed-core transformers finally made it technically and economically feasible to provide electric power for lighting in homes, businesses and public spaces. Bláthy had suggested the use of closed cores, Zipernowsky had suggested the use of parallel shunt connections, and Déri had performed the experiments

Transformers today are designed on the principles discovered by the three engineers. They also popularized the word 'transformer' to describe a device for altering the emf of an electric current, although the term had already been in use by 1882. In 1886, the ZBD engineers designed, and the Ganz factory supplied electrical equipment for, the world's first power station that used AC generators to power a parallel connected common electrical network, the steam-powered Rome-Cerchi power plant.

Although George Westinghouse had bought Gaulard and Gibbs' patents in 1885, the Edison Electric Light Company held an option on the US rights for the ZBD transformers, requiring Westinghouse to pursue alternative designs on the same principles. He assigned to William Stanley the task of developing a device for commercial use in United States. Stanley's first patented design was for induction coils with single cores of soft iron and adjustable gaps to regulate the emf present in the secondary winding. This design was first used commercially in the US in 1886 but Westinghouse was intent on improving the Stanley design to make it (unlike the ZBD type) easy and cheap to produce.

Westinghouse, Stanley and associates soon developed an easier to manufacture core, consisting of a stack of thin 'Eshaped' iron plates, insulated by thin sheets of paper or other insulating material. Prewound copper coils could then be slid into place, and straight iron plates laid in to create a closed magnetic circuit. Westinghouse applied for a patent for the new low-cost design in December 1886; it was granted in July 1887.

Other Early Transformers

In 1889, Russian-born engineer Mikhail Dolivo-Dobrovolsky developed the first three-phase transformer at the Allgemeine Elektricitäts-Gesellschaft ('General Electricity Company') in Germany.

In 1891, Nikola Tesla invented the Tesla coil, an air-cored, dual-tuned resonant transformer for generating very high voltages at high frequency.

TYPES OF TRANSFORMER

Various specific electrical application designs require a variety of transformer types. Although they all share the basic characteristic transformer principles, they are customize in construction or electrical properties for certain installation requirements or circuit conditions.

Autotransformer

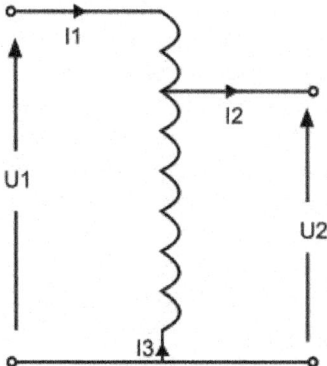

Fig. : Autotransformer.

An **autotransformer** (sometimes called *autostep down transformer*) is an electrical transformer with only one winding. The "auto" (Greek for "self") prefix refers to the single coil acting on itself and not to any kind of automatic mechanism. In an autotransformer, portions of the same winding act as both the primary and secondary sides of the transformer. The winding has at least three taps where electrical connections are made. Autotransformers have the advantages of often being smaller, lighter, and cheaper than typical dual-winding transformers, but the disadvantage of not providing electrical isolation. Other advantages of autotransformers include lower leakage reactance, lower losses, lower excitation current, and increased KVA rating.

Autotransformers are often used to step up or step down voltages in the 110-115-120 V range and voltages in the 220-230-240 volt range — for example. Providing 110 V or 120 V (with taps) from 230 V input, allowing equipment designed for 100 or 120 volts to be used with a 230 volt supply (as in using US electrical equipment with higher European voltages).

Operation

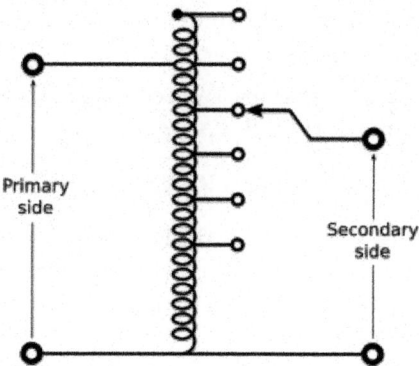

Fig. : Single-phase tapped autotransformer with output voltage range of 40%–115% of input.

An autotransformer has a single winding with two end terminals, and one or more terminals at intermediate tap points, or a transformer in which the primary and secondary coils have part or all of their turns in common. The primary voltage is applied across two of the terminals, and the secondary voltage taken from two terminals, almost always having one terminal in common with the primary voltage. The primary and secondary circuits therefore have a number of windings turns in common. Since the volts-per-turn is the same in both windings, each develops a voltage in proportion to its number of turns. In an autotransformer part of the current flows directly from the input to the output, and only part is transferred inductively, allowing a smaller, lighter, cheaper core to be used as well as requiring only a single winding. However the voltage and current ratio of autotransformers can be formulated the same as other two-winding transformers :

$$\frac{V_1}{V_2} = \frac{N_1}{N_2} = a$$

$$(0 < V_2 < V_1)$$

The ampere-turns provided by the upper half :

$$F_U = (N_1 - N_2)\, I_1 = \left(1 - \frac{1}{a}\right) N_1 I_1$$

The ampere-turns provided by the lower half :

$$F_L = N_2\, (I_2 - I_1) = \frac{N_1}{a}\, (I_2 - I_1)$$

For ampere-turn balance, $F_U = F_L$:

$$\left(1 - \frac{1}{a}\right) N_1 I_1 = \frac{N_1}{a}\, (I_2 - I_1)$$

Therefore :

$$\frac{I_1}{I_2} = \frac{1}{a}$$

One end of the winding is usually connected in common to both the voltage source and the electrical load. The other end of the source and load are connected to taps along the winding. Different taps on the winding correspond to different voltages, measured from the common end. In a step-down transformer the source is usually connected across the entire winding while the load is connected by a tap across only a portion of the winding. In a step-up transformer, conversely, the load is attached across the full winding while the source is connected to a tap across a portion of the winding.

As in a two-winding transformer, the ratio of secondary to primary voltages is equal to the ratio of the number of turns of the winding they connect to. For example, connecting the load between the middle and bottom of the autotransformer will reduce the voltage by 50%. Depending on the application, that portion of the winding used solely in the higher-voltage (lower current) portion may be wound with wire of a smaller gauge, though the entire winding is directly connected.

Limitations

An autotransformer does not provide electrical isolation between its windings as an ordinary transformer does; if the neutral side of the input is not at ground voltage, the neutral side of the output will not be either. A failure of the insulation of the windings of an autotransformer can result in full input voltage applied to the output. Also, a break in the part of the winding that is used as both primary and secondary will result in the transformer acting as an inductor in series with the load (which under light load conditions may result in near full input voltage being applied to the output). These are important safety considerations when deciding to use an autotransformer in a given application.

Because it requires both fewer windings and a smaller core, an autotransformer for power applications is typically lighter and less costly than a two-winding transformer, up to a voltage ratio of about 3 : 1; beyond that range, a two-winding transformer is usually more economical.

In three phase power transmission applications, autotransformers have the limitations of not suppressing harmonic currents and as acting as another source of ground fault currents. A large three-phase autotransformer may have a "buried" delta winding, not connected to the outside of the tank, to absorb some harmonic currents.

In practice, losses mean that both standard transformers and autotransformers are not perfectly reversible; one designed for stepping down a voltage will deliver slightly less voltage than required if it is used to step up. The difference is usually slight enough to allow reversal where the actual voltage level is not critical.

Like multiple-winding transformers, autotransformers use time-varying magnetic fields to transfer power. They require alternating currents to operate properly and will not function on direct current.

Applications

Power Transmission and Distribution

Autotransformers are frequently used in power applications to interconnect systems operating at different voltage classes, for example 138 kV to 66 kV for transmission. Another application in industry is to adapt machinery built (for example) for 480 V supplies to operate on a 600 V supply. They are also often used for providing conversions between the two common domestic mains voltage bands in the world (100 V - 130 V and 200 V - 250 V). The links between the UK 400 kV and 275 kV 'Super Grid' networks are normally three phase autotransformers with taps at the common neutral end.

On long rural power distribution lines, special autotransformers with automatic tap-changing equipment are inserted as voltage regulators, so that customers at the far end of the line receive the same average voltage as those closer to the source. The variable ratio of the autotransformer compensates for the voltage drop along the line.

A special form of autotransformer called a *zig zag* is used to provide grounding on three-phase systems that otherwise have no connection to ground. A zig-zag transformer provides a path for current that is common to all three phases (so-called *zero sequence* current).

Audio

In audio applications, tapped autotransformers are used to adapt speakers to constant-voltage audio distribution systems, and for impedance matching such as between a low-impedance microphone and a high-impedance amplifier input.

Railways

In UK railway applications, it is common to power the trains at 25 kV AC. To increase the distance between electricity supply Grid feeder points they can be arranged to supply a 25-0-25 kV supply with the third wire (opposite phase) out of reach of the train's overhead collector pantograph. The 0 V point of the supply is connected to the rail while one 25 kV point is connected to the overhead contact wire. At frequent (about 10 km) intervals, an autotransformer links the contact wire to rail and to the second (antiphase) supply conductor. This system increases usable transmission distance, reduces induced interference into external equipment and reduces cost. A variant is occasionally seen where the supply conductor is at a different voltage to the contact wire with the autotransformer ratio modified to suit.

Variable Autotransformers

As with two-winding transformers, autotransformers may be equipped with many taps and automatic switchgear to allow them to act as automatic voltage regulators, to maintain a steady voltage at the customers' service during a wide range of load conditions. They can also be used to simulate low line conditions for testing. Another application is a lighting dimmer that doesn't produce the EMI typical of most thyristor dimmers.

By exposing part of the winding coils and making the secondary connection through a sliding brush, a continuously variable turns ratio can be obtained, allowing for very smooth control of voltage. Applicable only for relatively low voltage designs, this device is known as a variable AC transformer (often referred to by the trademark name Variac). The output voltage is not limited to the discrete voltages represented by actual number of turns. The voltage can be smoothly varied between turns as the brush has a relatively high resistance (compared with a metal contact) and the actual output voltage is a function of the relative area of brush in contact with adjacent windings. The relatively high resistance of the brush also prevents it from acting as a short circuited turn when it contacts two adjacent turns. Typically the primary connection connects to only a part of the winding allowing the output voltage to be varied smoothly from zero to above the input voltage and thus allowing the device to be used for testing electrical equipment at the limits of its specified voltage range.

Variac Trademark

From 1934 to 2002, **Variac** was a U.S. trademark of General Radio for a variable autotransformer intended to conveniently vary the output voltage for a steady AC input voltage. In 2004, Instrument Service Equipment applied for and obtained the *Variac* trademark for the same type of product.

Capacitor Voltage Transformer

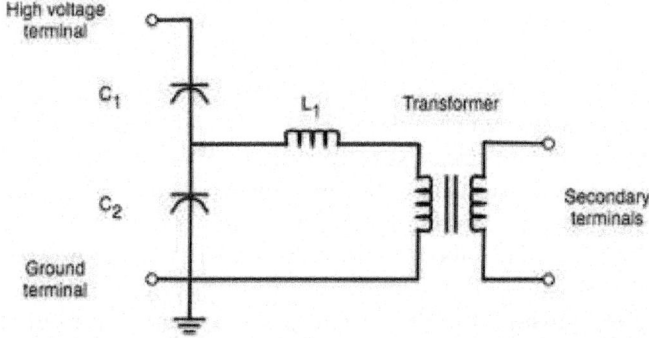

Fig. : The circuit diagram for a simple capacitor voltage transformer.

A **capacitor voltage transformer** (CVT), or **capacitance coupled voltage transformer** (CCVT) is a transformer used in power systems to step down extra high voltage signals and provide a low voltage signal, for measurement or to operate a protective relay. In its most basic form the device consists of three parts : two capacitors across which the transmission line signal is split, an inductive element to tune the device to the line frequency, and a transformer to isolate and further step down the voltage for the instrumentation or protective relay. The tuning of the divider to the line frequency makes the overall division ratio less sensitive to changes in the burden of the connected metering or protection devices. The device has at least four terminals : a terminal for connection to the high voltage signal, a ground terminal, and two secondary terminals which connect to the instrumentation or protective relay. CVTs are typically single-phase devices used for measuring voltages in excess of one hundred kilovolts where the use of wound primary voltage transformers would be uneconomical. In practice, capacitor C_1 is often constructed as a stack of smaller capacitors connected in series. This provides a large voltage drop across C_1 and a relatively small voltage drop across C_2.

The CVT is also useful in communication systems. CVTs in combination with wave traps are used for filtering high frequency communication signals from power frequency. This forms a carrier communication network throughout the transmission network.

Distribution Transformer

A **distribution transformer** is a transformer that provides the final voltage transformation in the electric power distribution system, stepping down the voltage used in the distribution lines to the level used by the customer. The invention of a practical efficient transformer made AC power distribution feasible; a system using distribution transformers was demonstrated as early as 1882.

If mounted on a utility pole, they are called **pole-mount transformers**. If the distribution lines are located at ground level or underground, distribution transformers are mounted on concrete pads and locked in steel cases, thus known as pad-mount transformers.

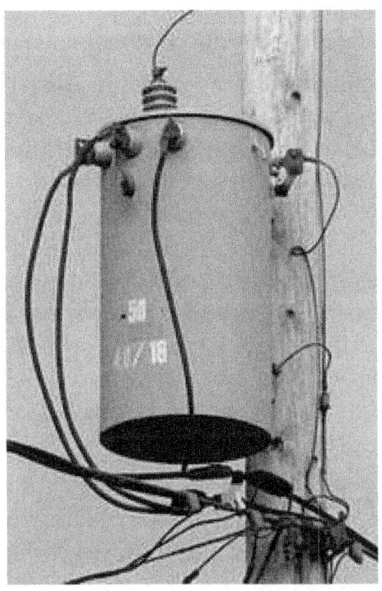

Fig. : Single-phase distribution transformer in Canada.

Distribution transformers normally have ratings up to 200 kVA, although national standard may describe units up to 5000 kVA as distribution transformers. Since distribution transformer are energized for 24 hours a day (even when they don't carry any load), reducing iron losses has an important role in their design. As they usually don't operate at full load, they are designed to have maximum efficiency at lower loads. To have a better efficiency, voltage regulation in these transformers should be kept minimum. Hence they are designed to have small leakage reactance.

Classification

Distribution transformers are classified into different categories based on certain factors such as :

- Mounting location - pole, pad, underground vault
- Type of insulation - liquid-immersed or dry-type
- Number of Phases - single-phase or three-phase
- Voltage class
- Basic impulse insulation level (BIL).

Use

Distribution transformers are normally located at a service drop, where wires run from a utility pole or underground power lines to a customer's premises. They are often used for the power supply of facilities outside settlements, such as isolated houses, farmyards or pumping stations at voltages below 30 kV. Another application is the power supply of the overhead wire of railways electrified with AC. In this case single phase distribution transformers are used.

The number of customers fed by a single distribution transformer varies depending on the number of customers in an area. Several homes may be fed off a single transfomer in urban areas; rural distribution may require one transfomer per customer. A large commercial or industrial complex will have multiple distribution transformers. Padmount transformers are used in urban areas and neighbourhoods where the primary distribution lines run underground. Many large buildings have electric service provided at primary distribution voltage. These buildings have customer-owned transformers in the basement for step-down purposes. In a secondary network system as used in urban areas, many distribution transformers may be connected in parallel, each equipped with its own network protector circuit breaker to isolate it from the secondary netwrok in case of a fault.

Distribution transfomers are also found in the power collector networks of wind farms, where they step up power from each wind turbine to connect to a substation that may be several miles (kilometres) distant.

Connections

Both pole-mount and pad-mount transformers convert the high 'primary' voltage of the overhead or underground distribution lines to the lower 'secondary' voltage of the distribution wires inside the building. The primaries use the three-phase system. Main distribution lines always have three wires, while smaller "laterals" (close to the customer) may include one or two phases, used to serve all customers with single-phase power. If three-phase service is desired, one must have a three-phase supply. Primaries provide power at the standard distribution voltages used in the area; these arange from as low as 2300 volts to about 35,000 volts depending on local distribution practice and standards; often 11,000 V (50 Hz systems)and 13,800 V (60 Hz) systems are used but many other voltages are standard.

Primary

The high voltage primary windings are brought out to bushings on the top of the case.

- Single phase transformers, generally used in the USA system, are attached to the overhead wires with two different types of connections :
  - If a primary neutral wire is available, a 'wye' or 'phase to neutral' transformer can be used. This usually has only one bushing on top, connected to one of the primary phases. The other end of the primary winding is 'grounded' to the transformer's case, which is connected to the neutral wire of the 3 phase system, and also earth ground. This type of distribution system, called 'grounded wye', is preferred because the transformers

present unbalanced loads on the line, causing currents in the neutral wire. With the 'delta' connection, this can cause variations in the voltages on the 3 phase wires.

o If no neutral wire is available, a 'delta' or 'phase to phase' transformer must be used. This has two bushings on top which are connected to two of the three primary wires, so the voltage across the primary winding is the phase-to-phase voltage. This type is used on long distribution lines where it is uneconomical to run a fourth neutral wire.

- Transformers providing three-phase secondary power, which are used for residential service in the European system, have three primary windings and are attached to all three primary phase wires. The windings are almost always connected in a 'wye' configuration, with the ends of the three windings connected and grounded.

The transformer is always connected to the primary distribution lines through protective fuses and disconnect switches. For pole-mounted transformers this usually takes the form of a 'fused cutout'. An electrical fault causes the fuse to melt, and the device drops open to give a visual indication of trouble. It can also be manually opened while the line is energized by lineworkers using insulated hot sticks.

Secondary

The low voltage secondary windings are attached to three or four terminals on the transformer's side.

- In the USA and countries using its system, the secondary is most often the split-phase 240/120 volt system. The 240 V secondary winding is center-tapped and the center neutral wire is grounded, making the two end conductors "hot" with respect to the center tap. These three wires run down the service drop to the electric meter and service panel inside the building. Connecting a load between either hot wire and the neutral gives 120 volts. Connecting between both hot wires gives 240 volts.

- In Europe and countries using its system, the secondary is often the three phase 400Y/230 system. There are three 230 V secondary windings, each receiving power from a primary winding attached to one of the primary phases. One end of each secondary winding is connected to a 'neutral' wire, which is grounded. The other end of the 3 secondary windings, along with the neutral, are brought down the service drop to the service panel. 230 V loads are connected between any of the three phase wires and the neutral.

Higher secondary voltages, such as 480 volts, are sometimes required for commercial and industrial uses. Some industrial customers require three-phase power at secondary voltages. To provide this, three-phase transformers can be used. In the US, which uses mostly single phase transformers, three identical single phase transformers are often wired in a *transformer bank* in either a wye or delta connection, to create a three phase transformer.

Construction

Distribution transformers are made using a core made from laminations of sheet steel stacked and either glued together with resin or banded together with steel straps. Where large numbers of transformers are made to standard designs, a wound C-shaped core is economic to manufacture. A steel strip is wrapped around a former, pressed into shape and then cut into two C-shaped halves, which are re-assembled on tp th ecopper windings.

The primary coils are wound from enamel coated copper or aluminum wire and the high current, low voltage secondaries are wound using a thick ribbon of aluminum or copper. The windings are insulated with resin-impregnated paper. The entire assembly is baked to cure the resin then submerged in a powder coated steel tank which is then filled with transformer oil (or other insulating liquid), which is inert and non-conductive. The transformer oil cools and insulates the windings, and protects the transformer winding from moisture, which will float on the surface of the oil. The tank is temporarily depressurized to remove any remaining moisture that would cause arcing and is sealed against the weather with a gasket at the top.

Formerly distribution transformers for indoor use would be filled with a polychlorinated biphenyl (PCB) liquid. Because these liquids persist in the environment and have adverse effects on animals, they have been banned. Other fire-resistant liquids such as silicones are used where a liquid-filled transformer must be used indoors. Certain vegetable oils have been applied as transformer oil; these have the advantage of a high fire point and are completely biodegradeable in the enviornment.

Pole-mounted transformers often include accessories such as surge arrestors or protective fuse links. A self-protected transformer includes an internal fuse and surge arrestor; other transformers have these components mounted separately outside the tank. Pole-mounted transformers may have lugs allownig direct mounting to a pole, or may be mounted on crossarms bolted to the pole. Aerial transformers, larger than around 75 kVA, may be mounted on a platform supported by one or more poles. A three-phase service may use three idential transformers, one per phase.

Transformers designed for below-grade installation can be designed for periodic submersion in water.

Distribution transformers may include an off-load tapchanger to allow slight adjust ment of the ratio between primary and secondary voltage, to bring the customer voltage within the desired range on long or heavily loaded lines.

Pad-mounted transformers have secure locked and bolted grounded metal enclosures to discourage unauhtorized access to live internal parts. The enclosure may also include fuses, isolating switches, load-break bushings, and other accessories as described in technical standards. Pad-mounted transformers for distribution systems typically range from around 100 to 2000 kVA, although some larger units are also used.

Quadrature Booster

A **phase angle regulating transformer**, **phase angle regulator** (**PAR**, American usage), **phase-shifting transformer**, **phase shifter** (West coast American usage), or **quadrature booster** (**quad booster**, British usage), is a specialised form of transformer used to control the flow of real power on three-phase electricity transmission networks.

For an alternating current transmission line, power flow through the line is proportional to the sine of the difference in the phase angle of the voltage between the transmitting end and the receiving end of the line. Where parallel circuits with different capacity exist between two points in a transmission grid (for example, an overhead line and an underground cable), direct manipulation of the phase angle allows control of the division of power flow between the paths, preventing overload. Quadrature boosters thus provide a means of relieving overloads on heavily laden circuits and re-routing power via more favourable paths.

Alternately, where an interchange partner is intentionally causing significant "inadvertent energy" to flow through an unwilling interchange partner's system, the unwilling partner may threaten to install a phase shifter to prevent such "inadvertent energy", with the unwilling partner's tactical objective being the improvement of his system's stability at the expense of the other system's stability. As power system reliability is really a regional or national strategic objective, the threat to install a phase shifter is usually sufficient to cause the other system to implement the required changes to his system to reduce or eliminate the "inadvertent energy".

The capital cost of a quadrature booster can be high : as much as four to six million GBP (6–9 million USD) for a unit rated over 2 GVA. However, the utility to transmission system operators in flexibility and speed of operation, and particularly savings in permitting more economical dispatch of generation, can soon recover the cost of ownership.

Method of Operation

By means of a voltage derived from the supply that is first phase-shifted by 90° (hence is in quadrature), and then re-applied to it, a phase angle is developed across the quadrature booster. It is this induced phase angle that affects the flow of power through specified circuits.

Arrangement

A quadrature booster typically consists of two separate transformers : a shunt unit and a series unit. The shunt unit has its windings connected across the phases, so it produces output voltages shifted by 90° with respect to the supply. Its output is then applied as input to the series unit, which, because its secondary winding is in series with the main circuit, adds the phase-shifted component. The overall output voltage is hence the vector sum of the supply voltage and the 90° quadrature component.

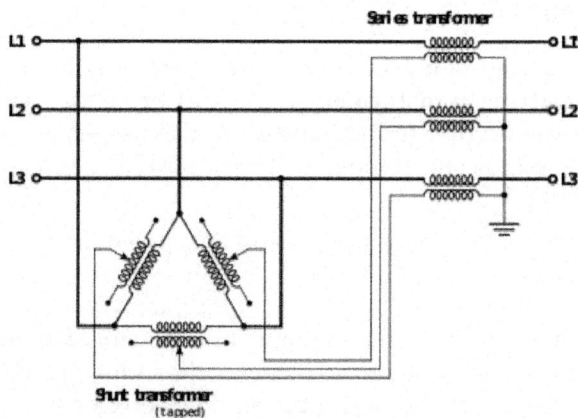

Fig. : Simplified circuit diagram of a three-phase quadrature booster. Arrows shown on shunt transformer secondary windings are movable taps; the windings have floating ends shown, and grounded centre taps (not shown).

Tap connections on the shunt unit allow the magnitude of the quadrature component to be controlled, and thus the magnitude of the phase shift across the quadrature booster. The flow on the circuit containing the quadrature booster may be increased (*boost tapping*) or reduced (*buck tapping*). Subject to system conditions, the flow may even be bucked enough to completely reverse from its neutral-tap direction.

Illustration of Effect

The one-line diagram below shows the effect of tapping a quadrature booster on a notional 100 MW generator-load system with two parallel transmission lines, one of which features a quadrature booster (shaded grey) with a tap range of 1 to 19.

In the left-hand image, the quadrature booster is at its centre tap position of 10 and has a phase angle of 0°. It thus does not affect the power flow through its circuit and both lines are equally loaded at 50 MW. The right-hand image shows the same network with the quadrature booster tapped down so to buck the power flow. The resulting negative phase angle has diverted 23 MW of loading onto the parallel circuit, while the total load supplied is unchanged at 100 MW. (Note that the values used here are hypothetical; the actual phase angle and transfer in load would depend upon the parameters of the quadrature booster and the transmission lines.)

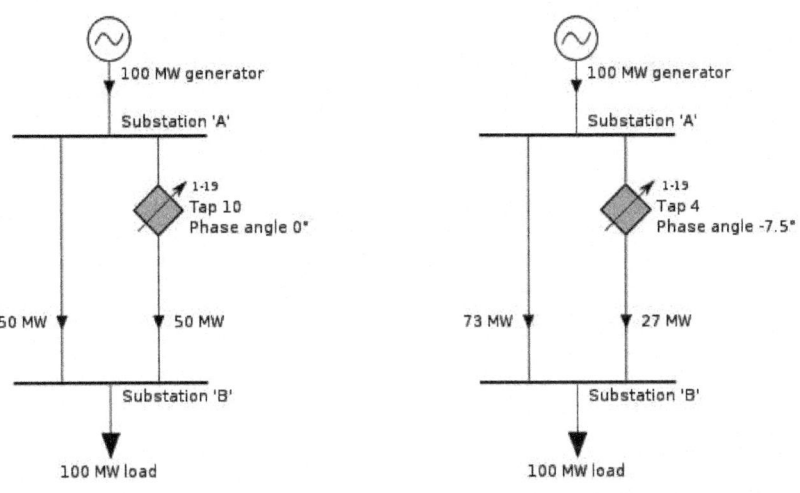

Fig. : Effect of tapping a quadrature booster.

The intended effect is opposite : equalizing power on lines where naturally one would be heavily loaded and one would be lightly loaded.

Scott-T Transformer

A **Scott-T transformer** (also called a **Scott connection**) is a type of circuit used to derive two-phase electric power (2-φ, 90-degree phase rotation) from a three-phase (3-φ, 120-degree phase rotation) source, or *vice-versa*. The Scott connection evenly distributes a balanced load between the phases of the source. The Scott three-phase transformer was invented by a Westinghouse engineer, Charles F. Scott, in the late 1890s to bypass Thomas Edison's more expensive rotary converter and thereby permit two-phase generator plants to drive Nikola Tesla's three-phase motors.

Interconnection

At this time two-phase motor loads also existed and the Scott connection allowed connecting them to newer three-phase supplies with the currents equal on the three phases. This was valuable for getting equal voltage drop and thus feasible regulation of the voltage from the electric generator (the phases cannot be varied separately in a three-phase machine). Nikola Tesla's original polyphase power system was based on simple to build two-phase four-wire components. However, as transmission distances increased, the more transmission line efficient three-phase system became more common. (Three phase power can be transmitted with only three wires, where the two-phase power systems required four wires, two per phase.) Both 2-φ and 3-φ components coexisted for a number of years and the Scott-T transformer connection allowed them to be interconnected.

Scott-T transformer converts 3-φ to 2-φ

Fig. : Standard Scott Connection 3-φ to 2-φ.

Technical Details

Assuming the desired voltage is the same on the two and three phase sides, the Scott-T transformer connection (shown right) consists of a centre-tapped 1 : 1 ratio main transformer, T1, and an 86.6% ($0.5\sqrt{3}$) ratio teaser transformer, T2. The centre-tapped side of T1 is connected between two of the phases on the three-phase side. Its centre tap then connects to one end of the lower turn count side of T2, the other end connects to the remaining phase. The other side of the transformers then connect directly to the two pairs of a two-phase four-wire system.

Unbalanced Loads

Two-phase motors draw constant power the same as three-phase motors, so a balanced two-phase load is converted to a balanced three-phase load. However if the two-phase load is not balanced (more power on one phase than the other), the Scott-T transformer cannot fix this. Unbalanced current on the two-phase side causes unbalanced current on the three-phase side. As the typical 2 phase load was a motor, equality of the current in the 2 phases was inherently presumed during the Scott development. In modern times people have tried to revive the Scott connection as a way to power single phase electric railways from 3 phase Utility supplies. This will not result in balanced current on the 3 phase side as it is unlikely that 2 different railway sections connected as the 2 phases will at all times conform to the Scott presumption of being equal. The instantaneous difference in loading on the 2 sections will be seen as an imbalance in the 3 phase supply, there is no ability to smooth it out.

Back to Back Arrangement

The Scott-T transformer connection may be also be used in a back to back T to T arrangement for a three-phase to 3 phase connection. This is a cost saving in the smaller kVA transformers due to the 2 coil T connected to a secondary 2 coil T in-lieu of the traditional three-coil primary to three-coil secondary transformer.

In this arrangement the X0 Neutral tap is part way up on the secondary teaser transformer. The voltage stability of this T to T arrangement as compared to the traditional 3 coil primary to three-coil secondary transformer is questioned.

3-phase

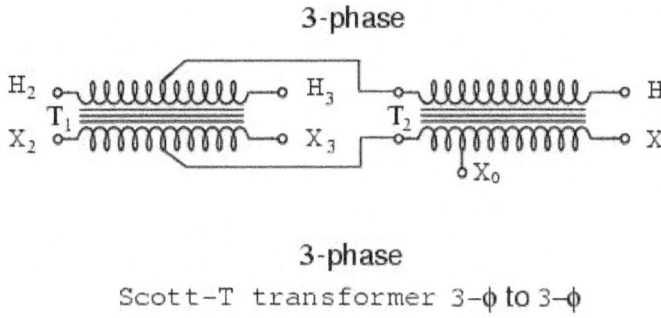

3-phase

Scott-T transformer 3-φ to 3-φ

Fig. : Scott Connection 3-φ to 3-φ.

Current Transformer

A **current transformer (CT)** is used for measurement of alternating electric currents. Current transformers, together with voltage transformers (VT) (potential transformers (PT)), are known as **instrument transformers**. When current in a circuit is too high to apply directly to measuring instruments, a current transformer produces a reduced current accurately proportional to the current in the circuit, which can be conveniently connected to measuring and recording instruments. A current transformer isolates the measuring instruments from what may be very high voltage in the monitored circuit. Current transformers are commonly used in metering and protective relays in the electrical power industry.

Design

Fig. : Basic operation of current transformer.

Like any other transformer, a current transformer has a primary winding, a magnetic core and a secondary winding. The alternating current flowing in the primary produces an alternating magnetic field in the core, which then induces an alternating current in the secondary winding circuit. An essential objective of current transformer design is to ensure the primary and secondary circuits are efficiently coupled, so the secondary current is linearly proportional to the primary current.

The most common design of CT consists of a length of wire wrapped many times around a silicon steel ring passed 'around' the circuit being measured. The CT's primary circuit therefore consists of a single 'turn' of conductor, with a secondary of many tens or hundreds of turns. The primary winding may be a permanent part of the current transformer, with a heavy copper bar to carry current through the magnetic core. Window-type current transformers (aka zero sequence current transformers, or ZSCT) are also common, which can have circuit cables run through the middle of an opening in the core to provide a single-turn primary winding. When conductors passing through a CT are not centered in the circular (or oval) opening, slight inaccuracies may occur.

Shapes and sizes can vary depending on the end user or switchgear manufacturer. Typical examples of low voltage single ratio metering current transformers are either ring type or plastic molded case. High-voltage current transformers are mounted on porcelain insulators to isolate them from ground. Some CT configurations slip around the bushing of a high-voltage transformer or circuit breaker, which automatically centers the conductor inside the CT window.

Current transformers can be mounted on the low voltage or high voltage leads of a power transformer; sometimes a section of bus bar is arranged to be easily removed for exchange of current transformers.

Usage

Current transformers are used extensively for measuring current and monitoring the operation of the power grid. Along with voltage leads, revenue-grade CTs drive the electrical utility's watt-hour meter on virtually every building with three-phase service and single-phase services greater than 200 amps.

The CT is typically described by its current ratio from primary to secondary. Often, multiple CTs are installed as a "stack" for various uses. For example, protection devices and revenue metering may use separate CTs to provide isolation between metering and protection circuits, and allows current transformers with different characteristics (accuracy, overload performance) to be used for the devices.

The primary circuit is largely unaffected by the insertion of the CT. The rated secondary current is commonly standardized at 1 or 5 amperes. For example, a 4000 : 5 CT secondary winding will supply an output current of 5 amperes when the primary winding current is 4000 amperes. The secondary winding can be single or multi-ratio, with five taps being common for multi-ratio CTs.

The load, or burden, of the CT should be a low resistance. If the voltage time integral area is higher than the core's design rating, the core goes into saturation toward the end of each cycle, distorting the waveform and affecting accuracy.

Safety Precautions

Care must be taken that the secondary of a current transformer is not disconnected from its load while current is flowing in the primary, as the transformer secondary will attempt to continue driving current across the effectively infinite impedance up to its core saturation voltage. This may produce a high voltage across the open secondary into the range of several kilovolts, causing arcing, compromising operator and equipment safety, or permanently affect the accuracy of the transformer.

Accuracy

The accuracy of a CT is directly related to a number of factors including :

- Burden
- Burden class/saturation class
- Rating factor
- Load
- External electromagnetic fields
- Temperature and
- Physical configuration.
- The selected tap, for multi-ratio CTs
- Phase change.

For the IEC standard, accuracy classes for various types of measurement are set out in IEC 60044-1, Classes 0.1, 0.2s, 0.2, 0.5, 0.5s, 1 and 3. The class designation is an approximate measure of the CT's accuracy. The ratio (primary to secondary current) error of a Class 1 CT is 1% at rated current; the ratio error of a Class 0.5 CT is 0.5% or less. Errors in phase are also important especially in power measuring circuits, and each class has an allowable maximum phase error for a specified load impedance.

Current transformers used for protective relaying also have accuracy requirements at overload currents in excess of the normal rating to ensure accurate performance of relays during system faults. A CT with a rating of 2.5L400 specifies with an output from its secondary winding of 20 times its rated secondary current (usually 5 A x 20 = 100 A) and 400 V (IZ drop) its output accuracy will be within 2.5 percent.

Burden

The secondary load of a current transformer is usually called the "burden" to distinguish it from the load of the circuit whose current is being measured.

The burden, in a CT metering circuit is the (largely resistive) impedance presented to its secondary winding. Typical burden ratings for IEC CTs are 1.5 VA, 3 VA, 5 VA, 10 VA, 15 VA, 20 VA, 30 VA, 45 VA and 60 VA. As for ANSI/ IEEE burden ratings are B-0.1, B-0.2, B-0.5, B-1.0, B-2.0 and B-4.0. This means a CT with a burden rating of B-0.2 can tolerate up to 0.2 Ω of impedance in the

metering circuit before its secondary accuracy falls outside of an accuracy specification. These specification diagrams show accuracy parallelograms on a grid
incorporating magnitude and phase angle error scales at the CT's rated burden.
Items that contribute to the burden of a current measurement circuit are switch-
blocks, meters and intermediate conductors. The most common source of excess
burden is the conductor between the meter and the CT. When sub-station meters
are located far from the meter cabinets, the excessive length of wire creates a large
resistance. This problem can be reduced by using CTs with 1 ampere secondaries,
which will produce less voltage drop between a CT and its metering devices.

Knee-point Core-saturation Voltage

The **knee-point voltage** of a current transformer is the magnitude of the secondary
voltage above which the output current ceases to linearly follow the input current
within declared accuracy. In testing, if a voltage is applied across the secondary
terminals the magnetizing current will increase in proportion to the applied
voltage, until the knee point is reached. The knee point is defined as the voltage
at which a 10% increase in applied voltage increases the magnetizing current by
50%. For voltages greater than the knee point, the magnetizing current increases
considerably even for small increments in voltage across the secondary terminals.
The knee-point voltage is less applicable for metering current transformers as their
accuracy is generally much higher, but constrained within a very small range of the
current transformer rating, typically 1.2 to 1.5 times rated current. However, the
concept of knee point voltage is very pertinent to protection current transformers,
since they are necessarily exposed to fault currents of 20 to 30 times rated current.

Rating Factor

Rating factor is a factor by which the nominal full load current of a CT can be
multiplied to determine its absolute maximum measurable primary current.
Conversely, the minimum primary current a CT can accurately measure is "light
load," or 10% of the nominal current (there are, however, special CTs designed
to measure accurately currents as small as 2% of the nominal current). The rating
factor of a CT is largely dependent upon ambient temperature. Most CTs have
rating factors for 35 degrees Celsius and 55 degrees Celsius. It is important to
be mindful of ambient temperatures and resultant rating factors when CTs are
installed inside padmount transformers or poorly ventilated mechanical rooms.
Recently, manufacturers have been moving towards lower nominal primary
currents with greater rating factors. This is made possible by the development of
more efficient ferrites and their corresponding hysteresis curves.

Phase Shift

Ideally the secondary current of a current transformer should be perfectly in phase
with the primary current. In practice, this is impossible to achieve, but phase
shifts as low as a few tenths of a degree for well constructed transformers up to
as much as six degrees for simpler designs may be encountered (for the normal

power frequencies). For the purposes of current measurement, any phase shift is immaterial as the indicating ammeter, only displays the magnitude of the current. However, if the current transformer is used in conjunction with the current circuit of a wattmeter, energy meter or power factor meter, any phase shift in the measured current can affect the accuracy of the target measurement. For power and energy measurement, this error is generally considered to be negligible at unity power factor but increases in significance as the power factor approaches zero. At true zero power factor, all the measured power is entirely due to the current transformer's phase error. In recent years the introduction of electronic based power and energy meters has allowed the phase error to be calibrated out.

Special Designs

Specially constructed *wideband current transformers* are also used (usually with an oscilloscope) to measure waveforms of high frequency or pulsed currents within pulsed power systems. One type of specially constructed wideband transformer provides a voltage output that is proportional to the measured current. Another type (called a Rogowski coil) requires an external integrator in order to provide a voltage output that is proportional to the measured current. Unlike CTs used for power circuitry, wideband CTs are rated in output volts per ampere of primary current. CT RATIO

Standards

Ultimately, depending on client requirements, there are two main standards to which current transformers are designed. IEC 60044-1 (BSEN 60044-1) & IEEE C57.13 (ANSI), although the Canadian and Australian standards are also recognised.

High Voltage Types

Current transformers are used for protection, measurement and control in high voltage electrical sub-stations and the electrical grid. Current transformers may be installed inside switchgear or in apparatus bushings, but very often free-standing outdoor current transformers are used. In a switchyard, *live tank* current transformers have a substantial part of their enclosure energized at the line voltage and must be mounted on insulators. *Dead tank* current transformers isolate the measured circuit from the enclosure. Live tank CTs are useful because the primary conductor is short, which gives better stability and a higher short-circuit current rating. The primary of the winding can be evenly distributed around the magnetic core, which gives better performance for overloads and transients. Since the major insulation of a live-tank current transformer is not exposed to the heat of the primary conductors, insulation life and thermal stability is improved.

A high-voltage current transformer may contain several cores, each with a secondary winding, for different purposes (such as metering circuits, control, or protection). A neutral current transformer is used as earth fault protection to measured any fault current flowing through the neutral line from the wye neutral point of a transformer.

Zigzag Transformer

A **zigzag transformer** is a special purpose transformer with a zigzag or 'interconnected star' winding connection. Its applications are for the derivation of a neutral connection from an ungrounded 3-phase system and the grounding of that neutral to an earth reference point and harmonics mitigation. It can cancel triplet (3rd, 9th, 15th, 21st, etc.) harmonic currents, to supply 3-phase power as an autotransformer (serving as the primary and secondary with no isolated circuits), and to supply non-standard phase-shifted 3-phase power.

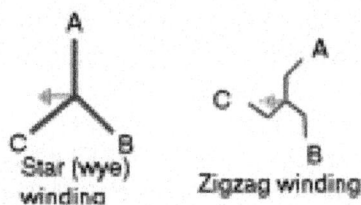

Fig. : 9-winding zigzag transformer.

Nine-winding three-phase transformers, typically having six identical secondary windings, can be used in zigzag winding connection as pictured. As with the conventional delta or wye winding configuration three-phase transformer, a standard stand-alone transformer containing only six windings on three cores can also be used in zigzag winding connection, such transformer sometimes being referred to as a **zigzag bank**. In all cases, six or nine winding, the first coil on each zigzag winding core is connected contrariwise to the second coil on the next core. The second coils are then all tied together to form the neutral and the phases are connected to the primary coils. Each phase, therefore, couples with each other phase and the voltages cancel out. As such, there would be negligible current through the neutral point, which can be tied to ground.

Each of the three "limbs" are split into two sections. The two halves of each limb have the equal number of turns and they're wound in opposite directions. With the neutral grounded, during a phase to ground short fault, a third of current returns to the fault current and the remainder must go through two of the three phases when used to derive a grounding point from a delta source.

If one phase, or more, faults to earth, the voltage applied to each phase of the transformer is no longer in balance; fluxes in the windings no longer oppose. (Using symmetrical components, this is $I_{a0} = I_{b0} = I_{c0}$.) Zero sequence (earth fault) current exists between the transformer's neutral to the faulting phase. The purpose of a zigzag transformer in this application is to provide a return path for earth faults on delta-connected systems. With negligible current in the neutral under normal conditions, undersized transformer maybe used; a short time rating is applied (*i.e.*, the transformer can only carry full rated current for, say, 60 s). Impedance should not be too low for desired fault current. Impedance can be added after the secondaries are summed (the $3I_0$ path).

An application example : A combination of Y (wye or star), delta, and zigzag windings may be used to achieve a vector phase shift. For example, an electrical network may have a transmission network of 110 kV/33 kV star/star transformers, with 33 kV/11 kV delta/star for the high voltage distribution network. If a transformation is required directly between the 110 kV/11 kV network an option is to use a 110 kV/11 kV star/delta transformer. The problem is that the 11 kV delta no longer has an earth reference point. Installing a zigzag transformer near the secondary side of the 110 kV/11 kV transformer provides the required earth reference point.

Balun

A **balun** is an electrical device that converts between a balanced signal (two signals working against each other where ground is irrelevant) and an unbalanced signal (a single signal working against ground or pseudo-ground). A balun can take many forms and may include devices that also transform impedances but need not do so. Transformer baluns can also be used to connect lines of differing impedance. The origin of the word balun is **bal**ance + **un**balance.

Baluns can take many forms and their presence is not always obvious. Sometimes, in the case of transformer baluns, they use magnetic coupling but need not do so. Common-mode chokes are also used as baluns and work by eliminating, rather than ignoring, common mode signals.

A variation of this device is the UNUN, which transfers signal from one unbalanced line to another.

Types of Balun

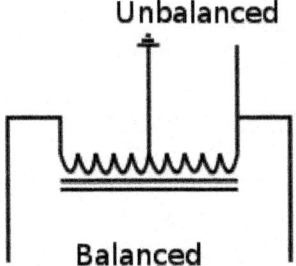

Fig. : Autotransformer 4 : 1 wideband balun using two windings on a ferrite rod.

Generally a balun consists of two wires (primary and secondary) and a toroid core : it converts the electrical energy of the primary wire into a magnetic field. Depending on how the secondary wire is done, the magnetic field is converted back to an electric field.

Autotransformer Type

In an autotransformer, two coils on a ferrite rod can be used as a balun by winding the individual strands of enameled wire comprising the coil very tightly together. This winding can take one of two forms : either the two windings must be wound such that the two form a single layer where each turn is touching each of the adjacent turns of the other winding; or the two wires are twisted together before being wound into the coil.

The two windings are joined to become a single coil. The end of one of the windings, on one side of the coil, is connected to the end of the other winding on the other side of the coil. This point then becomes the ground for the unbalanced circuit. One of the remaining ends is connected to the ungrounded side of the unbalanced circuit and one side of the balanced circuit. Finally, the other side of the balanced circuit is connected to the remaining end.

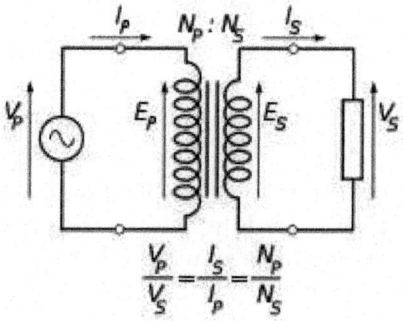

$$\frac{V_p}{V_s} = \frac{I_s}{I_p} = \frac{N_p}{N_s}$$

Fig. : Isolated transformer.

Classical Transformer Type

Isolated transformers have a real impedance at a resonance frequency where impedances from self-inductance and self-capacitance for each individual winding are equal at a given frequency.

Transmission-line Transformer Type

Baluns can be considered as simple forms of transmission line transformers.

A more complex (and subtle) type results when the transformer type (magnetic coupling) is combined with the transmission line type (electro-magnetic coupling). This is where transmission lines are used as windings, resulting in devices capable of very wideband operation. "Transmission line transformers" commonly use small ferrite cores in toroidal or "binocular" shapes. Something as simple as 10 turns of coaxial cable coiled up on a diameter about the size of a dinner plate makes an extremely effective choke balun for frequencies from about 10 MHz to beyond 30 MHz. The magnetic material may be "air", but it is a transmission line transformer.

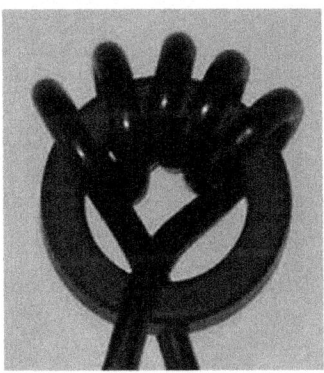

Fig. : Homemade 1 : 1 balun using a toroidal core and coaxial cable. This simple RF choke works as a balun by preventing signals passing along the outside of the braid. Such a device can be used to cure television interference by acting as a braid-breaker.

The Guanella transmission line transformer (Guanella 1944) is often combined with a balun to act as an impedance matching transformer. Putting balancing aside a 1 : 4 transformer of this type consists of a 75 Ω transmission line divided in parallel into two 150 Ω cables, which are then combined in series for 300 Ω. It is implemented as a specific wiring around the ferrite core of the balun.

Delay Line Type

A large class of baluns uses connected transmission lines of specific lengths, with no obvious "transformer" part. These are usually built for (narrow) frequency ranges where the lengths involved are some multiple of a quarter wavelength of the intended frequency in the transmission line medium. A common application is in making a coaxial connection to a balanced antenna, and designs include many types involving coaxial loops and variously connected "stubs".

One easy way to make a balun is a one-half wavelength ($\lambda/2$) length of co-axial cable. The inner core of the cable is linked at each end to one of the balanced connections for a feeder or dipole. One of these terminals should be connected to the inner core of the coaxial feeder. All three braids should be connected. This then forms a 4 : 1 balun which works at only one frequency.

Another narrow band design is to use a $\lambda/4$ length of metal pipe. The coaxial cable is placed inside the pipe; at one end the braid is wired to the pipe while at the other end no connection is made to the pipe. The balanced end of this balun is at the end where the pipe is wired to the braid. The $\lambda/4$ conductor acts as a transformer converting the infinite impedance at the unconnected end into a zero impedance at the end connected to the braid. Hence any current entering the balun through the connection, which goes to the braid at the end with the connection to the pipe, will flow into the pipe. This balun design is not good for low frequencies because of the long length of pipe that will be needed. An easy way to make such a balun is to paint the outside of the coax with conductive paint, then to connect this paint to the braid.

Balun Alternatives

An RF choke can be used in place of a balun. If a coil is made using coaxial cable near to the feed point of a balanced antenna, then the RF current that flows on the outer surface of the coaxial cable can be attenuated. One way of doing this would be to pass the cable through a ferrite toroid. (Straw 2005, 25-26)

Applications

A balun's function is generally to achieve compatibility between systems, and as such, finds extensive application in modern communications, particularly in realising frequency conversion mixers to make cellular phone and data transmission networks possible. They are also used to convert an E1 carrier signal from coaxial cable (BNC connector,1.0/2.3 connector,1.6/5.6 connector,Type 43 connectors) to UTP CAT-5 cable or IDC connector.

Radio and Television

In television, amateur radio, and other antenna installations and connections, baluns convert between impedances and symmetry of feedlines and antennas. For example, transformation of 300 Ω twin-lead or 450 Ω ladder line (balanced) and 75 Ω coaxial cable (unbalanced), or to directly connect a balanced antenna to unbalanced coaxial cable. To avoid feed line radiation, baluns are typically used as a form of common mode choke attached at the antenna feed point to prevent the coaxial cable from acting as an antenna and radiating power. This typically is needed when a balanced antenna (for instance, a dipole) is fed with coax; without a balun, the shield of the coax could couple with one side of the dipole, inducing common mode current, and becoming part of the antenna and unintentionally radiating.

When it comes to transmitting antennas the choice of the toroid core is crucial. A rule of thumb is : the more power the bigger the core.

In measuring the impedance or radiation pattern of a balanced antenna using a coaxial cable, it is important to place a balun between the cable and the antenna feed. Unbalanced currents that may otherwise flow on the cable will make the measured antenna impedance sensitive to the configuration of the feed cable, and the radiation pattern of small antennas may be distorted by radiation from the cable.

Baluns are present in radars, transmitters, satellites, in every telephone network, and probably in most wireless network modem/routers used in homes. It can be combined with transimpedance amplifiers to compose high-voltage amplifiers out of low-voltage components.

Video

Baseband video uses frequencies up to several megahertz. A balun can be used to couple video signals to twisted-pair cables instead of using coaxial cable. Many security cameras now have both a balanced unshielded twisted pair (UTP) output and an unbalanced coaxial one via an internal balun. A balun is also used on the video recorder end to convert back from the 100 Ω balanced to 75 Ω unbalanced. A balun of this type has a BNC connector with two screw terminals. VGA/DVI baluns are baluns with electronic circuitry used to connect VGA/DVI sources (laptop, DVD, etc.) to VGA/DVI display devices over long-runs of CAT-5/CAT-6 cable. Runs over 130 m (400 ft) may lose quality due to attenuation and variations in the arrival time of each signal. A skew control and special low skew or skew free cable is used for runs over 130 m (400 ft).

Audio

In audio applications, baluns convert between high impedance unbalanced and low impedance balanced lines. Another application is decoupling of devices (avoidance of earth loops).

A third application of baluns in audio systems is in the provision of balanced mains power to the equipment. Due to the common-mode rejection of interference characteristic of balanced mains power, a wide range of noise coming from the wall plug is eliminated, *e.g.* mains-borne interference from air conditioner/furnace/refrigerator motors, switching noise produced by fluorescent lighting and dimmer switches, digital noise from personal computers, and radio frequency signals picked up by the power lines/cords acting as antennae. This noise infiltrates the audio/video system through the power supplies and raises the noise floor of the entire system.

Except for the connections, the three devices in the image are electrically identical, but only the leftmost two can be used as baluns. The device on the left would normally be used to connect a high impedance source, such as a guitar, into a balanced microphone input, serving as a passive DI unit. The one in the centre is for connecting a low impedance balanced source, such as a microphone, into a guitar amplifier. The one at the right is not a balun, as it provides only impedance matching.

Other Applications

- In power line communications, baluns are used in coupling signals onto a power line.
- In electronic communications, baluns convert Twinax cables to Category 5 cables, and back.

Rotary Transformer

A **rotary (rotatory) transformer** is a specialized transformer used to couple electrical signals between two parts that rotate in relation to each other.

Slip rings can be used for the same purpose, but are subject to friction, wear, intermittent contact, and limitations on the rotational speed that can be accommodated without damage. Wear can be eliminated by using a pool of liquid mercury instead of a solid ring contact, but the toxicity and slow corrosion of mercury are problematic, and very high rotational speeds are again difficult to achieve. A rotary transformer has none of these limitations.

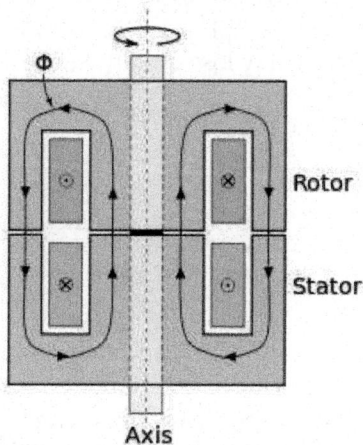

Fig. : Cross-section diagram of a simple rotary transformer.

Rotary transformers are constructed by winding the primary and secondary windings into separate halves of a *cup core*; these concentric halves face each other, with each half mounted to one of the rotating parts. Magnetic flux provides the coupling from one half of the cup core to the other across an air gap, providing the mutual inductance that couples energy from the transformer's primary to its secondary.

In brushless synchros, typical rotary transformers (in pairs) provide longer life than slip rings. These rotary transformers have a cylindrical, rather than a disc-shaped, air gap between windings. The rotor winding is a spool-shaped ferromagnetic core, with the winding placed like thread on a spool. The flanges are the pole pieces. The stator winding is a ferromagnetic cylinder with the winding inside, and end poles that are discs with holes, like washers.

Uses

Rotary transformers are most commonly used in videocassette recorders. Signals must be coupled from the electronics of the VCR to the fast-moving tape heads carried on the rotating head drum; a rotary transformer is ideal for this purpose. Most VCR designs require more than one signal to be coupled to the head drum. In this case, the cup core has more than one concentric winding, isolated by individual raised portions of the core. The transformer for the head drum shown to the right couples six individual channels.

Another use is to transmit the signals from rotary torque sensors installed on electric motors, to allow electronic control of motor speed and torque using feedback.

Rotary transformers cannot be used in most DC motors instead of commutators, as transformers can only transfer AC current.

The so-called "Brushless DC electric motors" as used in an increasing array of electronic devices, are actually AC motors. A constant voltage supply is made available to the motor/controller, which is then converted by the motor control module into a variable-frequency, variable-voltage AC signal which drives the windings of the motor.

Austin Transformer

An **Austin transformer** is a special type of an Isolation transformer used for feeding the air-traffic obstacle lamps and other devices on a mast radiator antenna insulated from ground. As the electrical potential difference between the antenna and ground is high (up to 300 kV), feeding the lamps directly is impossible. The transformer consists of two ring-like windings with a large air space between the winding and the magnetic core. The large spacing provides both isolation from high voltage and low inter-winding coupling capacitance.

The Austin transformer is named after its inventor, Arthur O. Austin, who graduated from Stanford University in 1903 and who obtained 225 patents in his career.

Condition Monitoring of Transformers

Condition Monitoring of Transformers is the process of acquisition and processing of data related to various parameters of transformers so as to predict and prevent the failure of a transformer. This is done by observing the deviation of the transformer parameters from their expected values. Transformers are the most critical assets of electrical transmission and distribution system. Transformer failures could cause power outages, personal and environmental hazards and expensive rerouting or purchase of power from other suppliers. Transformer failures can occur due to various causes. Transformer in-service interruptions and failures usually result from dielectric breakdown, winding distortion caused by short-circuit withstand, winding and magnetic circuit hot spot, electrical disturbances, deterioration of insulation, lightning, inadequate maintenance, loose connections, overloading, failure of accessories such as OLTCs, bushings, etc. Integrating the 'individual cause' monitoring allows for monitoring the overall condition of transformer. The important aspects of condition monitoring of transformers are :

1. Thermal Modelling
2. Dissolved Gas Analysis
3. Frequency Response Analysis
4. Partial Discharge Analysis.

Thermal Modelling

The useful life of a transformer is determined partially by the ability of transformer to dissipate the internally generated heat to its surroundings. The comparison of actual and predicted operating temperatures can provide a sensitive diagnosis of the transformer condition and might indicate abnormal operation. The consequences of temperature rise may not be sudden, but gradual as long as it is within break down limit. Among these consequences, insulation deterioration is economically important. Insulation being very costly, its deterioration is undesirable. Thermal modelling is the development of a mathematical model that predicts the temperature profile of the power transformer using the principle of thermal analysis. The thermal model is used to determine the top oil temperature and hot spot temperature (maximum temperature occurring in the winding insulation system)

Dissolved Gas Analysis

Gases are produced by degradation of the transformer oil and solid insulating materials. Gases are generated at a much more rapid rate whenever an electrical fault occurs. Normal causes of fault gases are classified into three categories: Corona or partial discharge, thermal heating and arcing. These faults can be detected by evaluating the quantities of hydrocarbon gases, hydrogen and oxides of carbon that are present in the transformer. Different gases can serve as markers for different types of faults. The concentration and the relation of individual gases allow a prediction of whether a fault has occurred and what type it is likely to be.

Frequency Response Analysis

When a transformer is subjected to high currents through fault currents, the mechanical structure and windings are subjected to severe mechanical stresses causing winding movement and deformations. It may also result in insulation damage and turn-to-turn faults. Frequency response analysis (FRA) is a non-intrusive very sensitive technique for detecting winding movement faults and deformation assessment caused by loss of clamping pressure or by short circuit forces. FRA technique involves measuring the impedance of the windings of the transformer with a low voltage sine input varying in a wide frequency range.

Partial Discharge Analysis

Partial discharge (PD) occurs when a local electric field exceeds a threshold value, resulting in a partial breakdown of the surrounding medium. Its cumulative effect leads to the degradation of insulation. PDs are initiated by the presence of defects during its manufacture, or the choice of higher stress dictated by design considerations. Measurements can be collected to detect these PDs and monitor the soundness of insulation. PDs manifest as sharp current pulses at transformer terminals, whose nature depends on the types of insulation, defects, measuring circuits and detectors used.

Energy Efficient Transformer

In a typical power distribution grid, electric transformer power loss typically contributes about 40-50% of the total transmission and distribution loss. Energy efficient transformers are therefore an important means to reduce transmission and distribution loss. With the improvement of electrical steel (silicon steel) properties, the losses of a transformer in 2010 can be half that of a similar transformer in the 1970s. With new magnetic materials, it is possible to achieve even higher efficiency. The amorphous metal transformer is a modern example.

Modulation Transformer

A **modulation transformer** is an audio-frequency transformer that forms a major part of most AM transmitters. The primary winding of a modulation transformer is fed by an audio amplifier that has about 1/2 of the rated input power of the transmitter's final amplifier stage. The secondary winding is in series with the power supply of that final radio-frequency amplifier stage, thereby lowering and raising the operating voltage of the power amplifier (PA) tube or transistor. Considering that the PA device is operated as a class-C amplifier, *i.e.* as a switch, the modulation transformer is responsible for the amplitude modulation (AM) of the transmitter.

Padmount Transformer

A **padmount** or **pad-mounted transformer** is a ground mounted electric power distribution transformer in a locked steel cabinet mounted on a concrete pad. Since all energized connection points are securely enclosed in a grounded metal housing, a padmount transformer can be installed in places that do not have room for a fenced enclosure. Padmount transformers are used with underground electric power distribution lines at service drops, to step down the primary voltage on the line to the lower secondary voltage supplied to utility customers. A single transformer may serve one large building, or many homes.

Pad-mounted transformers are made in power ratings from around 75 to around 5000 kVA and often include built-in fuses and switches. Primary power cables may be connected with elbow connectors, which can be operated when energized using a hot stick and allows for flexiblity in repair and maintenance.

Design

Pad-mount transformers are available in various electrical and mechanical configurations. Pad-mount transformers operate on medium-voltage distribution systems, up to about 35 kV. The low-voltage winding matches the customer requirement and may be single-phase or three-phase.

Pad-mount transformers are (nearly always) oil-filled units and so must be mounted outdoors only. The core and coils are enclosed in a steel oil-filled tank, with terminals for the transformer accessible in an adjacent lockable wiring cabinet. The wiring cabinet has high and low voltage wiring compartments. High and low voltage underground cables from below enter the terminal compartments directly. The top of the tank has a cover secured with carriage bolt-nut assemblies.

The wiring cabinet has sidewalls on two ends with doors that open sideways to expose the high and low voltage wiring compartments.

Pad-mount transformers have self-protecting fuses consisting of a bayonet mount fuse placed in a high voltage compartment, with a back-up high energy current limiting fuse in series to protect against secondary faults and transformer overload. The baoynet mount fuse protects against secondary faults and transformer overload and is a field replaceable device. The backup current limiting fuse operates only during transformer failure, therefore it is not field replaceable. These transformers also serve the conventional low voltage fusing requirements.

The use of polymeric cable and load break elbows enable switching and isolation to be carried out in the HV chamber in what is known as a "*dead front*" environment, *i.e.* all terminations are fully screened and watertight.

Single- and three-phase pad-mounted transformers are used in underground industrial and residential power distribution systems, where there is a need for safe, reliable and aesthetically appealing transformer design. Their enclosed construction allows the installation of pad-mount transformers in public areas without the need of protective fencing. In residential areas, pad-mount transformers are usually located on street easements and supply multiple households.

Three-phase pad-mounted transformers range in sizes from 75 kVA up to 5000 kVA with voltages ranging from 2,400 up to 34,500 delta or wye. Low-voltage pad-mounted range in size from 208y/120 through 24,940y/14,000.

While most traditional pad-mount transformer are fixed on a concrete 'pad', today small single-phase designs are also available with the transformer already mounted on a 'polypad' base so that they can be mounted on hard ground, connected, and switched on.

Tanks and Compartments

Pad-mount transformers have tanks and compartments assembled on top of a flat, rigid surface, usually a concrete pad. The pad-mounted transformer unit may be rolled, skidded or jacked into place, or raised using hooks. Underground cables are used to connect high voltage, which are connected to the bushings or to factory-installed auxiliary equipment. The high- and low-voltage compartments are separated by a metal barrier.

Terminations

Compartmental type pad-mounted transformers support underground entrance of primary and secondary conductors. Live and dead front primary termination on radial or loop feed service is provisioned for. Wet process electrical grade porcelain bushings with eyebolt terminals are provided for live front construction. For all voltage ratings, dead front construction with provisions for high voltage terminators is available. Secondary terminals are sealed into molded epoxy bushings and externally clamped to the tank wall.

Switch and Fuse

In a three-phase pad-mounted transformer, oil immersed switches that switch in three current ratings are available for radial and loop feed. Load break and latch operations are facilitated through spring load mechanism. A three-phase gang-operated switch is mounted near the core and coil assembly for low cable capacitance. Depending on the customers requirement, and transformer size and rating, the most common fuse options include weak link expulsion fuses; Bay-o-Net fuses; dry well canisters; arc strangler fused switch blades; clip-mounted, full-range, current-limiting fuses; and S & C disconnects with E-rated power fuses.

Installation

Pad-mount transformers are constructed particularly for installation in public areas, where there is a need for a tamper-proof design. Depending on the connection and access type, pad-mount transformers can be installed indoor and outdoor. Pad-mount transformers for indoors require dry-type construction.

Usually inspection authorities require pad-mount transformers to be located away from the main building. The transformer must have at least 8 feet of unobstructed space in front for the doors to the electrical equipment to open.

Upon installation, there should be a padlock on the transformer and a clearly visible warning sign indicating electrical voltage.

Application

Pad-mount transformers are used for power distribution for residential areas and for commercial and industrial buildings. Pad-mounted transformers are used for the collector systems of wind farms.

Chapter 6

DC Motor

Electrical motors are everywhere around us. Almost all the electro-mechanical movements we see around us are caused either by an A.C. or a **DC motor**. Here we will be exploring this kind of motors. This is a device that converts DC electrical energy to a mechanical energy.

PRINCIPLE OF DC MOTOR

This DC or **direct current motor** works on the principal, when a current carrying conductor is placed in a magnetic field, it experiences a torque and has a tendency to move. This is known as motoring action. If the direction of electric current in the wire is reversed, the direction of rotation also reverses. When magnetic field and electric field interact they produce a mechanical force, and based on that the working principle of **dc motor** established.

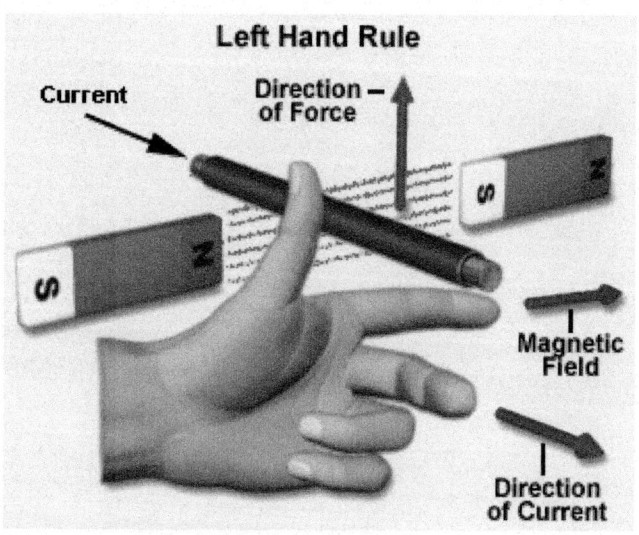

The direction of rotation of a this motor is given by Fleming's left hand rule, which states that if the index finger, middle finger and thumb of your left hand are extended mutually perpendicular to each other and if the index finger represents the direction of magnetic field, middle finger indicates the direction of electric current, then the thumb represents the direction in which force is experienced by the shaft of the **dc motor**.

Structurally and construction wise a direct current motor is exactly similar to a DC generator, but electrically it is just the opposite. Here we unlike a generator we supply electrical energy to the input port and derive mechanical energy from the output port. We can represent it by the block diagram shown below.

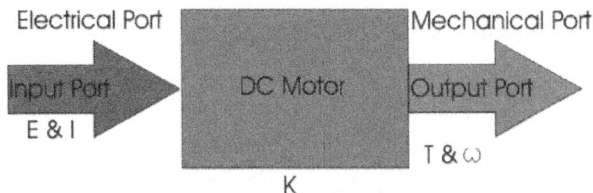

Here in a DC motor, the supply voltage E and electric current I is given to the electrical port or the input port and we derive the mechanical output *i.e.* torque T and speed ω from the mechanical port or output port.

The input and output port variables of the **direct current motor** are related by the parameter K.

$$T = KI \text{ and } E = K\omega$$

So from the picture above we can well understand that motor is just the opposite phenomena of a DC generator, and we can derive both motoring and generating operation from the same machine by simply reversing the ports.

DETAILED DESCRIPTION OF A DC MOTOR

To understand the DC motor in details lets consider the diagram below,

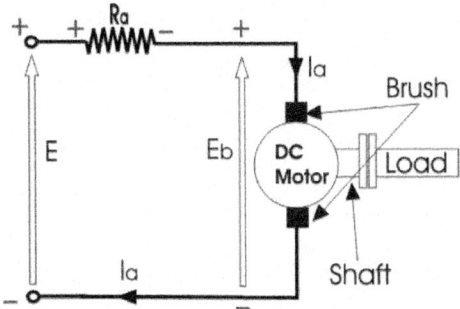

The direct current motor is represented by the circle in the center, on which is mounted the brushes, where we connect the external terminals, from where supply

voltage is given. On the mechanical terminal we have a shaft coming out of the Motor, and connected to the armature, and the armature-shaft is coupled to the mechanical load. On the supply terminals we represent the armature resistance R_a in series. Now, let the input voltage E, is applied across the brushes. Electric current which flows through the rotor armature via brushes, in presence of the magnetic field, produces a torque T_g. Due to this torque T_g the dc motor armature rotates. As the armature conductors are carrying currents and the armature rotates inside the stator magnetic field, it also produces an emf E_b in the manner very similar to that of a generator. The generated Emf E_b is directed opposite to the supplied voltage and is known as the back Emf, as it counters the forward voltage. The back emf like in case of a generator is represented by

$$E_b = \frac{P.\varphi.Z.N}{60.A}$$ (1)

Where, P = no of poles

     φ = flux per pole

     Z= No. of conductors

     A = No. of parallel paths

and

     N is the speed of the DC Motor.

So from the above equation we can see E_b is proportional to speed 'N'. That is whenever a direct current motor rotates, it results in the generation of back Emf. Now lets represent the rotor speed by ω in rad/sec. So E_b is proportional to ω.

So when the speed of the motor is reduced by the application of load, E_b decreases. Thus the voltage difference between supply voltage and back emf increases that means $E - E_b$ increases. Due to this increased voltage difference, armature current will increase and therefore torque and hence speed increases. Thus a DC Motor is capable of maintaining the same speed under variable load.

Now armature current I_a is represented by

$$I_a = \frac{E - E_b}{R_a}$$

Now at starting, speed $\omega = 0$ so at starting $E_b = 0$.

$$\therefore I_a = \frac{E}{R_a}$$ (2)

Now since the armature winding electrical resistance R_a is small, this motor has a very high starting current in the absence of back Emf. As a result we need to use a starter for starting a DC Motor.

Now as the motor continues to rotate, the back Emf starts being generated and gradually the current decreases as the motor picks up speed.

WORKING OR OPERATING PRINCIPLE OF DC MOTOR

A DC motor in simple words is a device that converts direct current (electrical energy) into mechanical energy. It's of vital importance for the industry today, and is equally important for engineers to look into the **working principle of DC motor** in details that has been discussed. In order to understand the **operating principle of dc motor** we need to first look into its constructional feature.

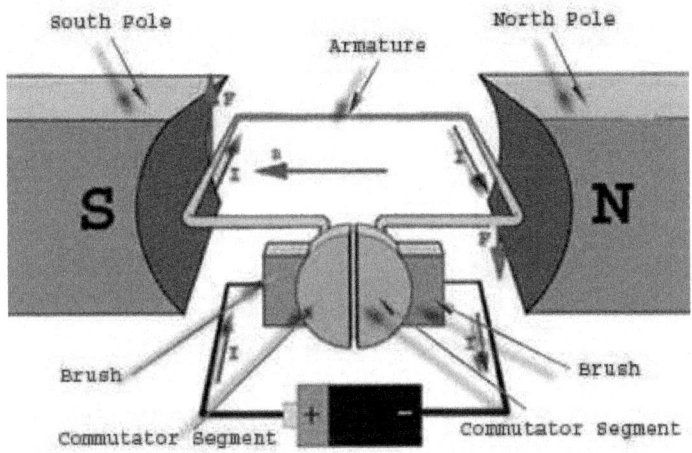

The very basic construction of a dc motor contains a current carrying armature which is connected to the supply end through commutator segments and brushes and placed within the north south poles of a permanent or an electro-magnet.

Now to go into the details of the **operating principle of DC motor** its important that we have a clear understanding of Fleming's left hand rule to determine the direction of force acting on the armature conductors of dc motor.

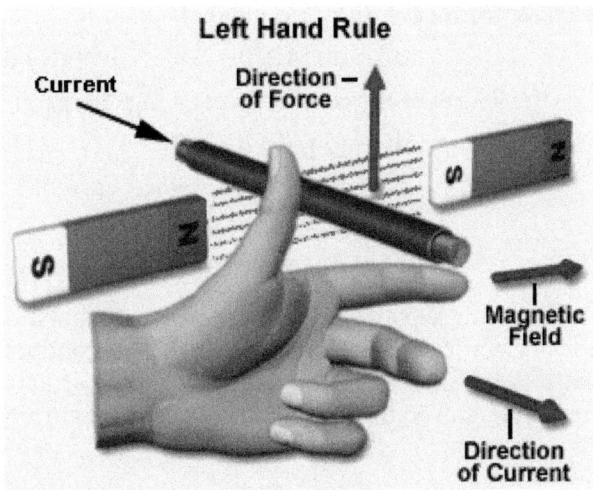

Fleming's left hand rule says that if we extend the index finger, middle finger and thumb of our left hand in such a way that the electric current carrying conductor is placed in a magnetic field (represented by the index finger) is perpendicular to the direction of current (represented by the middle finger), then the conductor experiences a force in the direction (represented by the thumb) mutually perpendicular to both the direction of field and the current in the conductor.

For clear understanding the **principle of DC motor** we have to determine the magnitude of the force.

We know that when an infinitely small charge dq is made to flow at a velocity 'v' under the influence of an electric field E, and a magnetic field B, then the Lorentz Force dF experienced by the charge is given by :-

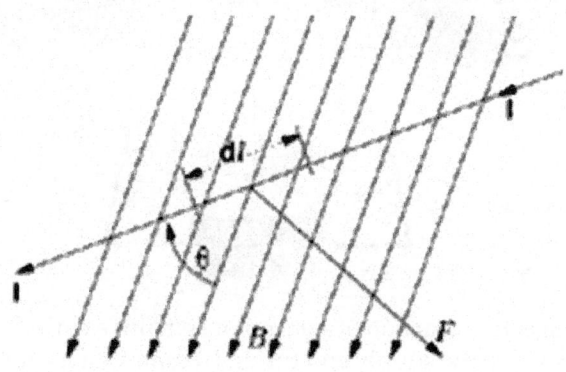

$$dF = dq(E + v \times B)$$

For the **operation of dc-motor**, considering $E = 0$

$$\therefore dF = dq \, v \times B$$

i.e. it's the cross product of dq v and magnetic field B.

or $dF = dq \, (dL/dt) \times B$ $[v = dL/dt]$

Where dL is the length of the conductor carrying charge q.

or $dF = (dq/dt) \, dL \times B$

or $dF = I \, dL \times B$ [Since, current $I = dq/dt$]

or $F = IL \times B = ILB \, Sin\theta$

or $F = BIL \, Sin\theta$

From the 1ˢᵗ diagram we can see that the construction of a DC motor is such that the direction of electric current through the armature conductor at all instance is perpendicular to the field. Hence the force acts on the armature conductor in the direction perpendicular to the both uniform field and current is constant.

i.e. $\theta = 90°$

So if we take the current in the left hand side of the armature conductor to be I, and current at right hand side of the armature conductor to be – I, because they are flowing in the opposite direction with respect to each other.

Then the force on the left hand side armature conductor, F_l = BIL Sin90° = BIL

Similarly force on the right hand side conductor F_r = B(– I)L.Sin90° = – BIL

∴ we can see that at that position the force on either side is equal in magnitude but opposite in direction. And since the two conductors are separated by some distance w = width of the armature turn, the two opposite forces produces a rotational force or a torque that results in the rotation of the armature conductor.

Now let's examine the expression of torque when the armature turn crate an angle of α with its initial position.

The torque produced is given by, Torque = force, tangential to the direction of armature rotation X distance.

or τ = Fcosα.w

or τ = BIL w cosα

Where α is the angle between the plane of the armature turn and the plane of reference or the initial position of the armature which is here along the direction of magnetic field.

The presence of the term cosα in the torque equation very well signifies that unlike force the torque at all position is not the same. It in fact varies with the variation of the angle α. To explain the variation of torque and the principle behind rotation of the motor let us do a step wise analysis.

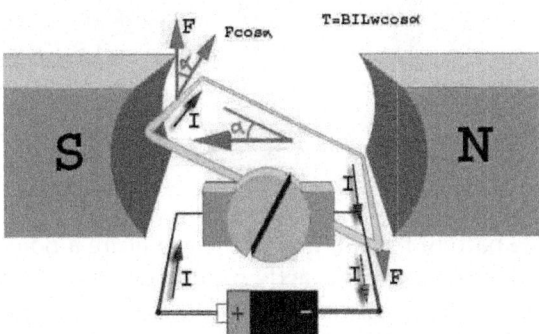

Step 1 : Initially considering the armature is in its starting point or reference position where the angle α = 0.

∴ τ = BIL w cos0 = BILw

Since α = 0, the term cos α = 1, or the maximum value, hence torque at this position is maximum given by τ = BILw. This high starting torque helps in overcoming the initial inertia of rest of the armature and sets it into rotation.

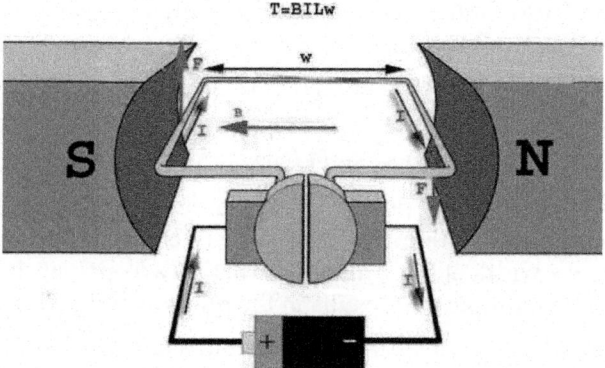

Step 2 : Once the armature is set in motion, the angle α between the actual position of the armature and its reference initial position goes on increasing in the path of its rotation until it becomes 90° from its initial position. Consequently the term cosα decreases and also the value of torque.

The torque in this case is given by τ = BILwcosα which is less than BIL w when α is greater than 0°.

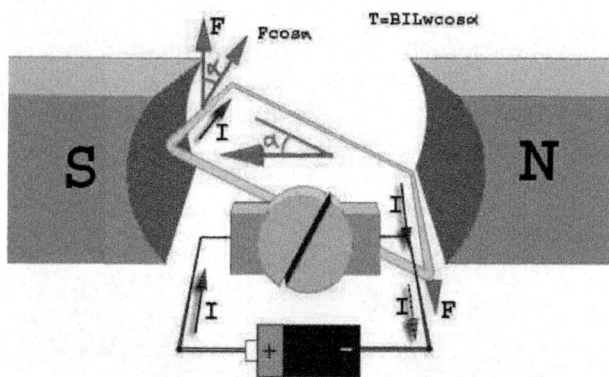

Step 3 : In the path of the rotation of the armature a point is reached where the actual position of the rotor is exactly perpendicular to its initial position, *i.e.* α = 90°, and as a result the term cosα = 0.

The torque acting on the conductor at this position is given by,

$$τ = BILwcos90° = 0$$

i.e. virtually no rotating torque acts on the armature at this instance. But still the armature does not come to a standstill, this is because of the fact that the operation of dc motor has been engineered in such a way that the inertia of motion at this point is just enough to overcome this point of null torque. Once the rotor crosses over this position the angle between the actual position of the armature and the initial plane again decreases and torque starts acting on it again.

$$T=BILw\cos 90 = 0$$

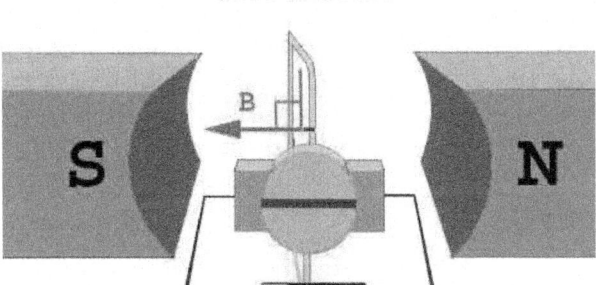

CONSTRUCTION OF DC MOTOR

A DC motor like we all know is a device that deals in the conversion of electrical energy to mechanical energy and this is essentially brought about by two major parts required for the **construction of dc motor**, namely.

1. Stator – The static part that houses the field windings and receives the supply and,

2. Rotor – The rotating part that brings about the mechanical rotations. Other than that there are several subsidiary parts namely the

3. Yoke of dc motor.

4. Poles of dc motor.

5. Field winding of dc motor.

6. Armature winding of dc motor.

7. Commutator of dc motor.

8. Brushes of dc motor.

All these parts put together configures the total **construction of a dc motor**.

Now let's do a detailed discussion about all the essential parts of dc motor.

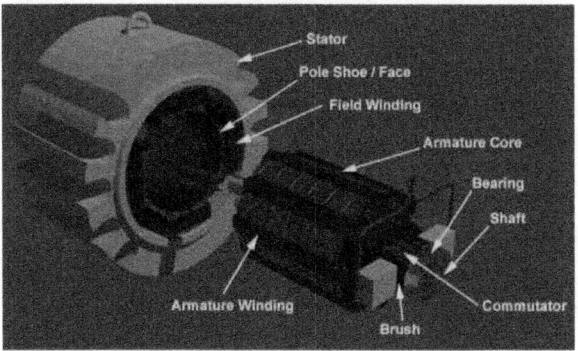

Fig. : Essential Parts of DC Machine.

Yoke of DC Motor

The magnetic frame or the **yoke of dc motor** made up of cast iron or steel and forms an integral part of the stator or the static part of the motor. Its main function is to form a protective covering over the inner sophisticated parts of the motor and provide support to the armature. It also supports the field system by housing the magnetic poles and field winding of the dc motor.

Poles of DC Motor

The magnetic **poles of DC motor** are structures fitted onto the inner wall of the yoke with screws. The construction of magnetic poles basically comprises of two parts namely, the pole core and the pole shoe stacked together under hydraulic pressure and then attached to the yoke. These two structures are assigned for different purposes, the pole core is of small cross sectional area and its function is to just hold the pole shoe over the yoke, whereas the pole shoe having a relatively larger cross-sectional area spreads the flux produced over the air gap between the stator and rotor to reduce the loss due to reluctance. The pole shoe also carries slots for the field windings that produce the field flux.

Field Winding of DC Motor

The **field winding of dc motor** are made with field coils (copper wire) wound over the slots of the pole shoes in such a manner that when field current flows through it, then adjacent poles have opposite polarity are produced. The field winding basically form an electromagnet, that produces field flux within which the rotor armature of the dc motor rotates, and results in the effective flux cutting.

Armature Winding of DC Motor

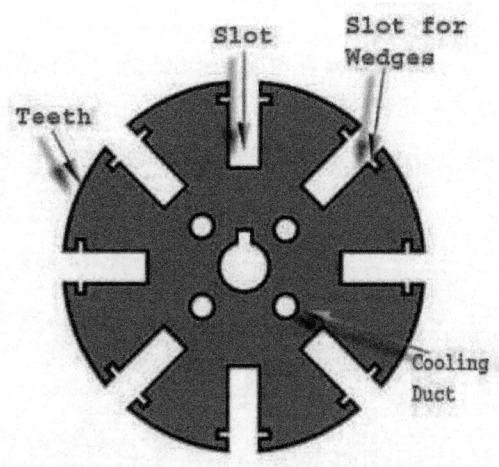

The **armature winding of dc motor** is attached to the rotor, or the rotating part of the machine, and as a result is subjected to altering magnetic field in the path of its rotation which directly results in magnetic losses. For this reason the rotor is made of armature core, that's made with several low-hysteresis silicon steel lamination, to reduce the magnetic losses like hysteresis and eddy current loss respectively. These laminated steel sheets are stacked together to form the cylindrical structure of the armature core.

The armature core are provided with slots made of the same material as the core to which the armature winding made with several turns of copper wire distributed uniformly over the entire periphery of the core. The slot openings a shut with fibrous wedges to prevent the conductor from plying out due to the high centrifugal force produced during the rotation of the armature, in presence of supply current and field.

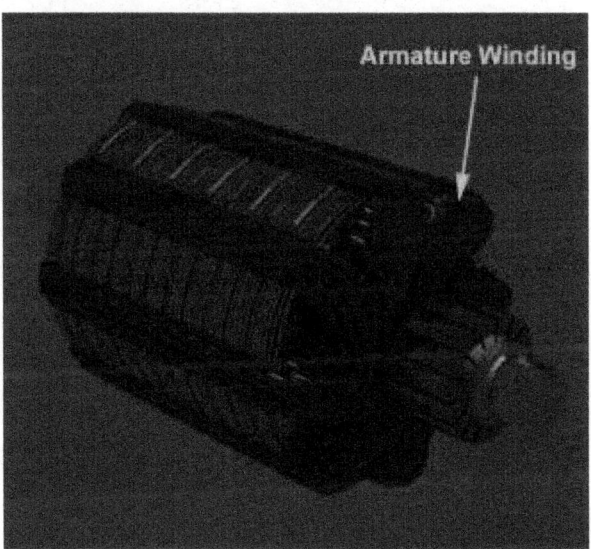

The construction of **armature winding of dc motor** can be of two types :-

Lap Winding

In this case the number of parallel paths between conductors A is equal to the number of poles P.

i.e. $A = P$

>

***An easy way of remembering it is by remembering the word LAP------→ L A=P

Wave Winding

Here in this case, the number of parallel paths between conductors A is always equal to 2 irrespective of the number of poles. Hence the machine designs are made accordingly.

Commutator of DC Motor

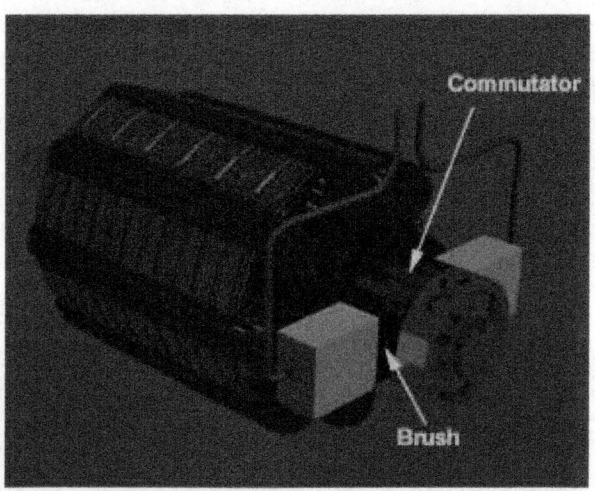

The **commutator of dc motor** is a cylindrical structure made up of copper segments stacked together, but insulated from each other by mica. Its main function as far as the dc motor is concerned is to commute or relay the supply current from the mains to the armature winding housed over a rotating structure through the **brushes of dc motor**.

Brushes of DC Motor

The **brushes of dc motor** are made with carbon or graphite structures, making sliding contact over the rotating commutator. The brushes are used to relay the electric current from external circuit to the rotating commutator form where it flows into the armature winding. So, the commutator and brush unit of the dc motor is concerned with transmitting the power from the static electrical circuit to the mechanically rotating region or the rotor.

TORQUE EQUATION OF DC MOTOR

The term torque as best explained by Dr. Huge d Young is the quantitative measure of the tendency of a force to cause a rotational motion, or to bring about a change in rotational motion. It is in fact the moment of a force that produces or changes a rotational motion.

The equation of torque is given by,

$$\tau = FR \sin\theta \tag{1}$$

Where F is force in linear direction.

R is radius of the object being rotated,

and θ is the angle, the force F is making with R vector

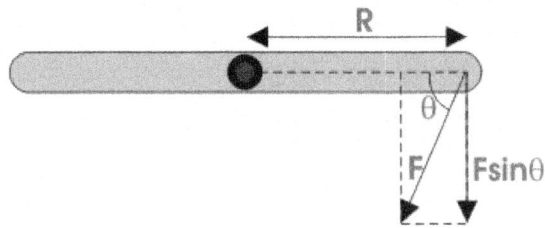

The dc motor as we all know is a rotational machine, and **torque of dc motor** is a very important parameter in this concern, and it's of utmost importance to understand the **torque equation of dc motor** for establishing its running characteristics.

To establish the torque equation, let us first consider the basic circuit diagram of a dc motor, and its voltage equation.

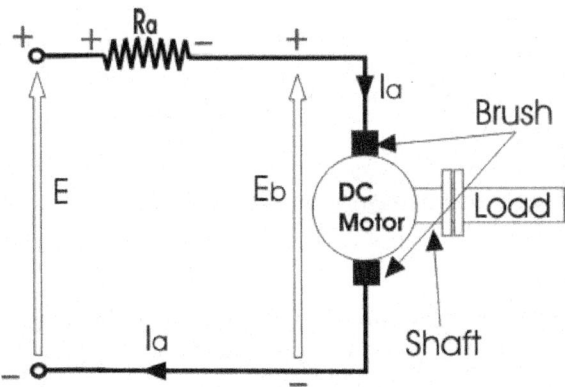

Referring to the diagram beside, we can see, that if E is the supply voltage, E_b is the back emf produced and I_a, R_a are the armature current and armature resistance respectively then the voltage equation is given by,

$$E = E_b + I_a R_a \tag{2}$$

But keeping in mind that our purpose is to derive the **torque equation of dc motor** we multiply both sides of equation (2) by I_a.

Therefore, $$EI_a = E_b I_a + I_a^2 R_a \tag{3}$$

Now $I_a^2 . R_a$ is the power loss due to heating of the armature coil, and the true effective mechanical power that is required to produce the desired torque of dc machine is given by,

$$P_m = E_b I_a \tag{4}$$

The mechanical power P_m is related to the electromagnetic torque T_g as,

$$P_m = T_g \omega \tag{5}$$

Where ω is speed in rad/sec.

Now equating equation (4) & (5) we get,

$$E_b I_a = T_g \omega$$

Now for simplifying the torque equation of dc motor we substitute.

$$E_b = \frac{P\varphi ZN}{60\,A} \tag{6}$$

Where, P is no of poles,

 φ is flux per pole,

 Z is no. of conductors,

 A is no. of parallel paths,

and N is the speed of the D.C. motor.

Hence, $$\omega = \frac{2\pi N}{60} \tag{7}$$

Substituting equation (6) and (7) in equation (4), we get :

$$T_g = \frac{P.Z.\,\varphi.\,I_a}{2\pi A}$$

The torque we so obtain, is known as the electromagnetic torque of dc motor, and subtracting the mechanical and rotational losses from it we get the mechanical torque.

Therefore, $T_m = T_g$ - mechanical losses.

This is the torque equation of dc motor. It can be further simplified as :

$$T_g = k_a \phi I_a$$

Where, $$k_a = \frac{P.Z}{2\pi A}$$

Which is constant for a particular machine and therefore the torque of dc motor varies with only flux φ and armature current I_a.

The Torque equation of a dc motor can also be explained considering the figure below.

Here we can see Area per pole Ar = $\dfrac{2\pi.r.L}{P}$

$$B = \frac{\varphi}{A_r}$$

$$B = \frac{P.\varphi}{2\pi rL}$$

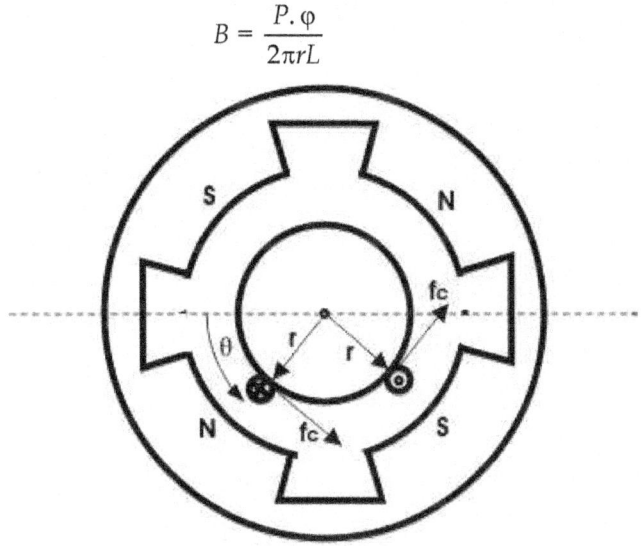

Current / conductor $I_c = I_a / A$

Therefore, force per conductor = f_c = BLI_a / A

Now torque $T_c = f_c.r = BLI_a.r / A$

$$\therefore T_C = \frac{\varphi P.I_a}{2\pi A}$$

Hence the total torque developed of a dc machine is,

$$T_g = \frac{P.Z.\varphi.I_a}{2\pi.A}$$

This torque equation of dc motor can be further simplified as :

$$T_g = k_a \varphi I_a$$

Where, $$k_a = \frac{P.Z}{2\pi.A}$$

Which is constant for a particular machine and therefore the torque of dc motor varies with only flux φ and armature current I_a

TYPES OF DC MOTOR

The direct current motor or the DC motor has a lot of application in today's field of engineering and technology. Starting from an electric shaver to parts of auto-mobiles, in all small or medium sized motoring applications DC motors come handy. And because of its wide range of application different functional **types of dc motor** are available in the market for specific requirements.

*The **types of DC motor** can be listed as follows :*

- DC motor
- Permanent Magnet DC Motor
- Separately Excited DC Motor
- Self-Excited DC Motor
  - Shunt Wound DC Motor
  - Series Wound DC Motor
- Compound Wound DC Motor
- Cumulative compound DC motor
  - Short shunt DC Motor
  - Long shunt DC Motor
- Differential Compound DC Motor
  - Short Shunt DC Motor
  - Long Shunt DC Motor

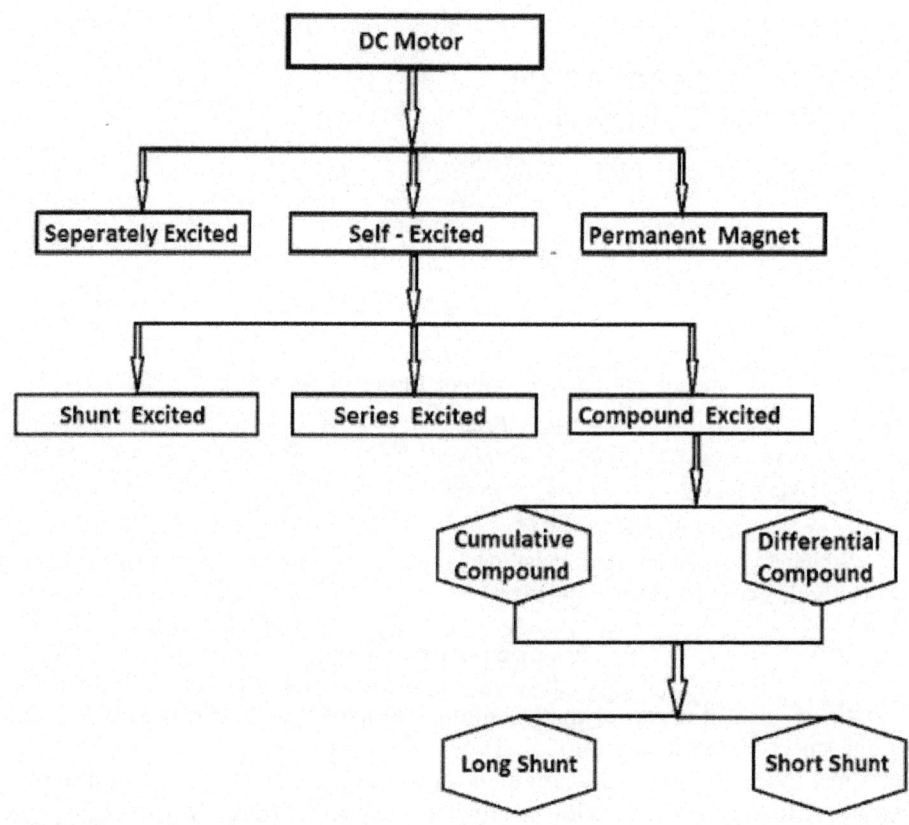

Now let's do a detailed discussion about all the essential types of dc motor.

Separately Excited DC Motor

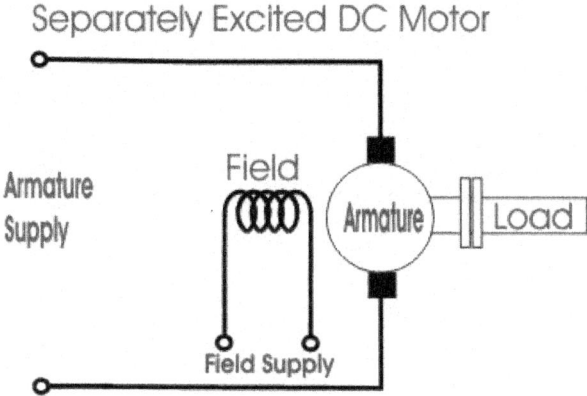

As the name suggests, in case of a separately excited DC motor the supply is given separately to the field and armature windings. The main distinguishing fact in these types of dc motor is that, the armature current does not flow through the field windings, as the field winding is energized from a separate external source of dc current as shown in the figure beside.

From the torque equation of dc motor we know $T_g = K_a \varphi I_a$. So the torque in this case can be varied by varying field flux φ, independent of the armature current I_a.

Permanent Magnet DC Motor

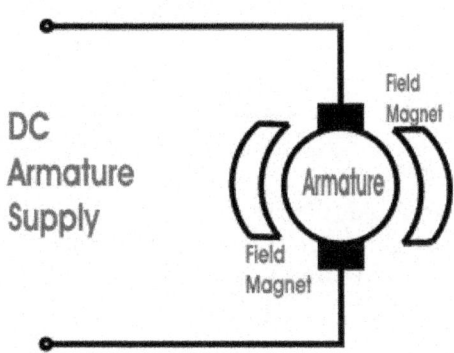

The **permanent magnet DC motor** consists of an armature winding as in case of an usual motor, but does not necessarily contain the field windings. The construction of these types of DC motor are such that, radially magnetized permanent magnets are mounted on the inner periphery of the stator core to produce the field flux.

The rotor on the other hand has a conventional dc armature with commutator segments and brushes.

The torque equation of dc motor suggests $T_g = K_a \varphi I_a$. Here φ is always constant, as permanent magnets of required flux density are chosen at the time of construction and can't be changed there after.

For a permanent magnet dc motor $T_g = K_{a1} I_a$

Where $K_{a1} = K_a . \varphi$ which is another constant. In this case the torque of DC Motor can only be changed by controlling armature supply.

Self-excited DC Motor

In case of self-excited dc motor, the field winding is connected either in series or in parallel or partly in series, partly in parallel to the armature winding, and on this basis its further classified as :-

i   **Shunt wound DC motor**

ii   **Series wound DC motor**

iii   **Compound wound DC motor**

Let's now go into the details of these types of self-excited dc motor.

Shunt Wound DC Motor

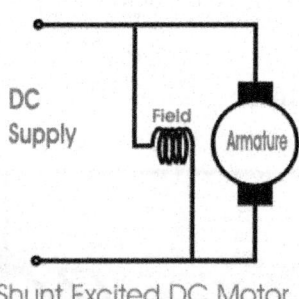

Shunt Excited DC Motor

In case of a **shunt wound dc motor** or more specifically shunt wound self-excited dc motor, the field windings are exposed to the entire terminal voltage as they are connected in parallel to the armature winding.

To understand the characteristic of these types of DC motor, lets consider the basic voltage equation given by,

$$E = E_b + I_a R_a \tag{1}$$

[Where E, E_b, I_a, R_a are the supply voltage, back emf, armature current and armature resistance respectively]

Now, $$E_b = k_a \phi \omega \tag{2}$$

[since back emf increases with flux ϕ and angular speed ω]

Now substituting E_b from equation (2) to equation (1) we get,

$$E = k_a \phi \omega + I_a R_a$$

$$\therefore \qquad \omega = \frac{E - I_a R_a}{k_a \varphi} \qquad (3)$$

The torque equation of a dc motor resembles,

$$T_g = K_a \phi I_a \qquad (4)$$

This is similar to the equation of a straight line, and we can graphically representing the torque speed characteristic of a shunt wound self-excited dc motor as

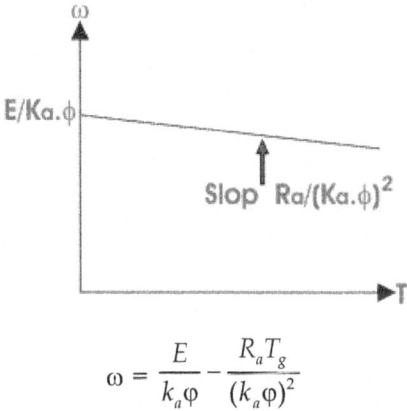

$$\omega = \frac{E}{k_a \varphi} - \frac{R_a T_g}{(k_a \varphi)^2}$$

The shunt wound dc motor is a constant speed motor, as the speed does not vary here with the variation of mechanical load on the output.

Series Wound DC Motor

In case of a series wound self-excited dc motor or simply **series wound dc motor**, the entire armature current flows through the field winding as its connected in series to the armature winding. The series wound self-excited dc motor is diagrammatically represented below for clear understanding.

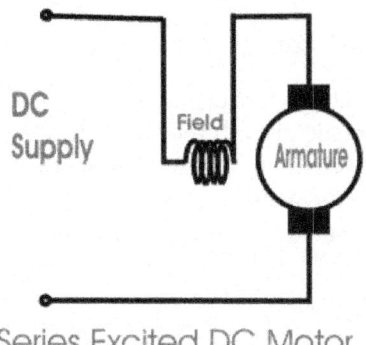

Series Excited DC Motor

Now to determint the torque speed characteristic of these types of DC motor, lets get to the torque speed equation.

From the circuit diagram we can see that the voltage equation gets modified to

$$E = E_b + I_a (R_a + R_s) \tag{5}$$

Where as back emf remains $E_b = k_a \varphi \omega$

Neglecting saturation we get,

$$\phi = K_1 I_f = K_1 I_a$$

[since field current = armature current]

Therefore, $\qquad E_b = k_a K_1 I_a \omega = K_S I_a \omega \tag{6}$

From equation (5) & (6)

$$\omega = \frac{E}{K_S I_a} - \frac{R_a + R_s}{K_S}$$

From this equation we obtain the torque speed characteristic as

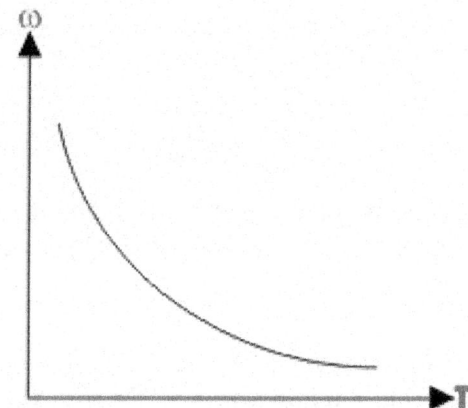

In a series wound dc motor, the speed varies with load. And operation wise this is its main difference from a shunt wound dc motor.

Compound Wound DC Motor

The compound excitation characteristic in a dc motor can be obtained by combining the operational characteristic of both the shunt and series excited dc motor. The compound wound self-excited dc motor or simply **compound wound dc motor** essentially contains the field winding connected both in series and in parallel to the armature winding as shown in the figure below :

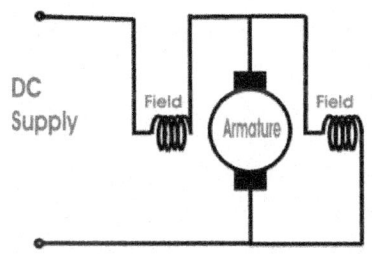

Cumulatively Compound Excited DC Motor

The excitation of compound wound dc motor can be of two types depending on the nature of compounding.

Cumulative Compound DC Motor

When the shunt field flux assists the main field flux, produced by the main field connected in series to the armature winding then its called cumulative compound dc motor.

$$\phi_{total} = \phi_{series} + \phi shunt$$

Differential Compound DC Motor

In case of a differentially compounded self-excited dc motor *i.e.* differential compound dc motor, the arrangement of shunt and series winding is such that the field flux produced by the shunt field winding diminishes the effect of flux by the main series field winding.

$$\phi_{total} = \phi_{series} + \phi shunt$$

The net flux produced in this case is lesser than the original flux and hence does not find much of a practical application.

The compounding characteristic of the self-excited dc motor is shown in the figure below.

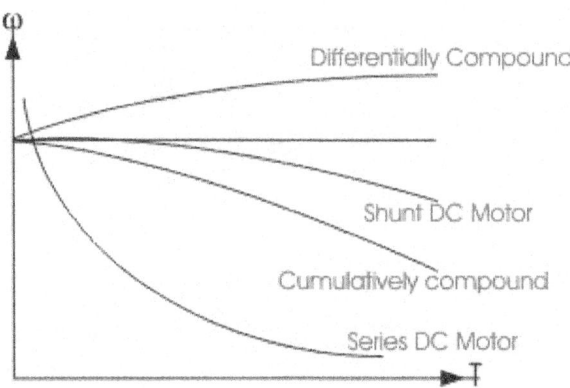

Both the cumulative compound and differential compound dc motor can either be of short shunt or long shunt type depending on the nature of arrangement.

Short Shunt DC Motor

If the shunt field winding is only parallel to the armature winding and not the series field winding then its known as short shunt dc motor or more specifically short shunt type compound wound dc motor.

Long Shunt DC Motor

If the shunt field winding is parallel to both the armature winding and the series field winding then it's known as long shunt type compounded wound dc motor or simply long shunt dc motor.

SHUNT WOUND DC MOTOR

The **shunt wound dc motor** falls under the category of self-excited dc motors, where the field windings are shunted to, or are connected in parallel to the armature winding of the motor, as its name is suggestive of. And for this reason both the armature winding and the field winding are exposed to the same supply voltage, though there are separate branches for the flow of armature current and the field current as shown in the figure of **dc shunt motor** below.

Voltage and Current Equation of a Shunt Wound DC Motor

Let us now consider the voltage and electric current being supplied from the electrical terminal to the motor be given by E and I_{total} respectively. This supply current in case of the **shunt wound dc motor** is split up into 2 parts. I_a, flowing through the armature winding of resistance R_a and I_{sh} flowing through the field winding of resistance R_{sh}. The voltage across both windings remains the same.

From there we can write $I_{total} = I_a + I_{sh}$

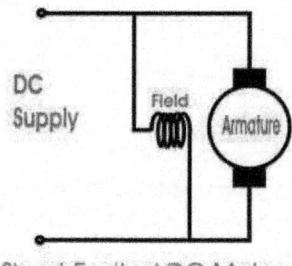

Shunt Excited DC Motor

Where
$$I_{sh} = \frac{E}{R_{sh}}$$

or
$$I_a = I_{total} - I_{sh} = \frac{E}{R_a}$$

Thus we put this value of armature current I_a to get general voltage equation of a **dc shunt motor**.

$$E = E_b + I_a R_a$$
Or
$$E = E_b + (I_{total} - I_{sh})R_a$$

Now in general practice, when the motor is in its running condition, and supply voltage is constant the shunt field current given by,

$$I_{sh} = \frac{E}{R_{sh}},\ \text{remains constant.}$$

But we know $I_{sh} \propto \Phi$

i.e. field flux Φ is proportional to filed current I_{sh}

Thus the field flux remains more or less constant and for this reason a shunt wound dc motor is called a constant flux motor.

Construction of a Shunt Wound DC Motor

The construction of a dc shunt motor is pretty similar to other types of dc motor, as shown in the figure below.

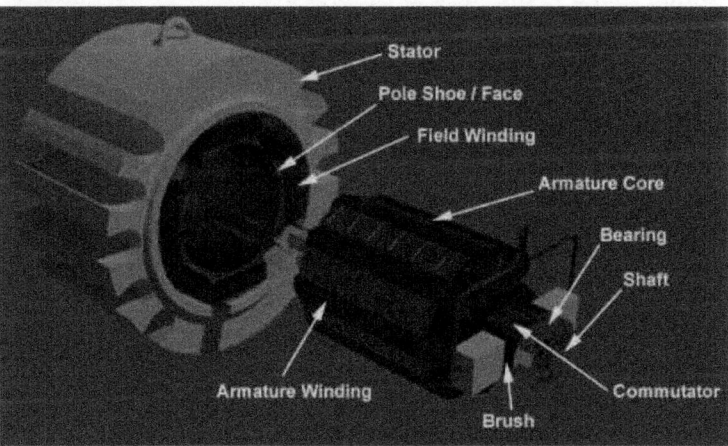

Fig. : Essential Parts of DC Machine.

Just that there is one distinguishable feature in its designing which can be explained by taking into consideration, the torque generated by the motor. To produce a high torque,

i   The armature winding must be exposed to an amount of electric current that's much higher than the field windings current, as the torque is proportional to the armature current.

ii   The field winding must be wound with many turns to increase the flux linkage, as flux linkage between the field and armature winding is also proportional to the torque.

Keeping these two above mentioned criterion in mind a dc shunt motor has been designed in a way, that the field winding possess much higher number of turns to increase net flux linkage and are lesser in diameter of conductor to increase resistance (reduce current flow) compared to the armature winding of the dc motor. And this is how a shunt wound dc motor is visibly distinguishable in static condition from the dc series motor (having thicker field coils) of the self-excited type motor's category.

Self-Speed Regulation of a Shunt Wound DC Motor

A very important and interesting fact about the dc shunt motor, is in its ability to self-regulate its speed on application of load to the shaft of the rotor terminals. This essentially means that on switching the motor running condition from no load to loaded, surprisingly there is no considerable change in speed of running, as would be expected in the absence of any speed regulating modifications from outside. Let us see how?

Let us do a step-wise analysis to understand it better.

1   Initially considering the motor to be running under no load or lightly loaded condition at a speed of N rpm.

2   On adding a load to the shaft, the motor does slow down initially, but this is where the concept of self regulation comes into the picture.

3   At the very onset of load introduction to a shunt wound dc motor, the speed definitely reduces, and along with speed also reduces the **back emf**, E_b. Since $E_b \propto N$, given by,

$$E_b = \frac{P. \varphi.Z.N}{60\,A}$$

This can be graphically explained below.

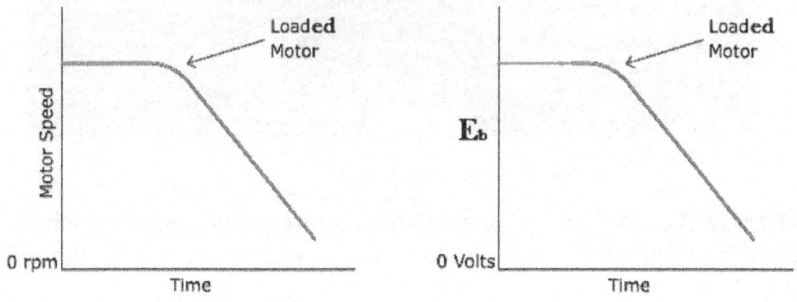

No load to loaded condition

4   This reduction in the counter emf or the back emf E_b results in the increase of the net voltage. As net voltage $E_{net} = E - E_b$. Since supply voltage E remains constant.

5   As a result of this increased amount of net voltage, the armature current increases and consequently the torque increases.

Since, $I_a \propto T$ given by

$$T = \frac{P.Z.\varphi.I_a}{2.\pi.A}$$

The change in armature current and torque on supplying load is graphically shown below.

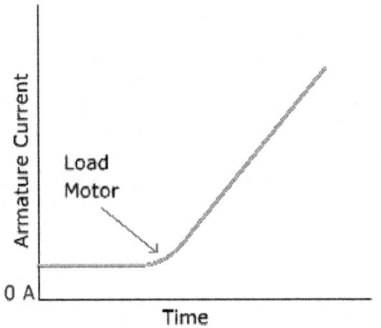   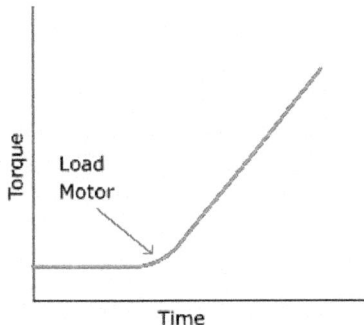

No load to loaded condition

6   This increase in the amount of torque increases the speed and thus compensating for the speed loss on loading. Thus the final speed characteristic of a dc shunt motor, looks like.

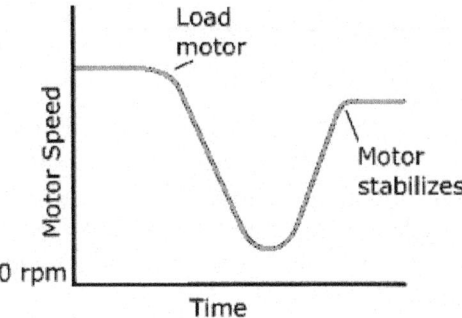

No load to loaded condition

From there we can well understand this special ability of the shunt wound dc motor to regulate its speed by itself on loading and thus its rightly called the constant flux or constant speed motor. Because of which it finds wide spread industrial application where ever constant speed operation is required.

DC SERIES MOTOR

Construction of Series DC Motor

Construction wise a this motor is similar to any other types of dc motors in almost all aspects. It consists of all the fundamental components like the stator housing the field winding or the rotor carrying the armature conductors, and the other vital parts like the commutator or the brush segments all attached in the proper sequence as in the case of a generic DC motor.

Yet if we are to take a close look into the wiring of the field and armature coils of this dc motor, its clearly distinguishable from the other members of this type. To understand that let us revert back into the above mentioned basic fact, that the this motor has field coil connected in series to the armature winding. For this reason relatively higher current flows through the field coils, and its designed accordingly as mentioned below.

i The field coils of dc series motor are wound with relatively fewer turns as the electric current through the field is its armature current and hence for required mmf less numbers of turns are required.

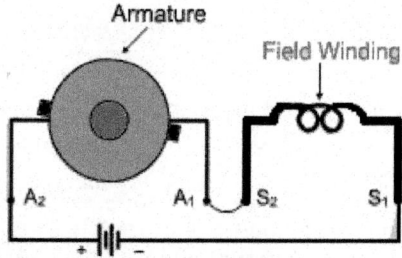

Fig. : Field winding with thicker diameter & fewer turns.

ii The wire is heavier, as the diameter is considerable increased to provide minimum electrical resistance to the flow of full armature current.

In spite of the above mentioned differences, about having fewer coil turns the running of this dc motor remains unaffected, as the electric current through the field is reasonably high to produce a field strong enough for generating the required amount of torque. To understand that better lets look into the voltage and current equation of dc series motor.

Voltage and Current Equation of Series dc Motor

The electrical layout of a typical series wound dc motor is shown in the diagram below.

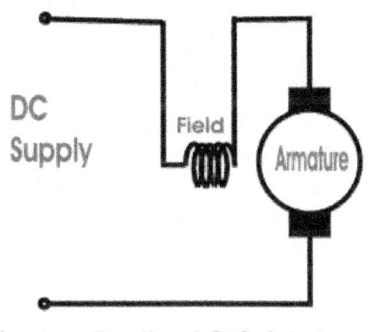

Series Excited DC Motor

Let the supply voltage and current given to the electrical port of the motor be given by E and I_{total} respectively.

Since the entire supply current flows through both the armature and field conductor.

Therefore, $$I_{total} = I_{se} = I_a$$

Where I_{se} is the series current in the field coil and I_a is the armature current.

Now form the basic voltage equation of the dc motor.

$$E = E_b + I_{se}R_{se} + I_aR_{se}$$

Where E_b is the back emf.

R_{se} is the series coil resistance & R_a is the armature resistance. .

Since $I_{se} = I_a$, we can write,

$$E = E_b + I_a(R_a + R_{se})$$

This is the basic voltage equation of a series wound dc motor.

Another interesting fact about the dc series motor worth noting is that, the field flux like in the case of any other dc motor is proportional to field current.

$$I_{se} \propto \phi$$

But since here $I_{se} = I_a = I_{total}$

$$\phi \propto I_{se} \propto I_a$$

i.e. the field flux is proportional to the entire armature current or the total supply current. And for this reason, the flux produced in this motor is strong enough to produce sufficient torque, even with the bare minimum number of turns it has in the field coil.

Speed & Torque of Series DC Motor

A series wound motors has linear relationship existing between the field current and the amount of torque produced. *i.e.* torque is directly proportional to electric current over the entire range of the graph. As in this case relatively higher current flows through the heavy series field winding with thicker diameter, the electromagnetic torque produced here is much higher than normal. This high electromagnetic torque produces motor speed, strong enough to lift heavy load overcoming its initial inertial of rest. And for this particular reason the motor becomes extremely essential as starter motors for most industrial applications dealing in heavy mechanical load like huge cranes or large metal chunks etc. Series motors are generally operated for a very small duration, about only a few seconds, just for the purpose of starting. Because if its run for too long, the high series current might burn out the series field coils thus leaving the motor useless.

Speed Regulation of Series Wound DC Motor

Unlike in the case of a DC shunt motor, the dc series motor has very poor speed regulation. *i.e.* the series motor is unable to maintain its speed on addition of external load to the shaft. Let us see why?

When mechanical load is added to the shaft at any instance, the speed automatically reduces whatever be the type of motor. But the term speed regulation refers to the ability of the motor to bring back the reduced speed to its original previous value within reasonable amount of time. But this motor is highly incapable of doing that as with reduction in speed N on addition of load, the back emf given by,

$$E_b = \frac{p\varphi ZN}{60\,A}$$

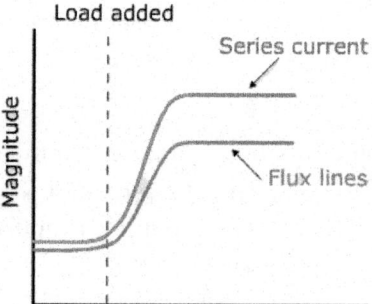

Fig. : Effect of load addition on dc series motor.

Also reduces as $E_b \propto N$

This decrease in back Emf E_b, increases the net voltage E- E_b, and consequently the series field current increases,

$$I_{se} = E - \frac{E_b}{R_a + R_{se}}$$

The value of series current through the field coil becomes so high that it tends to saturate of the magnetic core of the field. As a result the magnetic flux linking the coils increases at a much slower rate compared to the increase in current beyond the saturation region. The weak magnetic field produced as a consequence is unable to provide for the necessary amount of force to bring back the speed at its previous value before application of load. So keeping all the above mentioned facts in mind, a series wound dc motor is most applicable as a starting motor for industrial applications.

DC COMPOUND MOTOR

A **compound wound dc motor** or rather a **dc compound motor** falls under the category of self-excited motors, and is made up of both series the field coils $S_1 S_2$ and shunt field coils $F_1 F_2$ connected to the armature winding.

Both the field coils provide for the required amount of magnetic flux, that links with the armature coil and brings about the torque necessary to facilitate rotation at desired speed.

As we can understand, a **compound wound dc motor** is basically formed by the amalgamation of a shunt wound dc motor and series wound dc motor to achieve the better off properties of both these types. Like a shunt wound dc motor is bestowed with an extremely efficient speed regulation characteristic, whereas the dc series motor has high starting torque.

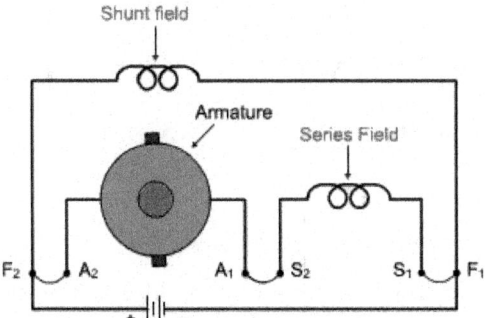

Fig. : Schematic diagram of dc compound motor.

So the compound wound dc motor reaches a compromise in terms of both this features and has a good combination of proper speed regulation and high starting touque. Though its staring torque is not as high as in case of dc motor, nor is its speed regulation as good as a shunt dc motor. Overall characteristics of dc shunt motor falls somewhere in between these 2 extreme limits.

Types of Compound Wound DC Motor

The compound wound dc motor can further be subdivided into 2 major types on the basis of its field winding connection with respect to the armature winding, and they are :

Long Shunt Compound Wound DC Motor

In case of long shunt compound wound dc motor, the shunt field winding is connected in parallel across the series combination of both the armature and series field coil, as shown in the diagram below.

Voltage and Current Equation of Long Shunt Compound Wound DC Motor

Let E and I_{total} be the total supply voltage and electric current supplied to the input terminals of the motor. And I_a, I_{se}, I_{sh} be the values of current flowing through armature resistance R_a, series winding resistance R_{se} and shunt winding resistance R_{sh} respectively.

Now we know in shunt motor, $I_{total} = I_a + I_{sh}$

And in series motor $I_a = I_{se}$

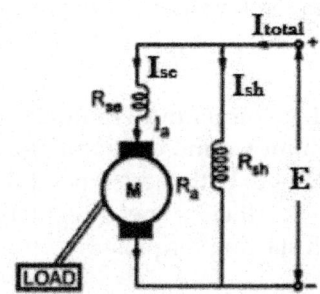

Therefore, the current equation of a compound wound dc motor is given by

$$I_{total} = I_{se} + I_{sh} \tag{1}$$

And its voltage equation is,

$$E = E_b + I_a(R_a + R_{se})$$

Short Shunt Compound Wound DC Motor

In case of short shunt compound wound dc motor, the shunt field winding is connected in parallel across the armature winding only. And series field coil is exposed to the entire supply current, before being split up into armature and shunt field current as shown in the diagram below.

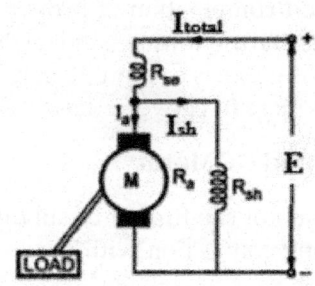

Voltage and Current Equation of Short Shunt Compound Wound DC Motor

Here also let, E and I_{total} be the total supply voltage and current supplied to the input terminals of the motor. And I_a , I_{se} , I_{sh} be the values of current flowing through armature resistance R_a , series winding resistance R_{se} and shunt winding resistance R_{sh} respectively.

But from the diagram above we can see,

$$I_{total} = I_{se} \qquad (2)$$

Since the entire supply current flows through the series field winding.

And like in the case of a dc shunt motor,

$$I_{total} = I_a + I_{sh} \qquad (3)$$

Equation (2) and (3) gives the current equation of a short shunt compound wound dc motor.

Now for equating the voltage equation, we apply Kirchoff's law to the circuit and get,

$$E = E_b + I_a R_a + I_{se} R_{se}$$

But since $I_{se} = I_{total}$

Thus the final voltage equation can be written as,

$$E = E_b + I_a R_a + I_{total} R_{se}$$

Apart from the above mentioned classification, a compound wound dc motor can further be sub-divided into 2 types depending upon excitation or the nature of compounding. *i.e.*

Cumulative Compounding of DC Motor

A compound wound dc motor is said to be cumulatively compounded when the shunt field flux produced by the shunt winding assists or enhances the effect of main field flux, produced by the series winding.

$$\phi_{total} = \phi_{series} + \phi_{shunt}$$

Differential Compounding of DC Motor

Similarly a compound wound dc motor is said to be differentially compounded when the flux due to the shunt field winding diminishes the effect of the main series winding. This particular trait is not really desirable, and hence does not find much of a practical application.

$$\phi_{total} = \phi_{series} - \phi_{shunt}$$

The net flux produced in this case is lesser than the original flux and hence does not find much of a practical application.

The compounding characteristic of the self-excited dc motor is shown in the figure below.

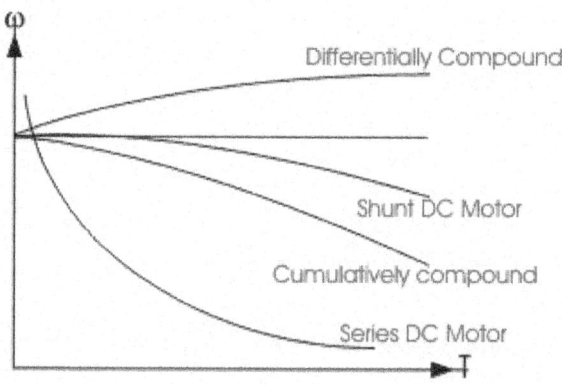

STARTING METHODS TO LIMIT STARTING CURRENT & TORQUE OF DC MOTOR

Starting of DC Motor

The **starting of DC motor** is somewhat different from the starting of all other types of electrical motors. This difference is credited to the fact that a dc motor unlike other types of motor has a very high starting electric current that has the potential of damaging the internal circuit of the armature winding of dc motor if not restricted to some limited value. This limitation to the **starting current of dc motor** is brought about by means of the starter. Thus the distinguishing fact about the **starting methods of dc motor** is that it is facilitated by means of a starter. Or rather a device containing a variable resistance connected in series to the armature winding so as to limit the starting current of dc motor to a desired optimum value taking into consideration the safety aspect of the motor.

Now the immediate question in why the DC motor has such high starting current?

To give an explanation to the above mentioned question let us take into consideration the basic operational voltage equation of the dc motor given by,

$$E = E_b + I_a R_a$$

Where E is the supply voltage, I_a is the armature current, R_a is the armature resistance. And the back emf is given by E_b.

Now the back emf, in case of a dc motor, is very similar to the generated emf of a dc generator as it's produced by the rotational motion of the electric current carrying armature conductor in presence of the field. This back emf of dc motor is given by

$$E_b = \frac{P.\phi.Z.N}{60\,A}$$

and has a major role to play in case of the **starting of dc motor**.

From this equation we can see that E_b is directly proportional to the speed N of the motor. Now since at starting N = 0, E_b is also zero, and under this circumstance the voltage equation is modified to

$$E = 0 + E_b R_a$$

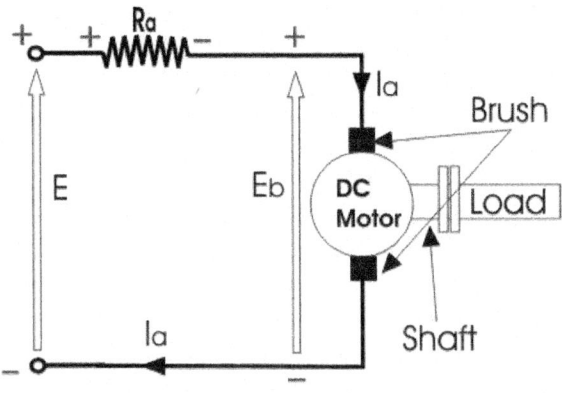

Therefore,
$$I_a = \frac{E}{R_a}$$

For all practical practices to obtain optimum operation of the motor the armature resistance is kept very small usually of the order of 0.5 Ω and the bare minimum supply voltage being 220 volts. Even under these circumstance the starting current, I_a is as high as 220/0.5 amp = 440 amp.

Such high starting current of dc motor creates two major problems :

1. Firstly, current of the order of 400 A has the potential of damaging the internal circuit of the armature winding of dc motor at the very onset.

2. Secondly, since the torque equation of dc motor is given by

Therefore,
$$I_a = \frac{E}{R_a}$$

Very high electromagnetic **starting torque of DC motor** is produced by virtue of the high starting current, which has the potential of producing huge centrifugal force capable of flying off the rotor winding from the slots.

Starting Methods of DC Motor

As a direct consequence of the two above mentioned facts i.e high starting current and high starting torque of DC motor, the entire motoring system can undergo a total disarray and lead towards into an engineering massacre and non-functionality. To prevent such an incidence from occurring several starting methods of dc motor has been adopted. The main principal of this being the addition of external electrical resistance R_{ext} to the armature winding, so as to increase the effective resistance to $R_a + R_{ext}$, thus limiting the armature current to the rated value. The new value of starting armature current is desirably low and is given by.

Therefore,
$$I_a = \frac{E}{R_a + R_{ext}}$$

Now as the motor continues to run and gather speed, the back emf successively develops and increases, countering the supply voltage, resulting in the decrease of the net working voltage. Thus now,

Therefore,
$$I_a = \frac{E - E_b}{R_a + R_{ext}}$$

At this moment to maintain the armature current to its rated value, R_{ext} is progressively decreased unless its made zero, when the back emf produced is at its maximum. This regulation of the external electrical resistance in case of the starting of dc motor is facilitated by means of the starter.

Starters can be of several types and requires a great deal of explanation and some intricate level understanding. But on a brief over-view the main types of starters used in the industry today can be illustrated as :-

1.   3 point starter.
2.   4 point starter.

Used for the starting of shunt wound DC motor and compound wound DC motor.

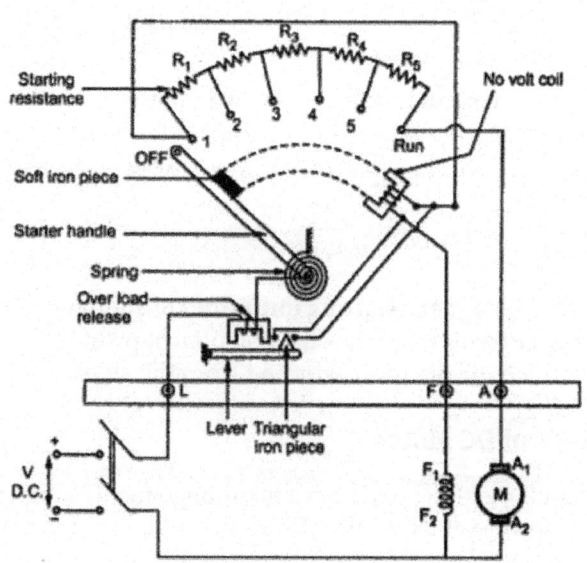

3 point Starter

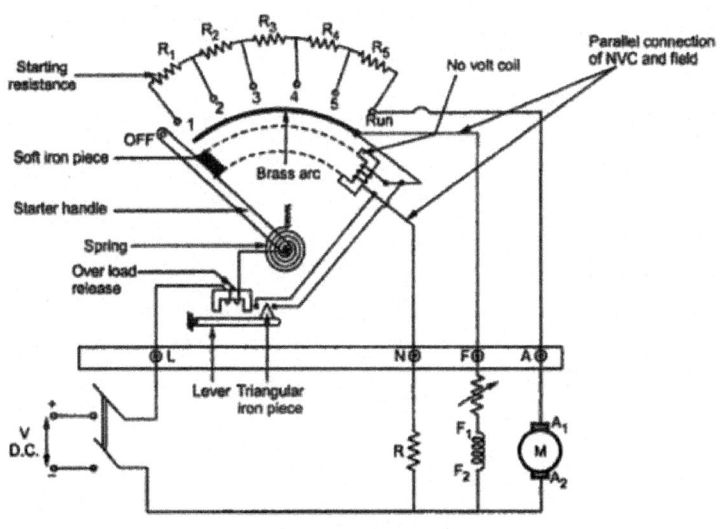

4 point Starter

3. Series wound DC motor's starter using no load release coil.

All of these play a very significant role in limiting starting current of DC motor for proper starting and running of the DC motor, and are described vividly under their respective sub-headings.

3 POINT STARTER

A **3 point starter** in simple words is a device that helps in the starting and running of a shunt wound DC motor or compound wound DC motor. Now the question is why these types of DC motors require the assistance of the starter in the first case. The only explanation to that is given by the presence of back emf E_b, which plays a critical role in governing the operation of the motor. The back emf, develops as the motor armature starts to rotate in presence of the magnetic field, by generating action and counters the supply voltage. This also essentially means, that the back emf at the starting is zero, and develops gradually as the motor gathers speed.

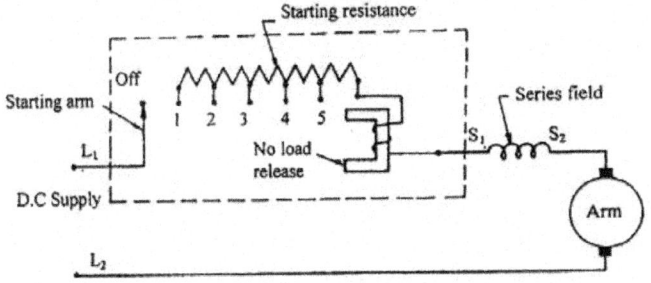

Series motor starter, no-load release.

The general motor emf equation $E = E_b + I_a.R_a$, at starting is modified to $E = I_a.R_a$ as at starting $E_b = 0$.

$$\therefore \qquad\qquad I_a = \frac{E}{R_a}$$

Thus we can well understand from the above equation that the electric current will be dangerously high at starting (as armature resistance R_a is small) and hence its important that we make use of a device like the **3 point starter** to limit the starting current to an allowable lower value.

Let us now look into the construction and **working of three point starter** to understand how the starting current is restricted to the desired value. For that let's consider the diagram given below showing all essential parts of the three point starter.

Construction of 3 Point Starter

Construction wise a starter is a variable resistance, integrated into number of sections as shown in the figure beside. The contact points of these sections are called studs and are shown separately as **OFF, 1, 2,3,4,5, RUN**. Other than that there are 3 main points, referred to as

1.   'L' Line terminal. (Connected to positive of supply.)
2.   'A' Armature terminal. (Connected to the armature winding.)
3.   'F' Field terminal. (Connected to the field winding.)

And from there it gets the name 3 point starter.

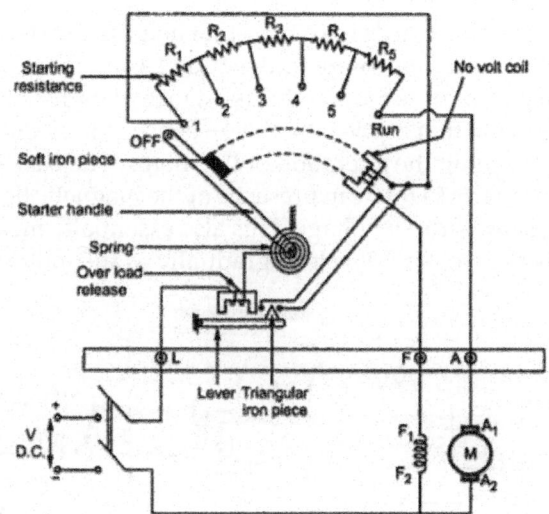

3 point Starter

Now studying the construction of 3 point starter in further details reveals that, the point 'L' is connected to an electromagnet called overload release (OLR) as shown in the figure. The other end of 'OLR' is connected to the lower end of conducting lever of starter handle where a spring is also attached with it and the starter handle contains also a soft iron piece housed on it. This handle is free to move to the other side RUN against the force of the spring. This spring brings back the handle to its original OFF position under the influence of its own force. Another parallel path is derived from the stud '1', given to the another electro-magnet called No Volt Coil (NVC) which is further connected to terminal 'F'. The starting resistance at starting is entirely in series with the armature. The OLR and NVC acts as the two protecting devices of the starter.

Working of Three Point Starter

Having studied its construction, let us now go into the **working of the 3 point starter**. To start with the handle is in the OFF position when the supply to the DC motor is switched on. Then handle is slowly moved against the spring force to make a contact with stud No. 1. At this point, field winding of the shunt or the compound motor gets supply through the parallel path provided to starting resistance, through No Voltage Coil. While entire starting resistance comes in series with the armature. The high starting armature current thus gets limited as the current equation at this stage becomes $I_a = E/(R_a + R_{st})$. As the handle is moved further, it goes on making contact with studs 2, 3, 4 etc., thus gradually cutting off the series resistance from the armature circuit as the motor gathers speed. Finally when the starter handle is in 'RUN' position, the entire starting resistance is eliminated and the motor runs with normal speed.

This is because back emf is developed consequently with speed to counter the supply voltage and reduce the armature current. So the external electrical resistance is not required anymore, and is removed for optimum operation. The handle is moved manually from OFF to the RUN position with development of speed. Now the obvious question is once the handle is taken to the RUN position how is it supposed to stay there, as long as motor is running ?

To find the answer to this question let us look into the working of No Voltage Coil.

Working of No Voltage Coil of 3 Point Starter

The supply to the field winding is derived through no voltage coil. So when field current flows, the NVC is magnetized. Now when the handle is in the 'RUN' position, soft iron piece connected to the handle and gets attracted by the magnetic force produced by NVC, because of flow of electric current through it. The NVC is designed in such a way that it holds the handle in 'RUN' position against the force of the spring as long as supply is given to the motor. Thus NVC holds the handle in the 'RUN' position and hence also called **hold on coil**.

Now when there is any kind of supply failure, the electric current flow through NVC is affected and it immediately looses its magnetic property and is unable to keep the soft iron piece on the handle, attracted. At this point under the action of the spring force, the handle comes back to OFF position, opening the circuit and thus switching off the motor. So due to the combination of NVC and the spring, the starter handle always comes back to OFF position whenever there is any supply problems. Thus it also acts as a protective device safeguarding the motor from any kind of abnormality.

4 POINT STARTER

Working Principle of Four Point Starter

The **4 point starter** like in the case of a 3 point starter also acts as a protective device that helps in safeguarding the armature of the shunt or compound excited dc motor against the high starting current produced in the absence of back emf at starting.

The 4 point starter has a lot of constructional and functional similarity to a three point starter, but this special device has an additional point and a coil in its construction, which naturally brings about some difference in its functionality, though the basic operational characteristic remains the same.

Now to go into the details of **operation of 4 point starter**, lets have a look at its constructional diagram, and figure out its point of difference with a 3 point starter.

Construction and Operation of Four Point Starter

A 4 point starter as the name suggests has 4 main operational points, namely

1. 'L' Line terminal. (Connected to positive of supply.)
2. 'A' Armature terminal. (Connected to the armature winding.)
3. 'F' Field terminal. (Connected to the field winding.)
   Like in the case of the 3 point starter, and in addition to it there is,
4. A 4th point N. (Connected to the No Voltage Coil)

The remarkable difference in case of a 4 point starter is that the No Voltage Coil is connected independently across the supply through the fourth terminal called 'N' in addition to the 'L', 'F' and 'A'. As a direct consequence of that, any change in the field supply electric current does not bring about any difference in the performance of the NVC. Thus it must be ensured that no voltage coil always produce a force which is strong enough to hold the handle in its 'RUN' position, against force of the spring, under all the operational conditions. Such a current is adjusted through No Voltage Coil with the help of fixed resistance R connected in series with the NVC using fourth point 'N'.

Apart from this above mentioned fact, the 4 point and 3 point starters are similar in all other ways like possessing is a variable resistance, integrated into number of sections as shown in the figure above. The contact points of these sections are called studs and are shown separately as OFF, 1, 2, 3, 4, 5, RUN, over which the handle is free to be maneuvered manually to regulate the starting current with gathering speed.

Now to understand its way of operating lets have a closer look at the diagram given above. Considering that supply is given and the handle is taken stud No.1, then the circuit is complete and line current that starts flowing through the starter. In this situation we can see that the electric current will be divided into 3 parts, flowing through 3 different points.

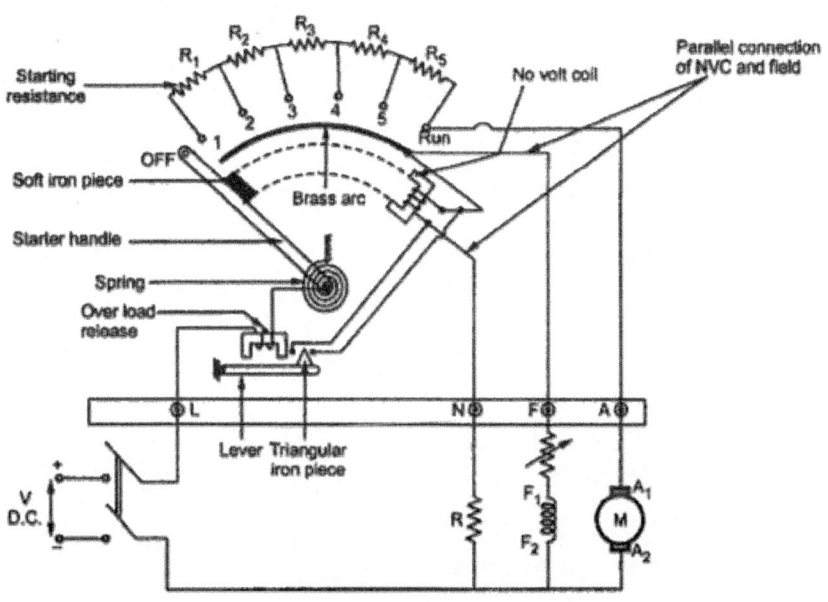

4 point Starter

i.   1 part flows through the starting resistance (R_1+ R_2+ R_3.....) and then to the armature.

ii. A 2nd part flowing through the field winding F.

iii. And a 3rd part flowing through the no voltage coil in series with the protective resistance R.

So the point to be noted here is that with this particular arrangement any change in the shunt field circuit does not bring about any change in the no voltage coil as the two circuits are independent of each other. This essentially means that the electromagnet pull subjected upon the soft iron bar of the handle by the no voltage coil at all points of time should be high enough to keep the handle at its RUN position, or rather prevent the spring force from restoring the handle at its original OFF position, irrespective of how the field rheostat is adjusted.

This marks the operational difference between a 4 point starter and a 3 point starter. As otherwise both are almost similar and are used for limiting the starting current to a shunt wound DC motor or compound wound DC motor, and thus act as a protective device.

SPEED REGULATION OF DC MOTOR

On application of load the speed of a dc motor decreases gradually. This is not at all desirable. So the difference between no load and full load speed should be very less. The motor capable of maintaining a nearly constant speed for varying load is said to have good speed regulation *i.e.* the difference between no load and full load speed is quite less. The speed regulation of a permanent magnet DC motor is good ranging from 10-15% whereas for dc shunt motor it is somewhat less than 10 %. DC series motor has poor value of regulation. In case of compound DC motor the speed regulation for dc cumulative compound is around 25 % while differential compound has its excellent value of 5 %.

Speed of a DC Motor

The emf equation of DC motor is given by

$$E = \frac{NP\Phi Z}{60\,A}$$

Here N = speed of rotation in rpm.

P = number of poles.

A = number of parallel paths.

Z = total no. conductors in armature.

Thus, speed of rotation $\quad N = \frac{60\,A}{P\,Z} \times \frac{E}{\Phi}$

$$\Rightarrow N = \frac{E}{k\phi} \text{ Where } k = \frac{PZ}{60\,A} \text{ is a constant}$$

Hence, speed of a DC motor is directly proportional to emf of rotation (E) and inversely proportional to flux per pole (φ).

Speed Regulation of a DC Motor

The speed regulation is defined as the change in speed from no load to full load, expressed as a fraction or percentage of full load speed.

Therefore, as per definition per unit (p.u) **speed regulation of DC motor** is given as,

$$\text{Speed Regulation}_{pu} = \frac{N_{no\ load} - N_{full\ load}}{N_{full\ load}}$$

Similarly, percentage (%) **speed regulation** is given as,

$$\text{Speed Regulation}(\%) = \frac{N_{no\ load} - N_{full\ load}}{N_{full\ load}} \times 100\%$$

Where $N_{no\ load}$ = no load speed and $N_{full\ lod}$ = full load speed of DC motor.

Therefore, Per cent speed regulation = Per unit (p.u.) speed regulation X 100 %.

A motor which has nearly constant speed at all load below full rated load, have good speed regulation.

SPEED CONTROL OF DC MOTOR

Speed control means intentional change of the drive speed to a value required for performing the specific work process. Speed control is a different concept from speed regulation where there is natural change in speed due change in load on the shaft. Speed control is either done manually by the operator or by means of some automatic control device.

One of the important features of dc motor is that its speed can be controlled with relative ease. We know that the expression of speed control dc motor is given as,

$$N = KV - \frac{I_a(R_a + R)}{\phi}$$

Therefore speed (N) of 3 types of dc motor – SERIES, SHUNT AND COMPOUND can be controlled by changing the quantities on RHS of the expression. So speed can be varied by changing

(i) terminal voltage of the armature V,

(ii) external resistance in armature circuit R and

(iii) flux per pole ϕ .

The first two cases involve change that affects armature circuit and the third one involves change in magnetic field. Therefore speed control of dc motor is classified as

1. armature control methods and

2. field control methods.

Speed Control of DC Series Motor

Speed control of dc series motor can be done either by armature control or by field control.

Armature Control of DC Series Motor

Speed adjustment of dc series motor by armature control may be done by any one of the methods that follow,

1. Armature resistance control method : This is the most common method employed. Here the controlling resistance is connected directly in series with the supply to the motor as shown in the fig. diagram

    The power loss in the control resistance of dc series motor can be neglected because this control method is utilized for a large portion of time for reducing the speed under light load condition. This method of speed control is most economical for constant torque. This method of speed control is employed for dc series motor driving cranes, hoists, trains etc.

2. Shunted armature control : The combination of a rheostat shunting the armature and a rheostat in series with the armature is involved in this method of speed control. The voltage applied to the armature is varies by varying series rheostat R_1. The exciting current can be varied by varying the armature shunting resistance R_2. This method of speed control is not economical due to considerable power losses in speed controlling resistances. Here speed control is obtained over wide range but below normal speed.

3. Armature terminal voltage control : The speed control of dc series motor can be accomplished by supplying the power to the motor from a separate variable voltage supply. This method involves high cost so it rarely used.

Field Control of DC Series Motor

The speed of dc motor can be controlled by this method by any one of the following ways –

1. Field diverter method : This method uses a diverter. Here the field flux can be reduced by shunting a portion of motor current around the series field. Lesser the diverter resistance less is the field current, less flux therefore more speed. This method gives speed above normal and the method is used in electric drives in which speed should rise sharply as soon as load is decreased. diagram

2. Tapped Field control : This is another method of increasing the speed by reducing the flux and it is done by lowering number of turns of field winding through which current flows. In this method a number of tapping from field winding are brought outside . This method is employed in electric traction.

Speed Control of DC Shunt Motor

Field Control of DC Shunt Motor

By this method speed control is obtained by any one of the following means :

1. Field rheostat control of DC Shunt Motor : In this method, speed variation is accomplished by means of a variable resistance inserted in series with the

shunt field. An increase in controlling resistances reduces the field current with a reduction in flux and an increase in speed. This method of speed control is independent of load on the motor. Power wasted in controlling resistance is very less as field current is a small value. This method of speed control is also used in DC compound motor.

Limitations of this method of speed control :

A. Creeping speeds cannot be obtained.

B. Top speeds only obtained at reduced torque.

C. The speed is maximum at minimum value of flux, which is governed by the demagnetizing effect of armature reaction on the field.

2. Field voltage control : This method requires a variable voltage supply for the field circuit which is separated from the main power supply to which the armature is connected. Such a variable supply can be obtained by an electronic rectifier.

Armature Control of DC Shunt Motor

Speed control by this method involves two ways . These are :

1. Armature resistance control : In this method armature circuit is provided with a variable resistance. Field is directly connected across the supply so flux is not changed due to variation of series resistance. This is applied for dc shunt motor. This method is used in printing press, cranes, hoists where speeds lower than rated is used for a short period only.

2. Armature voltage control : This method of speed control needs a variable source of voltage separated from the source supplying the field current. This method avoids disadvantages of poor speed regulation and low efficiency of armature-resistance control methods. The basic adjustable armature voltage control method of speed d control is accomplished by means of an adjustable voltage generator is called Ward Leonard system. This method involves using a motor –generator (M-G) set. This method is best suited for steel rolling mills, paper machines, elevators, mine hoists, etc.

Advantages of this method :

A. Very fine speed control over whole range in both directions

B. Uniform acceleration is obtained

C. Good speed regulation

Disadvantages :

A. Costly arrangement is needed , floor space required is more

B. Low efficiency at light loads.

LAP WINDING SIMPLEX AND DUPLEX LAP WINDING

Armature windings are mainly of two types – **lap winding** and wave winding. Here we are going to discuss about **lap winding**.

Lap winding is the winding in which successive coils overlap each other. It is named "Lap" winding because it doubles or laps back with its succeeding coils.

In this winding the finishing end of one coil is connected to one commutator segment and the starting end of the next coil situated under the same pole and connected with same commutator segment.

Here we can see, the finishing end of coil-1 and starting end of coil-2 are both connected to the commutator segment-2 and both coils are under the same magnetic pole that is N pole here.

Simplex Lap Winding

A winding in which the number of parallel path between the brushes is equal to the number of poles is called **simplex lap winding**.

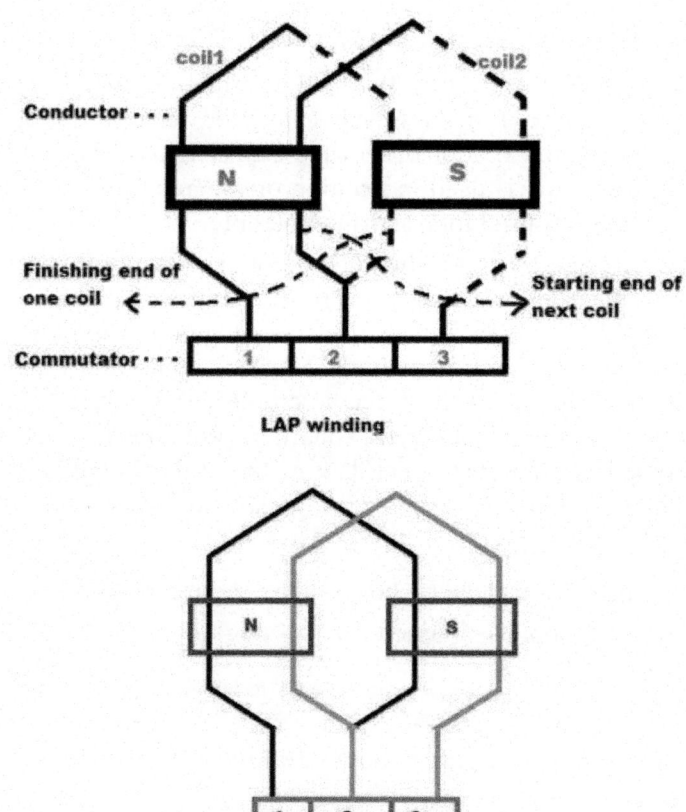

LAP winding

Simplex lap winding

Duplex Lap Winding

A winding in which the number of parallel path between the brushes is twice the number of poles is called **duplex lap winding.**

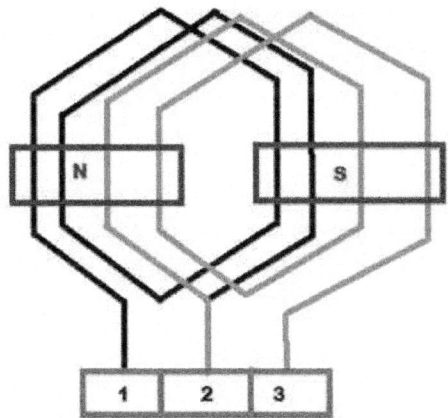

Duplex lap winding

Some important points to remember while designing the Lap winding :
If, Z= the number conductors

        P = number of poles

        Y_B = Back pitch

        Y_F = Front pitch

        Y_C = Commutator pitch

        Y_A = Average pole pitch

        Y_P = Pole pitch

        Y_R = Resultant pitch

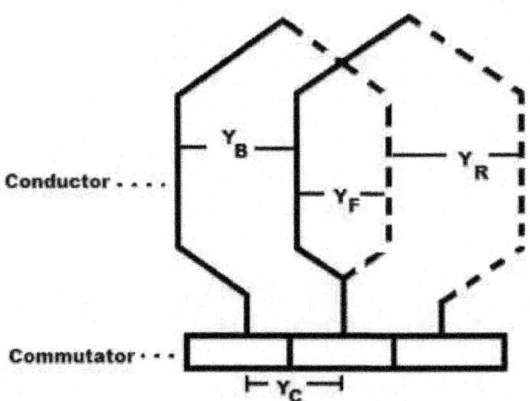

Then, the back and front pitches are of opposite sign and they cannot be equal.

$Y_B = Y_F \pm 2m$

m = multiplicity of the winding.

m = 1 for Simplex Lap winding

m = 2 for Duplex Lap winding

When,

$Y_B > Y_F$, it is called progressive winding.

$Y_B < Y_F$, it is called retrogressive winding.

Back pitch and front pitch must be odd.

Resultant pitch $(Y_R) = Y_B - Y_F = 2m$

Y_R is even because it is the difference between two odd numbers.

$$\text{Average pitch } (Y_A) = \frac{Y_B + Y_F}{2} = \text{pole pitch } (Y_P) = \frac{Z}{P}$$

$$\text{Average pitch } (Y_A) \approx \frac{Z}{P}$$

Commutator pitch $(Y_C) = \pm \ m$

Number of parallel path in the Lap winding = mP

Let us start from 1st conductor,

Back connections	Front connections
1 to $(1+Y_B)=(1+5)=6$	6 to $(6-Y_F)=(6-3)=3$
3 to $(3+5)=8$	8 to $(8-3)=5$
5 to $(5+5)=10$	10 to $(10-3)=7$
7 to $(7+5)=12$	12 to $(12-3)=9$
9 to $(9+5)=14$	14 to $(14-3)=11$
11 to $(11+5))=16$	16 to $(16-3)=13$
13 to $(13+5)=18=(18-16)=2$	2 to $(18-3)=15$
15 to $(15+5)=20=(20-16)=4$	4 to$(20-3)=17=(17-16)=1$

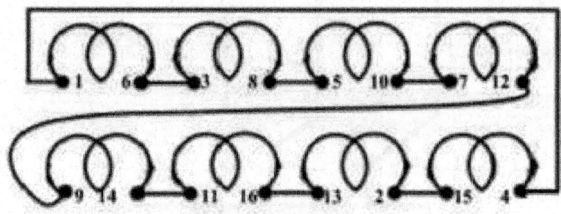

Simplified connection

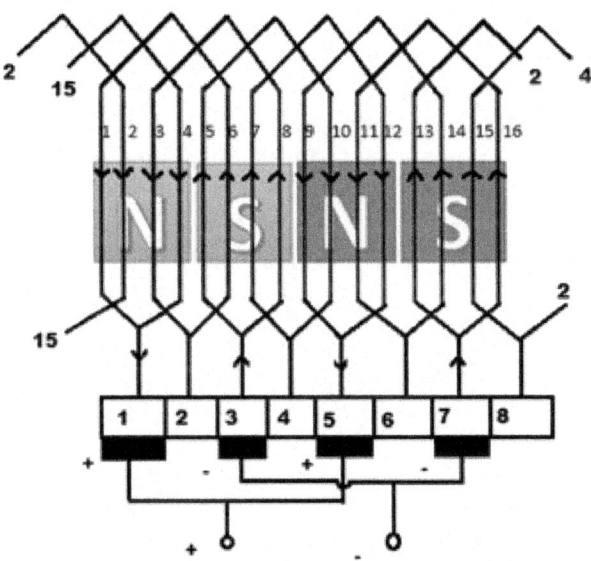

Advantages of Lap Winding

1. This winding is necessarily required for large current application because it has more parallel paths.
2. It is suitable for low voltage and high current generators.

Disadvantages of Lap Winding

1. It gives less emf compared to wave winding. This winding is required more no. of conductors for giving the same emf, it results high winding cost.
2. It has less efficient utilization of space in the armature slots.

PERMANENT MAGNET DC MOTOR OR PMDC MOTOR

In a dc motor, an armature rotates inside a magnetic field. Basic working principle of DC motor is based on the fact that whenever a current carrying conductor is placed inside a magnetic field, there will be mechanical force experienced by that conductor. All kinds of DC motors work in this principle only. Hence for constructing a dc motor it is essential to establish a magnetic field. The magnetic field is obviously established by means of magnet. The magnet can by any types *i.e.* it may be electromagnet or it can be permanent magnet. When permanent magnet is used to create magnetic field in a DC motor, the motor is referred as **permanent magnet dc motor** or **PMDC motor**. Have you ever uncovered any battery operated toy, if you did, you had obviously found a battery operated motor inside it. This battery operated motor is nothing but a **permanent magnet dc motor** or **PMDC motor**. These types of motor are essentially simple in construction. These motors are commonly used as starter motor in automobiles, windshield wipers, washer, for blowers used in heaters and air conditioners, to raise and lower windows, it also extensively used in toys. As the magnetic field strength of a permanent magnet is fixed it cannot be controlled externally, field control of this type of dc motor cannot be possible. Thus permanent magnet dc motor is used where there is no need of speed control of motor by means of controlling its field. Small fractional and sub fractional kW motors now constructed with permanent magnet.

Construction of Permanent Magnet DC Motor or PMDC Motor

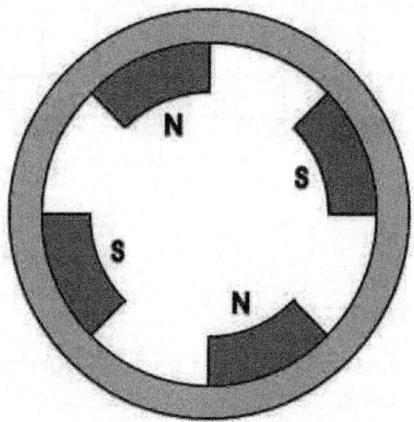

Fig. : Stator of Permanent Magnet DC Motor.

As it is indicated in name of permanent magnet dc motor, the field poles of this motor are essentially made of permanent magnet. A PMDC motor mainly consists of two parts. A stator and an armature. Here the stator which is a steel cylinder. The magnets are mounted in the inner periphery of this cylinder. The permanent magnets are mounted in such a way that the N – pole and S – pole of each magnet are alternatively faced towards armature. That means, if N – pole of one magnet is faced towards armature then S – pole of very next magnet is faced towards armature.

In addition to holding the magnet on its inner periphery, the steel cylindrical stator also serves as low reluctance return path for the magnetic flux. Although field coil is not required in permanent magnet dc motor but still it is sometimes found that they are used along with permanent magnet. This is because if permanent magnets lose their strength, these lost magnetic strengths can be compensated by field excitation through these field coils. Generally, rare earth hard magnetic materials are used for these permanent magnet.

Rotor : The rotor of pmdc motor is similar to other DC motor. The rotor or armature of permanent magnet dc motor also consists of core, windings and commutator. Armature core is made of number of varnish insulated, slotted circular lamination of steel sheets. By fixing these circular steel sheets one by one, a cylindrical shaped slotted armature core is formed. The varnish insulated laminated steel sheets are used to reduce eddy current loss in armature of permanent magnet dc motor. These slots on the outer periphery of the armature core are used for housing armature conductors in them. The armature conductors are connected in a suitable manner which gives rise to armature winding. The end terminals of the winding are connected to the commutator segments placed on the motor shaft. Like other dc motor, carbon or graphite brushes are placed with spring pressure on the commutator segments to supply current to the armature.

Working Principle of Permanent Magnet DC Motor or PMDC Motor

As we said earlier the working principle of PMDC motor is just similar to the general working principle of DC motor. That is when a carrying conductor comes inside a magnetic field, a mechanical force will be experienced by the conductor and the direction of this force is governed by Fleming's left hand rule. As in a permanent magnet dc motor, the armature is placed inside the magnetic field of permanent magnet; the armature rotates in the direction of the generated force. Here each conductor of the armature experiences the mechanical force $F = B.I.L$ Newton where B is the magnetic field strength in Tesla (weber / m^2), I is the current in Ampere flowing through that conductor and L is length of the conductor in metre comes under the magnetic field. Each conductor of the armature experiences a force and the compilation of those forces produces a torque, which tends to rotate the armature.

Equivalent Circuit of Permanent Magnet DC Motor or PMDC Motor

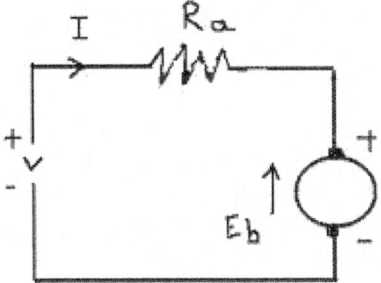

As in PMDC motor the field is produced by permanent magnet, there is no need of drawing field coils in the equivalent circuit of permanent magnet dc motor. The supply voltage to the armature will have armature resistance drop and rest of the supply voltage is countered by back emf of the motor. Hence voltage equation of the motor is given by,
$$V = I_R + E_b$$
Where I, is armature current and R is armature resistance of the motor.

E_b is the back emf and V is the supply voltage.

Advantages of Permanent Magnet DC Motor or PMDC Motor

PMDC motor have some advantages over other types of dc motors. They are :

- No need of field excitation arrangement.
- No input power in consumed for excitation which improve efficiency of dc motor.
- No field coil hence space for field coil is saved which reduces the overall size of the motor.
- Cheaper and economical for fractional kW rated applications.

Disadvantages of Permanent Magnet DC Motor or PMDC Motor

- In this case, the armature reaction of DC motor cannot be compensated hence the magnetic strength of the field may get weak due to demagnetizing effect armature reaction.
- There is also a chance of getting the poles permanently demagnetized (partial) due to excessive armature current during starting, reversal and overloading condition of the motor.
- Another major disadvantage of PMDC motor is that, the field in the air gap is fixed and limited and it cannot be controlled externally. Therefore, very efficient speed control of DC motor in this type of motor is difficult.

Applications of Permanent Magnet DC Motor or PMDC Motor

PMDC motor is extensively used where small dc motors are required and also very effective control is not required, such as in automobiles starter, toys, wipers, washers, hot blowers, air conditioners, computer disc drives and in many more.

WARD LEONARD METHOD OF SPEED CONTROL

Ward Leonard control system is introduced by Henry Ward Leonard in 1891. Ward Leonard method of speed control is used for controlling the speed of a DC motor. It is a basic armature control method. This control system is consisting of a dc motor M_1 and powered by a DC generator G. In this method the speed of the dc motor (M_1) is controlled by applying variable voltage across its armature. This variable voltage is obtained using a motor-generator set which consists of a motor M_2 (either ac or dc motor) directly coupled with the generator G. It is a very widely used method of speed control of DC motor.

Principle of Ward Leonard Method

The speed of motor M_1 is to be controlled which is powered by the generator G. The shunt field of the motor M_1 is connected across the dc supply lines. Now, generator G is driven by the motor M_2. The speed of the motor M_2 is constant. When the output voltage of the generator is fed to the motor M_1 then the motor starts to rotate. When the output voltage of the generator varies then the speed of the motor also varies. Now controlling the output voltage of the generator the speed of motor can also be controlled. For this purpose of controlling the output voltage, a field regulator is connected across the generator with the dc supply lines to control the field excitation. The direction of rotation of the motor M_1 can be reversed by excitation current of the generator and it can be done with the help of the reversing switch R.S. But the motor-generator set must run in the same direction.

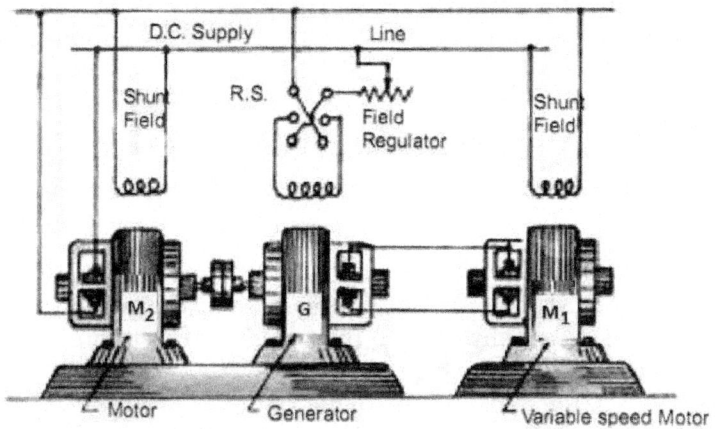

WARD LEONARD SYSTEM OF SPEED CONTROL

Advantages of Ward Leonard System

1. It is a very smooth speed control system over a very wide range (from zero to normal speed of the motor).
2. The speed can be controlled in both the direction of rotation of the motor easily.
3. The motor can run with a uniform acceleration.
4. Speed regulation of DC motor in this ward Leonard system is very good.

Disadvantages of Ward Leonard System

1. The system is very costly because two extra machines (motor-generator set) are required.
2. Overall efficiency of the system is not sufficient especially it is lightly loaded.

Application of Ward Leonard System

This Ward Leonard method of speed control system is used where a very wide and very sensitive speed control is of a DC motor in both the direction of rotation is required. This speed control system is mainly used in colliery winders, cranes, electric excavators, mine hoists, elevators, steel rolling mills and paper machines etc.

Chapter 7

SYNCHRONOUS MOTOR

A **synchronous electric motor** is an AC motor in which, at steady state, the rotation of the shaft is synchronized with the frequency of the supply current; the rotation period is exactly equal to an integral number of AC cycles. Synchronous motors contain electromagnets on the stator of the motor that create a magnetic field which rotates in time with the oscillations of the line current. The rotor turns in step with this field, at the same rate.

The synchronous motor and induction motor are the most widely used types of AC motor. The difference between the two types is that the synchronous motor rotates in exact synchronism with the line frequency. In contrast the induction motor requires *"slip"*, the rotor must rotate slightly slower than the AC current alternations, to develop torque. Therefore small synchronous motors are used in timing applications such as in synchronous clocks, timers in appliances, tape recorders and precision servomechanisms in which the motor must operate at a precise speed.

Synchronous motors are available in sub-fractional **self-excited** sizes to high-horsepower industrial sizes. In the fractional horsepower range, most synchronous motors are used where precise constant speed is required. In high-horsepower industrial sizes, the synchronous motor provides two important functions. First, it is a highly efficient means of converting AC energy to work. Second, it can operate at leading or unity power factor and thereby provide power-factor correction.

These machines are commonly used in analog electric clocks, timers and other devices where correct time is required.

TYPE

Synchronous motors fall under the more general category of *synchronous machines* which also includes the synchronous generator. Generator action will be observed if the field poles are "driven ahead of the resultant air-gap flux by the forward motion of the prime mover". Motor action will be observed if the field poles are "dragged behind the resultant air-gap flux by the retarding torque of a shaft load".

There are two major types of synchronous motors depending on how the rotor is magnetized : *non-excited* and *direct-current excited*.

Non-excited Motors

In non-excited motors, the rotor is made of steel. At synchronous speed it rotates in step with the rotating magnetic field of the stator, so it has an almost-constant magnetic field through it. The external stator field magnetizes the rotor, inducing the magnetic poles needed to turn it. The rotor is made of a high-retentivity steel such as cobalt steel, These are manufactured in permanent magnet, reluctance and hysteresis designs :

Reluctance Motors

These have a rotor consisting of a solid steel casting with projecting (salient) toothed poles, typically less than the stator poles to minimize torque ripple and prevents the poles from all aligning simultaneously—a position which cannot generate torque. The size of the air gap in the magnetic circuit and thus the reluctance is minimum when the poles are aligned with the (rotating) magnetic field of the stator, and increases with the angle between them. This creates a torque pulling the rotor into alignment with the nearest pole of the stator field. Thus at synchronous speed the rotor is "locked" to the rotating stator field. This cannot start the motor, so the rotor poles usually have squirrel-cage windings embedded in them. The machine starts as an induction motor until it approaches synchronous speed, when the rotor "pulls in" and locks to the rotating stator field.

Reluctance motor designs have ratings that range from fractional horsepower (a few watts) to about 22 kW. Very small reluctance motors have low torque, and are generally used for instrumentation applications. Moderate torque, integral horsepower motors use squirrel cage construction with toothed rotors. When used with an adjustable frequency power supply, all motors in the drive system can be controlled at exactly the same speed. The power supply frequency determines motor operating speed.

Hysteresis Motors

These have a solid smooth cylindrical rotor, cast of a high coercivity magnetically "hard" cobalt steel. This material has a wide hysteresis loop (high retentivity), meaning once it is magnetized in a given direction, it requires a large reverse magnetic field to reverse the magnetization. The rotating stator field causes each small volume of the rotor to experience a reversing magnetic field. Because of hysteresis the phase of the magnetization lags behind the phase of the applied field. The result of this is that the axis of the magnetic field induced in the rotor lags behind the axis of the stator field by a constant angle δ, producing a torque as the rotor tries to "catch up" with the stator field. As long as the rotor is below synchronous speed, each particle of the rotor experiences a reversing magnetic field at the "slip" frequency which drives it around its hysteresis loop, causing the rotor field to lag and create torque. There is a 2-pole low reluctance bar structure in the rotor. As the rotor approaches synchronous speed and slip goes to zero, this magnetizes and aligns with the stator field, causing the rotor to "lock" to the rotating stator field.

A major advantage of the hysteresis motor is that since the lag angle δ is independent of speed, it develops constant torque from startup to synchronous speed. Therefore, it is self-starting and doesn't need an induction winding to start it, although many designs do have a squirrel-cage conductive winding structure embedded in the rotor to provide extra torque at start-up.

Hysteresis motors are manufactured in sub-fractional horsepower ratings, primarily as servomotors and timing motors. More expensive than the reluctance type, hysteresis motors are used where precise constant speed is required.

Permanent Magnet Motors

These have permanent magnets embedded in the steel rotor to create a constant magnetic field. At synchronous speed these poles lock to the rotating magnetic field. They are not self-starting. Because of the constant magnetic field in the rotor these cannot use induction windings for starting, but it not necessary

DC-excited Motors

Made in sizes larger than 735 W, these motors require direct current supplied to the rotor for excitation. This is most straightforwardly supplied through slip rings, but a brushless AC induction and rectifier arrangement may also be used. The direct current may be supplied from a separate DC source or from a DC generator directly connected to the motor shaft. Made in sizes larger than 735 W, these motors require direct current supplied to the rotor for excitation. This is most straightforwardly supplied through slip rings and brushes, but a brushless AC induction and rectifier arrangement may also be used.

Synchronous Speed

The synchronous speed N_s (in RPM) of a synchronous motor is given by

$$N_s = \frac{120f}{p}$$

where,

f is the frequency of the AC supply current in Hz.

p is the number of poles per phase.

Synchronous speed can also be expressed in terms of angular speed,

$$\omega_s = \frac{4\pi f}{p}$$

Here,

ω_s is the angular speed expressed in rad·s^{-1}. and in difference with the previous formula for number of turns per a minute, **p** is the number of pair of poles per phase.

Example

A 3-phase synchronous motor is running with 12 poles (6 pole pairs) is operating at 50 Hz. The synchronous speed will be,

$$N_s = \frac{120 \times 50}{12} = 500 \text{ rpm}$$

Construction

The principal components of a synchonous motor are the stator and the rotor. The stator of synchronous motor and stator of induction motor are similar in construction. The stator frame contains *wrapper plate. Circumferential ribs* and *keybars* are attached to the wrapper plate. To carry the weight of the machine, *frame mounts* and *footings* are required. When the field winding is excited by DC excitation, brushes and slip rings are required to connect to the exctiation supply. The field winding can also be excited by a brushless exciter. Cylindrical, round rotors, (also known as non-salient pole rotor) are used for up to six poles. In some machines or when a large number of poles are needed, a salient pole rotor is used. The construction of synchronous motor is similar to that of a synchronous alternator.

Operation

The operation of a synchronous motor is due to the interaction of the magnetic fields of the stator and the rotor. Synchronous motor is a doubly excited machine *i.e.* two electrical inputs are provided to it. Its stator winding which consists of a 3 phase winding is provided with 3 phase supply and rotor is provided with DC supply. The 3 phase stator winding carrying 3 phase currents produces 3 phase rotating magnetic flux (and therefore rotating magnetic field). The rotor locks in with the rotating magnetic field and rotates along with it. Once the rotor locks in with the rotating magnetic field, the motor is said to be in synchronization. A single-phase (or two-phase derived from single phase) stator winding is possible, but in this case the direction of rotation is not defined and the machine may start in either direction unless prevented from doing so by the starting arrangements.

Once the motor is in operation, the speed of the motor is dependent only on the supply frequency. When the motor load is increased beyond the breakdown load, the motor falls out of synchronization and the field winding no longer follows the rotating magnetic field. Since the motor cannot produce (synchronous) torque if it falls out of synchronization, practical synchronous motors have a partial or complete squirrel-cage damper (amortisseur) winding to stabilize operation and facilitate starting. Because this winding is smaller than that of an equivalent induction motor and can overheat on long operation, and because large slip-frequency voltages are induced in the rotor excitation winding, synchronous motor protection devices sense this condition and interrupt the power supply (out of step protection).

Starting Methods

Above a certain size, synchronous motors are not self-starting motors. This property is due to the inertia of the rotor; it cannot instantly follow the rotation of the magnetic field of the stator. Since a synchronous motor produces no inherent average torque at standstill, it cannot accelerate to synchronous speed without some supplemental mechanism.

Large motors operating on commercial power frequency include a "squirrel cage" induction winding which provides sufficient torque for acceleration and which also serves to damp oscillations in motor speed in operation. Once the rotor nears the synchronous speed, the field winding is excited, and the motor pulls into synchronization. Very large motor systems may include a "pony" motor that accelerates the unloaded synchronous machine before load is applied. Motors that are electronically controlled can be accelerated from zero speed by changing the frequency of the stator current.

Very small synchronous motors are commonly used in line-powered electric mechanical clocks or timers that use the powerline frequency to run the gear mechanism at the correct speed. Such small synchronous motors are able to start without assistance if the moment of inertia of the rotor and its mechanical load is sufficiently small [because the motor] will be accelerated from slip speed up to synchronous speed during an accelerating half cycle of the reluctance torque." Single-phase synchronous motors such as in electric wall clocks can freely rotate in either direction unlike a shaded-pole type.

Applications, Special Properties, and Advantages

Use as Synchronous Condenser

By varying the excitation of a synchronous motor, it can be made to operate at lagging, leading and unity power factor. Excitation at which the power factor is unity is termed *normal excitation voltage*. The magnitude of current at this excitation is minimum. Excitation voltage more than normal excitation is called over excitation voltage, excitation voltage less than normal excitation is called under excitation. When the motor is over excited, the back emf will be greater than the motor terminal voltage. This causes a demagnetizing effect due to armature reaction.

The V curve of a synchronous machine shows armature current as a function of field current. With increasing field current armature current at first decreases, then reaches a minimum, then increases. The mimimum point is also the point at which power factor is unity.

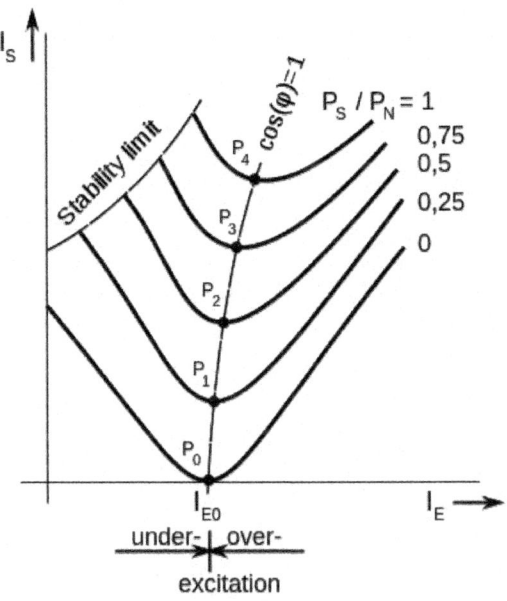

Fig. : V-curve of a synchronous machine.

This ability to selectively control power factor can be exploited for power factor correction of the power system to which the motor is connected. Since most power systems of any significant size have a net lagging power factor, the presence of overexcited synchonous motors moves the system's net power factor closer to unity, improving efficiency. Such power-factor correction is usually a side effect of motors already present in the system to provide mechanical work, although motors can be run without mechanical load simply to provide power-factor correction. In large industrial plants such as factories the interaction between synchronous motors and other, lagging, loads may be an explicit consideration in the plant's electrical design.

Steady State Stability Limit

$$T = T_{max} \sin \delta$$

where,

T is the torque

δ is the torque angle

T_{max} is the maximum torque

here,

$$T_{max} = \frac{3VE}{X_s \, \omega_s}$$

When load is applied, torque angle δ increases. When $\delta = 90°$ the torque will be maximum. If load is applied further then the motor will lose its synchronism, since motor torque will be less than load torque. The maximum load torque that can be applied to a motor without losing its synchronism is called steady state stability limit of a synchronous motor.

Other

Synchronous motors are especially useful in applications requiring precise speed and/or position control.

- Speed is independent of the load over the operating range of the motor.
- Speed and position may be accurately controlled using open loop controls, *e.g.* stepper motors.
- Low-power applications include positioning machines, where high precision is required, and robot actuators.
- They will hold their position when a DC current is applied to both the stator and the rotor windings.
- A clock driven by a synchronous motor is in principle as accurate as the line frequency of its power source. (Although small frequency drifts will occur over any given several hours, grid operators actively adjust line frequency in later periods to compensate, thereby keeping motor-driven clocks accurate
- Record player turntables
- Increased efficiency in low-speed applications (*e.g.* ball mills).

Chapter 8

INDUCTION MOTOR

One of the most common electrical motor used in most applications which is known as **induction motor**. This motor is also called as asynchronous motor because it runs at a speed less than synchronous speed. In this, we need to define what is synchronous speed. Synchronous speed is the speed of rotation of the magnetic field in a rotary machine and it depends upon the frequency and number poles of the machine. An **induction motor** always runs at a speed less than synchronous speed because the rotating magnetic field which is produced in the stator will generate flux in the rotor which will make the rotor to rotate, but due to the lagging of flux current in the rotor with flux current in the stator, the rotor will never reach to its rotating magnetic field speed *i.e.* the synchronous speed. There are basically two **types of induction motor** that depend upon the input supply - single phase induction motor and three phase induction motor. Single phase induction motor is not a self-starting motor which we will discuss later and three phase induction motor is a self-starting motor. Now in general we need to give two supply *i.e.* double excitation to make a machine to rotate. For example if we consider a DC motor, we will give one supply to the stator and another to the rotor through brush arrangement.

WORKING PRINCIPLE OF INDUCTION MOTOR

But in induction motor we give only one supply, so it is really interesting to know that how it works. It is very simple, from the name itself we can understand that there is induction process occurred. Actually when we are giving the supply to the stator winding, flux will generate in the coil due to flow of current in the coil. Now the rotor winding is arranged in such a way that it becomes short circuited in the rotor itself. The flux from the stator will cut the coil in the rotor and since the rotor coils are short circuited, according to Faraday's law of electromagnetic induction, electric current will start flowing in the coil of the rotor. When the current will flow, another flux will get generated in the rotor. Now there will be two flux, one is stator flux and another is rotor flux and the rotor flux will be lagging to the stator flux. Due to this, the rotor will feel a torque which will make the rotor to rotate in the direction of rotating magnetic flux. So the speed of the rotor will be depending upon the ac supply and the speed can be controlled by varying the input supply. This is the **working principle of an induction motor** of either type.

TYPES INDUCTION MOTOR

Single phase induction motor :
* Split phase induction motor
* Capacitor start induction motor
* Capacitor start capacitor run induction motor
* Shaded pole induction motor

Three phase induction motor :
* Squirrel cage induction motor
* Slip ring induction motor.

We had mentioned above that single phase induction motor is not a self-starting and three phase induction motor is self-starting. So what is self-starting? When the machine starts running automatically without any external force to the machine, then it is called as self-starting. For example, we see that when we press the key the fan starts to rotate automatically, so it is self-starting. Point to be note that fan used in home appliances is single phase induction motor but it is self-starting. How? We will discuss it how.

WHY IS THREE PHASE INDUCTION MOTOR SELF-STARTING?

In three phase system, there are three single phase line with 120° phase difference. So the rotating magnetic field is having the same phase difference which will make the rotor to move. If we consider three phases a, b and c, when phase a is magnetized, the rotor will move towards the phase a winding, in the next moment phase b will get magnetized and it will attract the rotor and than phase c. So the rotor will continue to rotate.

WORKING PRINCIPLE OF THREE PHASE INDUCTION MOTOR

An electrical motor is such an electromechanical device which converts electrical energy into a mechanical energy. In case of three phase AC operation, most widely used motor is **Three phase induction motor** as this type of motor does not require any starting device or we can say they are self-starting induction motor.

For better understanding the **principle of three phase induction motor**, the basic constructional feature of this motor must be known to us. This Motor consists of two major parts :

Stator : **Stator of three phase induction motor** is made up of numbers of slots to construct a 3 phase winding circuit which is connected to 3 phase AC source.

The three phase winding are arranged in such a manner in the slots that they produce a rotating magnetic field after AC is given to them.

Rotor : **Rotor of three phase induction motor** consists of cylindrical laminated core with parallel slots that can carry conductors. Conductors are heavy copper or aluminum bars which fits in each slots & they are short circuited by the end rings. The slots are not exactly made parallel to the axis of the shaft but are slotted a little skewed because this arrangement reduces magnetic humming noise & can avoid stalling of motor.

Working of Three Phase Induction Motor

Production of Rotating Magnetic Field

The stator of the motor consists of overlapping winding offset by an electrical angle of 120°. When the primary winding or the stator is connected to a 3 phase AC source, it establishes a rotating magnetic field which rotates at the synchronous speed.

Secrets Behind the Rotation :

According to Faraday's law an emf induced in any circuit is due to the rate of change of magnetic flux linkage through the circuit. As the rotor winding in an induction motor are either closed through an external resistance or directly shorted by end ring, and cut the stator rotating magnetic field, an emf is induced in the rotor copper bar and due to this emf a current flows through the rotor conductor.

Here the relative velocity between the rotating flux and static rotor conductor is the cause of electric current generation; hence as per Lenz's law the rotor will rotate in the same direction to reduce the cause *i.e.* the relative velocity.

Thus from the **working principle of three phase induction motor** it may observed that the rotor speed should not reach the synchronous speed produced by the stator. If the speeds equals, there would be no such relative velocity, so no emf induction in the rotor, & no current would be flowing, and therefore no torque would be generated. Consequently the rotor can not reach at the synchronous speed. The difference between the stator (synchronous speed) and rotor speeds is called the slip. The rotation of the magnetic field in an induction motor has the advantage that no electrical connections need to be made to the rotor.

Thus the **three phase induction motor** *is :*

- Self-starting.
- Less armature reaction and brush sparking because of the absence of commutators and brushes that may cause sparks.
- Robust in construction.
- Economical.
- Easier to maintain.

SINGLE PHASE INDUCTION MOTOR

For lightning and general purposes in homes, offices, shops, small factories single phase system is widely used as compared to three phase system as the single phase system is more economical and the power requirement in most of the houses, shops, offices are small, which can be easily met by single phase system. The single phase motors are simple in construction, cheap in cost, reliable and easy to repair and maintain. Due to all these advantages the single phase motor finds its application in vacuum cleaner, fans, washing machine, centrifugal pump, blowers, washing machine, small toys etc.

The single phase ac motors are further classified as :

- **Single phase induction motors** or **asynchronous motors**.
- Single phase synchronous motors.
- Commutator motors.

Construction of Single Phase Induction Motor

Like any other electrical motor asynchronous motor also have two main parts namely rotor and stator.

Stator : As its name indicates stator is a stationary part of induction motor. A single phase ac supply is given to the stator of single phase induction motor.

Rotor : The rotor is a rotating part of induction motor. The rotor is connected to the mechanical load through the shaft. The rotor in single phase induction motor is of squirrel cage rotor type.

The **construction of single phase induction motor** is almost similar to the squirrel cage three phase motor except that in case of asynchronous motor the stator have two windings instead of one as compare to the single stator winding in three phase induction motor.

Stator of Single Phase Induction Motor

The stator of the single phase induction motor has laminated stamping to reduce eddy current losses on its periphery. The slots are provided on its stamping to carry stator or main winding. In order to reduce the hysteresis losses, stamping are made up of silicon steel. When the stator winding is given a single phase ac supply, the magnetic field is produced and the motor rotates at a speed slightly less than the synchronous speed N_s which is given by

$$N_s = \frac{120 f}{P}$$

The construction of the stator of asynchronous motor is similar to that of three phase induction motor except there are two dissimilarity in the winding part of the single phase induction motor.

- Firstly the single phase induction motors are mostly provided with concentric coils. As the number of turns per coil can be easily adjusted with the help of concentric coils, the mmf distribution is almost sinusoidal.
- Except for shaded pole motor, the asynchronous motor has two stator windings namely the main winding and the auxiliary winding. These two windings are placed in space quadrature with respect to each other.

Rotor of Single Phase Induction Motor

The construction of the rotor of the single phase induction motor is similar to the squirrel cage three phase induction motor. The rotor is cylindrical in shape and has slots all over its periphery. The slots are not made parallel to each other but are bit skewed as the skewing prevents magnetic locking of stator and rotor teeth and makes the working of induction motor more smooth and quieter. The squirrel cage rotor consists of aluminium, brass or copper bars. These aluminium or copper bars are called rotor conductors and are placed in the slots on the periphery of the rotor. The rotor conductors are permanently shorted by the copper or aluminium rings called the end rings. In order to provide mechanical strength these rotor conductor are braced to the end ring and hence form a complete closed circuit resembling like a cage and hence got its name as "squirrel cage induction motor". As the bars are permanently shorted by end rings, the rotor electrical resistance is very small and it is not possible to add external resistance as the bars are permanently shorted. The absence of slip ring and brushes make the construction of single phase induction motor very simple and robust.

Working Principle of Single Phase Induction Motor

NOTE : We know that for the working of any electrical motor whether its ac or dc motor, we require two fluxes as, the interact of these two fluxes produced the required torque, which is desired parameter for any motor to rotate.

When single phase ac supply is given to the stator winding of single phase induction motor, the alternating current starts flowing through the stator or main winding. This alternating current produces an alternating flux called main flux. This main flux also links with the rotor conductors and hence cut the rotor conductors. According to the Faraday's law of electromagnetic induction, emf gets induced in the rotor. As the rotor circuit is closed one so, the current starts flowing in the rotor. This current is called the rotor current. This rotor current produces its own flux called rotor flux. Since this flux is produced due to induction principle so, the motor working on this principle got its name as induction motor. Now there are two fluxes one is main flux and another is called rotor flux. These two fluxes produce the desired torque which is required by the motor to rotate.

Why Single Phase Induction Motor is not Self-starting?

According to double field revolving theory, any alternating quantity can be resolved into two components, each component have magnitude equal to the half of the maximum magnitude of the alternating quantity and both these component rotates in opposite direction to each other. For example - a flux, φ can be resolved into two components

$$\frac{\phi_m}{2} \text{ and } \frac{\phi_m}{2}$$

Each of these components rotates in opposite direction *i.e.* if one $\varphi_m / 2$ is rotating in clockwise direction then the other $\varphi_m / 2$ rotates in anticlockwise direction.

When a single phase ac supply is given to the stator winding of single phase induction motor, it produces its flux of magnitude, φ_m. According to the double field revolving theory, this alternating flux, φ_m is divided into two components of magnitude $\varphi_m / 2$. Each of these components will rotate in opposite direction, with the synchronous speed, N_s. Let us call these two components of flux as forward component of flux, φ_f and backward component of flux, φ_b. The resultant of these two component of flux at any instant of time, gives the value of instantaneous stator flux at that particular instant.

i.e. $$\phi_r = \frac{\phi_m}{2} + \frac{\phi_m}{2} \text{ or } \phi_r = \phi_f + \phi_b$$

Now at starting, both the forward and backward components of flux are exactly opposite to each other. Also both of these components of flux are equal in magnitude. So, they cancel each other and hence the net torque experienced by the rotor at starting is zero. So, the single phase induction motors are not self-starting motors. Methods for Making Single Phase Induction as Self-starting Motor

We can easily conclude that the single phase induction motors are not self-starting because the produced stator flux is alternating in nature and at the starting the two components of this flux cancel each other and hence there is no net torque. The solution to this problem is that if the stator flux is made rotating type, rather than alternating type, which rotates in one particular direction only. Then the induction motor will become self-starting. Now for producing this rotating magnetic field we require two alternating flux, having some phase difference angle between them. When these two fluxes interact with each other they will produce a resultant flux. This resultant flux is rotating in nature and rotates in space in one particular direction only. Once the motor starts running, the additional flux can be removed. The motor will continue to run under the influence of the main flux only. Depending upon the methods for making asynchronous motor as Self-starting Motor, there are mainly four types of single phase induction motor namely,

- Split phase induction motor,
- Capacitor start inductor motor,
- Capacitor start capacitor run induction motor,
- Shaded pole induction motor.

Comparison between Single Phase and Three Phase Induction Motors

1. Single phase induction motors are simple in construction, reliable and economical for small power rating as compared to three phase induction motors.
2. The electrical power factor of Single phase induction motors is low as compared to three phase induction motors.
3. For same size, the single phase induction motors develop about 50% of the output as that of three phase induction motors.
4. The starting torque is also low for asynchronous motors.
5. The efficiency of single phase induction motors is less as compare it to the three phase induction motors.

TYPES OF THREE PHASE INDUCTION MOTOR

Induction motor is also called asynchronous motor as it runs at a speed other than the synchronous speed. Like any other electrical motor, induction motor have two main parts namely rotor and stator.

Fig. : Three Phase Induction Motor.

Stator

As its name indicate stator is a stationary part of induction motor. A three phase supply is given to the stator of induction motor.

Rotor

The rotor is a rotating part of induction motor. The rotor is connected to the mechanical load through the shaft. The rotor of the three phase induction motor are further classified as :

- Squirrel cage rotor,
- Slip ring rotor or wound rotor or phase wound rotor.

Depending upon the type of rotor used the three phase induction motor are classified as :

- Squirrel cage induction motor
- Slip ring induction motor or wound induction motor or phase wound induction motor.

The construction of stator for both the kind of three phase induction motor remains the same and is discussed in brief in next paragraph.

Stator of Three Phase Induction Motor

The stator of the three phase induction motor consists of three main parts :

- Stator frame
- Stator core
- Stator winding or field winding.

1. **Stator frame :** It is the outer most part of the three phase induction motor. Its main function is to support the stator core and the field winding. It acts as a covering and provide protection and mechanical strength to all the inner parts of the machine. The frame is either made up of die cast or fabricated steel. The frame of three phase induction motor should be very strong and rigid as the air gap length of three phase induction motor is very small, otherwise rotor will not remain concentric with stator which will give rise to unbalanced magnetic pull.

2. **Stator core :** The main function of the stator core is to carry alternating flux. In order to reduce the eddy current losses the stator core is laminated. This laminated type of structure are made up of stamping which is about 0.4 to 0.5 mm thick. All the stamping are stamped together to form stator core, which is then housed in stator frame. The stamping are generally made up of silicon steel, which reduces the hysteresis loss.

3. **Stator winding or field winding :** The slots on the periphery of stator core of the three phase induction motor carries three phase windings. This three phase winding is supplied by three phase ac supply. The three phases of the winding are connected either in star or delta depending upon which type of starting method is used. The squirrel cage motor is mostly started by star – delta stater and hence the stator of squirrel cage motor are delta connected. The slip ring three phase induction motor are started by inserting resistances so, the stator winding can be connected either in star or delta. The winding wound on the stator of three phase induction motor is also called field winding and when this winding is excited by three phase ac supply it produces rotating magnetic field.

TORQUE EQUATION OF THREE PHASE INDUCTION MOTOR

The torque produced by three phase induction motor depends upon the following three factors :

Firstly the magnitude of rotor current, secondly the flux which interact with the rotor of three phase induction motor and is responsible for producing emf in the rotor part of induction motor, lastly the power factor of rotor of the three phase induction motor.

Combining all these factors together we get the equation of torque as :

$$T \propto \phi I_2 \cos \theta 2$$

Where, T is the torque produced by induction motor,

φ is flux responsible of producing induced emf,

I_2 is rotor current,

$\cos\theta_2$ is the power factor of rotor circuit.

The flux φ produced by the stator is proportional to stator emf E_1.

i.e. $\varphi \propto E_1$

We know that transformation ratio K is defined as the ratio of secondary voltage (rotor voltage) to that of primary voltage (stator voltage).

$$K = \frac{E_2}{E_1} \text{ or } K = \frac{E_2}{\phi} \text{ or } E_2 = \phi$$

Rotor current I_2 is defined as the ratio of rotor induced emf under running condition, sE_2 to total impedance, Z_2 of rotor side,

i.e. $$I_2 = \frac{sE_2}{Z_2}$$

and total impedance Z_2 on rotor side is given by,

$$Z_2 = \sqrt{R_2^2 + (sX_2)^2}$$

Putting this value in above equation we get,

$$I_2 = \frac{sE_2}{\sqrt{R_2^2 + (sX_2)^2}}$$

We know that power factor is defined as ratio of resistance to that of impedance. The power factor of the rotor circuit is

$$\cos \theta_2 = \frac{R_2}{Z_2} = \frac{R_2}{\sqrt{R_2^2 + (sX_2)^2}}$$

Putting the value of flux φ, rotor current I_2, power factor $\cos\theta_2$ in the equation of torque we get,

$$T \propto E_2 \frac{sE_2}{\sqrt{R_2^2 + (sX_2)^2}} \times \frac{R_2}{\sqrt{R_2^2 + (sX_2)^2}}$$

Combining similar term we get,

$$T \propto sE_2^2 \frac{R_2}{\sqrt{R_2^2 + (sX_2)^2}}$$

Removing proportionality constant we get,

$$T = K\, sE_2^2 \frac{R_2}{\sqrt{R_2^2 + (sX_2)^2}}$$

This constant $K = \dfrac{3}{2\pi n_s}$

Where n_s is synchronous speed in r. p. s, $n_s = N_s / 60$. So, finally the equation of torque becomes,

$$T = sE_2^2 \times \frac{R_2}{R_2^2 + (sX_2)^2} \times \frac{3}{2\pi n_s} N - m$$

Derivation of K in torque equation.

In case of three phase induction motor, there occur copper losses in rotor. These rotor copper losses are expressed as

$$P_c = 3I_2^2 R^2$$

We know that rotor current,

$$I_2 = \frac{sE_2}{\sqrt{R_2^2 + (sX_2)^2}}$$

Substitute this value of I_2 in the equation of rotor copper losses, P_c. So, we get

$$P_c = 3R_2 \left(\frac{sE_2}{\sqrt{R_2^2 + (sX_2)^2}} \right)^2$$

On simplifying $P_c = \dfrac{3R_2 s^2 E_2^2}{R_2^2 + (sX_2)^2}$

The ratio of $P_2 : P_c : P_m = 1 : s : (1 - s)$

Where P_2 is the rotor input,

P_c is the rotor copper losses,

P_m is the mechanical power developed.

or $\dfrac{P_c}{P_m} = \dfrac{s}{1-s}$

or
$$P_m = \frac{(1-s)P_c}{s}$$

Substitute the value of P_c in above equation we get,

$$P_m = \frac{1}{s} \times \frac{(1-s) \, 3R_2 s^2 \, E_2^2}{R_2^2 + (sX_2)^2}$$

On simplifying we get,

$$P_m = \frac{(1-s) \, 3R_2 s E_2^2}{R_2^2 + (sX_2)^2}$$

The mechanical power developed $P_m = T\omega$,

Where,
$$\omega = \frac{2\pi N}{60}$$

or
$$P_m = T \frac{2\pi N}{60}$$

Substituting the value of P_m

$$\frac{1}{s} \times \frac{(1-s)3R_2 \, s^2 E_2^2}{R_2^2 + (sX_2)^2} = T \frac{2\pi N}{60}$$

or
$$T = \frac{1}{s} \times \frac{(1-s)3R_2 E_2^2}{R_2^2 + (sX_2)^2} X \frac{60}{2\pi N}$$

We know that the rotor speed $N = N_s (1 - s)$

Substituting this value of rotor speed in above equation we get,

$$T = \frac{1}{s} \times \frac{(1-s)3R_2 s^2 \, E_2^2}{R_2^2 + (sX_2)^2} \times \frac{60}{2\pi N_s \, (1-s)}$$

N_s is speed in revolution per minute (rpm) and n_s is speed in revolution per sec (rps) and the relation between the two is

$$\frac{N_s}{60} = n_s$$

Substitute this value of N_s in above equation and simplifying it we get

Torque,
$$T = \frac{sE_2^2 R_2}{R_2^2 + (sX_2)^2} \times \frac{3}{2\pi n_s}$$

or,
$$T = K_s E_2^2 \frac{R_2}{R_2^2 + (sX_2)^2}$$

or

Comparing both the equations, we get, constant $K = 3 \, / \, 2\pi n_s$

Equation of Starting Torque of Three Phase Induction Motor

Starting torque is the torque produced by induction motor when it is started. We know that at start the rotor speed, N is zero.

So, slip $\qquad S = \dfrac{N_s - N}{N_s}$ becomes 1

So, the equation of starting torque is easily obtained by simply putting the value of s = 1 in the equation of torque of the three phase induction motor,

$$T = \frac{E_2^2 R_2}{R_2^2 + X_2^2} \times \frac{3}{2\pi n_s} \, N - m$$

Maximum Torque Condition for Three Phase Induction Motor

In the equation of torque,

$$T = \frac{s E_2^2 R_2}{R_2^2 + (s X_2)^2} \times \frac{3}{2\pi n_s}$$

The rotor resistance, rotor inductive reactance and synchronous speed of induction motor remains constant. The supply voltage to the three phase induction motor is usually rated and remains constant so the stator emf also remains the constant. The transformation ratio is defined as the ratio of rotor emf to that of stator emf. So if stator emf remains constant then rotor emf also remains constant.

If we want to find the maximum value of some quantity then we have to differentiate that quantity with respect to some variable parameter and then put it equal to zero. In this case we have to find the condition for maximum torque so we have to differentiate torque with respect to some variable quantity which is slip, s in this case as all other parameters in the equation of torque remains constant.

So, for torque to be maximum

$$\frac{dT}{ds} = 0$$

$$T = K s E_2^2 \frac{R_2}{R_2^2 + (s X_2)^2}$$

Now differentiate the above equation by using division rule of differentiation. On differentiating and after putting the terms equal to zero we get,

$$s^2 = \frac{R_2^2}{X_2^2}$$

Neglecting the negative value of slip we get

$$s = \frac{R_2}{X_2}$$

So, when slip s = R_2 / X_2, the torque will be maximum and this slip is called maximum slip Sm and it is defined as the ratio of rotor resistance to that of rotor reactance.

NOTE : At starting S = 1, so the maximum starting torque occur when rotor resistance is equal to rotor reactance.

Equation of Maximum Torque

The equation of torque is

$$T = \frac{sE_2^2 R_2}{R_2^2 + (sX_2)^2}$$

The torque will be maximum when slip $s = R_2 / X_2$

Substituting the value of this slip in above equation we get the maximum value of torque as,

$$T_{max} = K\frac{E_2^2}{2X_2} N - m$$

CONCLUSION : *From the above equation it is concluded that :*

- The maximum torque is directly proportional to square of rotor induced emf at the standstill.
- The maximum torque is inversely proportional to rotor reactance.
- The maximum torque is independent of rotor resistance.
- The slip at which maximum torque occur depends upon rotor resistance, R_2. So, by varying the rotor resistance, maximum torque can be obtained at any required slip.

SPEED CONTROL OF THREE PHASE INDUCTION MOTOR

A three phase induction motor is basically a constant speed motor so it's somewhat difficult to control its speed. The speed control of induction motor is done at the cost of decrease in efficiency and low electrical power factor. Before discussing the methods to **control the speed of three phase induction motor** one should know the basic formulas of speed and torque of three phase induction motor as the methods of speed control depends upon these formulas.

- Synchronous speed

$$N_s = \frac{120f}{P}$$

Where f = frequency and P is the number of poles

- The speed of induction motor is given by,

$$N = N_s (1 - s)$$

Where N is the speed of rotor of induction motor,

N_s is the synchronous speed,

S is the slip.

- The torque produced by three phase induction motor is given by,

$$T = \frac{3}{2\pi N_s} X \frac{sE_2^2 R_2}{R_2^2 + (sX_2)^2}$$

When rotor is at sandstill slip, s is one.

So the equation of torque is,

$$T = \frac{3}{2\pi N_s} X \frac{E_2^2 R_2}{R_2^2 + X_2^2}$$

Where E_2 is the rotor emf

N_s is the synchronous speed

R_2 is the rotor resistance

X_2 is the rotor inductive reactance.

The Speed of Induction Motor is Changed from Both Stator and Rotor Side

The speed control of three phase induction motor from stator side are further classified as :

1. V / f control or frequency control
2. changing the number of stator poles
3. controlling supply voltage
4. adding rheostat in the stator circuit.

The speed controls of three phase induction motor from rotor side are further classified as :

1. Adding external resistance on rotor side
2. Cascade control method
3. Injecting slip frequency emf into rotor side

Speed Control from Stator Side

1. *V/f* control or frequency control - Whenever three phase sup ply is given to three phase induction motor rotating magnetic field is produced which rotates at synchronous speed given by

$$N_s = \frac{120f}{P}$$

In three phase induction motor emf is induced by induction similar to that of transformer which is given by

$$E \text{ or } V = 4.44\phi \, K. \, T. \, f \, .or \, \phi = \frac{V}{4.44 KT \, f}$$

Where K is the winding constant, T is the number of turns per phase and f is frequency. Now if we change frequency synchronous speed changes but with decrease in frequency flux will increase and this change in value of flux causes saturation of rotor and stator cores which will further cause increase in no load current of the motor. So, its important to maintain flux, φ constant and it is only possible if we change voltage. *i.e* if we decrease frequency flux increases but at the same time if we decrease voltage flux will also decease causing no change in flux and hence it remains constant. So, here we are keeping the ratio of V/f as constant. Hence its name is V/f method. For controlling the speed of three phase induction motor by V/f method we have to supply variable voltage and frequency which is easily obtained by using converter and inverter set.

2. Controlling supply voltage : The torque produced by running three phase induction motor is given by

$$T \propto \frac{sE_2^2 R_2}{R_2^2 + (sX_2)^2}$$

In low slip region $(sX)^2$ is very very small as com pared to R_2. So, it can be neglected. So torque becomes

$$T \propto \frac{sE_2^2}{R_2}$$

Since rotor resistance, R_2 is constant so the equation of torque further reduces to
$$T \propto sE^2{}_2$$

We know that rotor induced emf $E_2 \propto V$. So, $T \propto sV^2$.

From the equation above it is clear that if we decrease supply voltage torque will also decrease. But for supplying the same load, the torque must remains the same and it is only possible if we increase the slip and if the slip increases the motor will run at reduced speed. This method of speed control is rarely used because small change in speed requires large reduction in voltage, and hence the current drawn by motor increases, which cause over heating of induction motor.

3. Changing the number of stator poles : The stator poles can be changed by two methods

- Multiple stator winding method
- Pole amplitude modulation method (PAM)
- Multiple stator winding method – In this method of speed control of three phase induction motor, the stator is provided by two separate winding. These two stator windings are electrically isolated from each other and are wound for two different pole numbers. Using switching arrangement, at a time, supply is

given to one winding only and hence speed control is possible. Disadvantages of this method is that the smooth speed control is not possible. This method is more costly and less efficient as two different stator winding are required. This method of speed control can only be applied for squirrel cage motor.

- Pole amplitude modulation method (PAM) – In this method of speed control of three phase induction motor the original sinusoidal mmf wave is modulated by another sinusoidal mmf wave having different number of poles.

    Let $f_1(\theta)$ be the original mmf wave of induction motor whose speed is to be controlled $f_2(\theta)$ be the modulation mmf wave

    P_1 be the number of poles of induction motor whose speed is to be controlled

    P_2 be the number of poles of modulation wave

    $$f_1(\theta) = F_1 \sin\left(\frac{P_1\theta}{2}\right)$$

    $$f_2(\theta) = F_2 \sin\left(\frac{P_2\theta}{2}\right)$$

    After modulation resultant mmf wave

    $$f_r(\theta) = F_1 F_2 \sin\left(\frac{P_1\theta}{2}\right) \sin\left(\frac{P_2\theta}{2}\right)$$

    Apply formulae for $2 \sin A \sin B = \cos\dfrac{A-B}{2} - \cos\dfrac{A+B}{2}$

    So we get, resultant mmf wave

    $$F_r(\theta) = F_1 F_2 \frac{\cos\dfrac{(P_1 - P_2)\theta}{2} - \cos\dfrac{(P_1 + P_2)\theta}{2}}{2}$$

    Therefore the resultant mmf wave will have two different number of poles

    i.e. $P_{11} = P_1 - P_2$ and $P_{12} = P_1 + P_2$

    Therefore by changing the number of poles we can easily change the speed of three phase induction motor.

4. Adding rheostat in the stator circuit - In this method of speed control of three phase induction motor rheostat is added in the stator circuit due to this voltage gets dropped. In case of three phase induction motor torque produced is given by $T \propto sV_2^2$. If we decrease supply voltage torque will also decrease. But for supplying the same load, the torque must remains the same and it is only possible if we increase the slip and if the slip increase motor will run reduced speed.

Speed Control from Rotor Side

1. Adding external resistance on rotor side – In this method of speed control of three phase induction motor external resistance are added on rotor side. The equation of torque for three phase induction motor is

$$T \propto \frac{sE_2^2 R_2}{R_2^2 + (xX)^2}$$

The three phase induction motor operates in low slip region. In low slip region term $(sX)^2$ becomes very very small as compared to R_2. So, it can be neglected. and also E_2 is constant. So the equation of torque after simplification becomes,

$$T \propto \frac{s}{R_2}$$

Now if we increase rotor resistance, R_2 torque decreases but to supply the same load torque must remains constant. So, we increase slip, which will further results in decrease in rotor speed. Thus by adding additional resistance in rotor circuit we can decrease the speed of three phase induction motor. The main advantage of this method is that with addition of external resistance starting torque increases but this method of speed control of three phase induction motor also suffers from some disadvantages :

*   The speed above the normal value is not possible
*   Large speed change requires large value of resistance and if such large value of resistance is added in the circuit it will cause large copper loss and hence reduction in efficiency
*   Presence of resistance causes more losses
*   This method cannot be used for squirrel cage induction motor

2. Cascade control method – In this method of speed control of three phase induction motor, the two three phase induction motor are connected on common shaft and hence called cascaded motor. One motor is the called the main motor and another motor is called the auxiliary motor. The three phase supply is given to the stator of the main motor while the auxiliary motor is derived at a slip frequency from the slip ring of main motor

Let N_{S1} be the synchronous speed of main motor

N_{S2} be the synchronous speed of auxiliary motor

P_1 be the number of poles of the main motor

P_2 be the number of poles of the auxiliary motor

F is the supply frequency

F_1 is the frequency of rotor induced emf of main motor

N is the speed of set and it remains same for both the main and aux iliary motor as both the motors are mounted on common shaft

S_1 is the slip of main motor

$$S_1 = \frac{N_{S1} - N}{N_{S1}}$$

$$F_1 = S_1 F$$

The auxiliary motor is supplied with same frequency as the main motor *i.e.*

$$F_1 = F_2$$

$$N_{S2} = \frac{120 F_2}{P_2} = \frac{120 F_1}{P_2}$$

$$N_{S2} = \frac{120 S_1 F}{P_2}$$

Now put the value of

$$S_1 = \frac{N_{S1} - N}{N_{S1}}$$

We get,

$$N_{S2} = \frac{120 F (N_{S1} - N)}{P_2 N_{S1}}$$

Now at no load, the speed of auxiliary rotor is almost same as its synchronous speed *i.e* $N = N_{S2}$

$$N = \frac{120 F (N_{S1} - N)}{P_2 N_{S1}}$$

Now rearrange the above equation and find out the value of N, we get,

$$N = \frac{120 F}{P_1 - P_2}$$

This cascaded set of two motors will now run at new speed having number of poles $(P_1 + P_2)$. In the above method the torque produced by the main and auxiliary motor will act in same direction, resulting in number of poles $(P_1 + P_2)$. Such type of cascading is called cumulative cascading. There is one more type of cascading in which the torque produced by the main motor is in opposite direction to that of auxiliary motor. Such type of cascading is called differential cascading; resulting in speed corresponds to number of poles $(P_1 - P_2)$.

In this method of speed control of three phase induction motor, four different speeds can be obtained

1. when only main induction motor work, having speed corresponds to $N_{S1} = 120 \, F/P_1$

2. when only auxiliary induction motor work, having speed corresponds to $N_{S2} = 120 \, F/P_2$

3. when cumulative cascading is done, then the complete set runs at a speed of $N = 120 F/(P_1 + P_2)$

4. when differential cascading is done, then the complete set runs at a speed of $N = 120F/(P_1 P_2)$

3. Injecting slip frequency emf into rotor side - when the speed control of three phase induction motor is done by adding resistance in rotor circuit, some part of power called, the slip power is lost as I^2R losses. Therefore the efficiency of three phase induction motor is reduced by this method of speed control. This slip power loss can be recovered and supplied back in order to improve the overall efficiency of three phase induction motor and this scheme of recovering the power is called slip power recovery scheme and this is done by connecting an external source of emf of slip frequency to the rotor circuit. The injected emf can either oppose the rotor induced emf or aids the rotor induced emf. If it oppose the rotor induced emf, the total rotor resistance increases and hence speed decreases and if the injected emf aids the main rotor emf the total resistance decreases and hence speed increases. Therefore by injecting induced emf in rotor circuit the speed can be easily controlled. The main advantage of this type of speed control of three phase induction motor is that wide range of speed control is possible whether its above normal or below normal speed.

LOSSES AND EFFICIENCY OF INDUCTION MOTOR

There are two types of losses occur in three phase induction motor. These losses are,

1. Vonstant or fixed losses,
2. Variable losses.

Constant or Fixed Losses

Constant losses are those losses which are considered to remain constant over normal working range of induction motor. The fixed losses can be easily obtained by performing no-load test on the three phase induction motor. These losses are further classified as :

*   Iron or core losses,
*   Mechanical losses,
*   Brush friction losses.

Iron or Core Losses

Iron or core losses are further divided into hysteresis and eddy current losses. Eddy current losses are minimized by using lamination. Since by laminating the core, area decreases and hence resistance increases, which results in decrease in eddy currents. Hysteresis losses are minimized by using high grade silicon steel. The core losses depend upon frequency. The frequency of stator is always supply frequency, f and the frequency of rotor is slip times the supply frequency, (sf) which is always less than the stator frequency. Hence the rotor core loss is very small as compared to stator core loss and is usually neglected in running conditions.

Mechanical and Brush Friction Losses

Mechanical losses occur at the bearing and brush friction loss occurs in wound rotor induction motor. These losses occurs with the change in speed. In three phase induction motor the speed usually remains constant. hence these losses almost remains constant.

Variable Losses

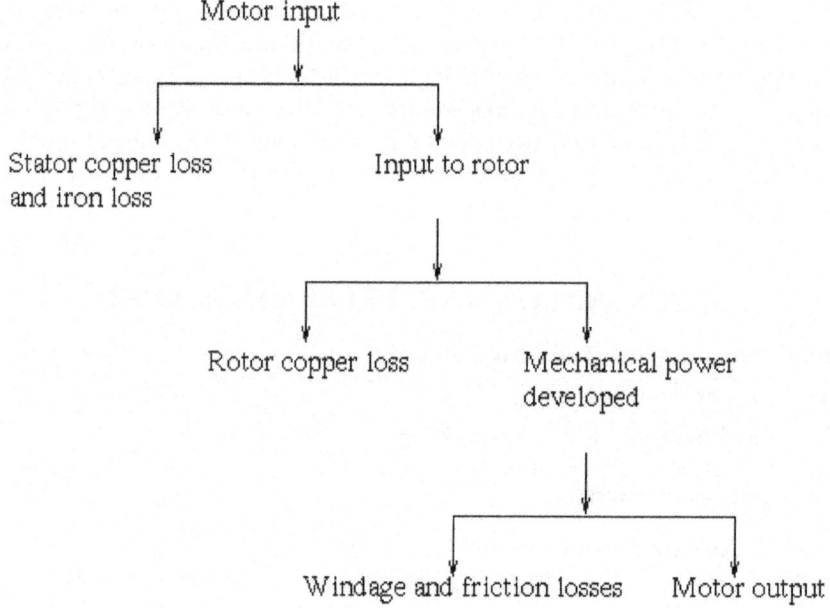

These losses are also called copper losses. These losses occur due to current flowing in stator and rotor windings. As the load changes, the current flowing in rotor and stator winding also changes and hence these losses also changes. Therefore these losses are called variable losses. The copper losses are obtained by performing blocked rotor test on three phase induction motor.

The main function of induction motor is to convert an electrical power into mechanical power. During this conversion of electrical energy into mechanical energy the power flows through different stages. This power flowing through different stages is shown by power flow diagram. As we all know the input to the three phase induction motor is three phase supply. So, the three phase supply is given to the stator of three phase induction motor.

Let, P_{in} = electrical power supplied to the stator of three phase induction motor,

V_L = line voltage supplied to the stator of three phase induction motor,

I_L = line current,

Cosφ = power factor of the three phase induction motor.

Electrical power input to the stator, P_{in} = $\sqrt{3} V_L I_L \cos\varphi$

A part of this power input is used to supply stator losses which are stator iron loss and stator copper loss. The remaining power *i.e.* (input electrical power – stator losses) are supplied to rotor as rotor input.

So, rotor input $P_2 = P_{in}$ – stator losses (stator copper loss and stator iron loss).

Now, the rotor has to convert this rotor input into mechanical energy but this complete input cannot be converted into mechanical output as it has to supply rotor losses. As explained earlier the rotor losses are of two types rotor iron loss and rotor copper loss. Since the iron loss depends upon the rotor frequency, which is very small when the rotor rotates, so it is usually neglected. So, the rotor has only rotor copper loss. Therefore the rotor input has to supply these rotor copper losses. After supplying the rotor copper losses, the remaining part of Rotor input, P_2 is converted into mechanical power, P_m.

Let P_c be the rotor copper loss,

I_2 be the rotor current under running condition,

R_2 is the rotor resistance,

P_m is the gross mechanical power developed.

$P_c = 3I_2^2 R_2$

$P_m = P_2 - P_c$

Now this mechanical power developed is given to the load by the shaft but there occur some mechanical losses like friction and windage losses. So, the gross mechanical power developed has to supplied these losses. Therefore, the net output power developed at the shaft, which is finally given to the load is P_{out}.

$P_{out} = P_m$ – Mechanical losses (friction and windage losses).

Efficiency of Three Phase Induction Motor

Efficiency is defined as the ratio of the output to that of input,

Efficiency, $\qquad \eta = \dfrac{output}{input}$

Rotor efficiency of the three phase induction motor,

$$= \dfrac{rotor\ output}{rotor\ input}$$

$$= \text{gross mechanical power developed} / \text{rotor input}$$

$$= \dfrac{P_m}{P_2}$$

Three phase induction motor efficiency

$$= \frac{power\ developed\ at\ shaft}{electrical\ input\ to\ the\ motor}$$

Three phase induction motor efficiency

$$\eta = \frac{P_{out}}{P_{in}}$$

CONSTRUCTION OF THREE PHASE INDUCTION MOTOR

The three phase induction motor is the most widely used electrical motor. Almost 80% of the mechanical power used by industries is provided by three phase induction motors because of its simple and rugged construction, low cost, good operating characteristics, absence of commutator and good speed regulation. In three phase induction motor the power is transferred from stator to rotor winding through induction. The Induction motor is also called asynchronous motor as it runs at a speed other than the synchronous speed.

Like any other electrical motor induction motor also have two main parts namely rotor and stator

- Stator : As its name indicates stator is a stationary part of induction motor. A stator winding is placed in the stator of induction motor and the three phase supply is given to it.
- Rotor : The rotor is a rotating part of induction motor. The rotor is connected to the mechanical load through the shaft.

    The rotor of the three phase induction motor are further classified as :

- Squirrel cage rotor,
- Slip ring rotor or wound rotor or phase wound rotor.

    Depending upon the type of rotor construction used the three phase induction motor are classified as :

- Squirrel cage induction motor,
- Slip ring induction motor or wound induction motor or phase wound induction motor.

    The construction of stator for both the kinds of three phase induction motor remains the same and is discussed in brief in next paragraph.

    The other parts, which are required to complete the induction motor, are :

- Shaft for transmitting the torque to the load. This shaft is made up of steel.
- Bearings for supporting the rotating shaft.
- One of the problems with electrical motor is the production of heat during its rotation. In order to overcome this problem we need fan for cooling.
- For receiving external electrical connection Terminal box is needed.
- There is a small distance between rotor and stator which usually varies from 0.4 mm to 4 mm. Such a distance is called air gap.

Stator of Three Phase Induction Motor

The stator of the three phase induction motor consists of three main parts :

- Stator frame,
- Stator core,
- Stator winding or field winding.

Stator Frame

It is the outer most part of the three phase induction motor. Its main function is to support the stator core and the field winding. It acts as a covering and it provide protection and mechanical strength to all the inner parts of the induction motor. The frame is either made up of die cast or fabricated steel. The frame of three phase induction motor should be very strong and rigid as the air gap length of three phase induction motor is very small, otherwise rotor will not remain concentric with stator, which will give rise to unbalanced magnetic pull.

Fig. : Stator Lamination.

Stator Core

The main function of the stator core is to carry the alternating flux. In order to reduce the eddy current loss, the stator core is laminated. These laminated types of structure are made up of stamping which is about 0.4 to 0.5 mm thick. All the stamping are stamped together to form stator core, which is then housed in stator frame. The stamping is generally made up of silicon steel, which helps to reduce the hysteresis loss occurring in motor.

Stator Winding or Field Winding

The slots on the periphery of stator core of the three phase induction motor carries three phase windings. This three phase winding is supplied by three phase ac supply. The three phases of the winding are connected either in star or delta depending upon which type of starting method is used. The squirrel cage motor is mostly started by star – delta stater and hence the stator of squirrel cage motor is delta connected. The slip ring three phase induction motor are started by inserting resistances so, the stator winding of slip ring induction motor can be connected either in star or delta. The winding wound on the stator of three phase induction motor is also called field winding and when this winding is excited by three phase ac supply it produces a rotating magnetic field.

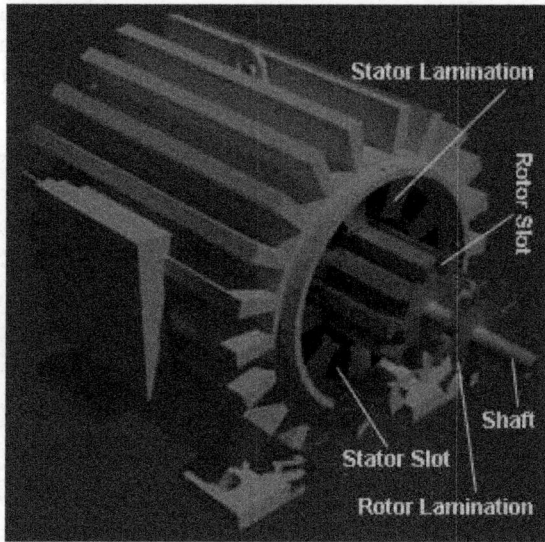

Fig. : Basic Construction of Induction Motor.

Types of Three Phase Induction Motor

1. **Squirrel cage three phase induction motor :** The rotor of the squirrel cage three phase induction motor is cylindrical in shape and have slots on its periphery. The slots are not made parallel to each other but are bit skewed (skewing is not shown in the figure of squirrel cadge rotor beside) as the skewing prevents magnetic locking of stator and rotor teeth and makes the working of motor more smooth and quieter. The squirrel cage rotor consists of aluminum, brass or copper bars (copper bras rotor is shown in the figure beside). These aluminum, brass or copper bars are called rotor conductors and are placed in the slots on the periphery of the rotor. The rotor conductors are permanently shorted by the copper or aluminum rings called the end rings. In order to provide mechanical strength these rotor conductor are braced to the end ring and hence form a complete closed circuit resembling like a cage and

hence got its name as 'squirrel cage induction motor". The squirrel cage rotor winding is made symmetrical. As the bars are permanently shorted by end rings, the rotor resistance is very small and it is not possible to add external resistance as the bars are permanently shorted. The absence of slip ring and brushes make the construction of Squirrel cage three phase induction motor very simple and robust and hence widely used three phase induction motor. These motors have the advantage of adapting any number of pole pairs.

Advantages of Squirrel Cage Induction Rotor

1. Its construction is very simple and rugged.
2. As there are no brushes and slip ring, these motors requires less maintenance.

Applications

Squirrel cage induction motor is used in lathes, drilling machine, fan, blower printing machines etc.

2. **Slip ring or wound three phase induction motor** : In this type of three phase induction motor the rotor is wound for the same number of poles as that of stator but it has less number of slots and has less turns per phase of a heavier conductor.The rotor also carries star or delta winding similar to that of stator winding. The rotor consists of numbers of slots and rotor winding are placed inside these slots. The three end terminals are connected together to form star connection. As its name indicates three phase slip ring induction motor consists of slip rings connected on same shaft as that of rotor. The three ends of three phase windings are permanently connected to these slip rings. The external resistance can be easily connected through the brushes and slip rings and hence used for speed control and improving the starting torque of three phase induction motor. The brushes are used to carry current to and from the rotor winding. These brushes are further connected to three phase star connected resistances. At starting, the resistance are connected in rotor circuit and is gradually cut out as the rotor pick up its speed. When the motor is running the slip ring are shorted by connecting a metal collar, which connect all slip ring together and the brushes are also removed. This reduces wear and tear of the brushes. Due to presence of slip rings and brushes the rotor construction becomes somewhat complicated therefore it is less used as compare to squirrel cage induction motor.

Advantages of Slip Ring Induction Motor

1. It has high starting torque and low starting current.
2. Possibility of adding additional resistance to control speed

Application :

Slip ring induction motor are used where high starting torque is required *i.e.* in hoists, cranes, elevator etc.

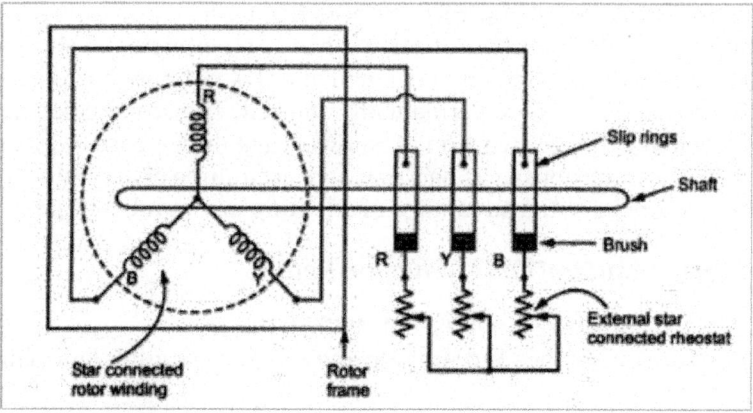

Fig. : Slip Ring Three Phase Induction Motor.

Difference between Slip Ring and Squirrel Cage Induction Motor

Slip ring or phase wound Induction motor	Squirrel cage induction motor
Construction is complicated due to presence of slip ring and brushes	Construction is very simple
The rotor winding is similar to the stator winding	The rotor consists of rotor bars which are permanently shorted with the help of end rings
We can easily add rotor resistance by using slip ring and brushes	Since the rotor bars are permanently shorted, its not possible to add external resistance
Due to presence of external resistance high starting torque can be obtained	Staring torque is low and cannot be improved
Slip ring and brushes are present	Slip ring and brushes are absent
Frequent maintenance is required due to presence of brushes	Less maintenance is required
The construction is complicated and the presence of brushes and slip ring makes the motor more costly	The construction is simple and robust and it is cheap as compared to slip ring induction motor
This motor is rarely used only 10% industry uses slip ring induction motor	Due to its simple construction and low cost. The squirrel cage induction motor is widely used
Rotor copper losses are high and hence less efficiency	Less rotor copper losses and hence high efficiency
Speed control by rotor resistance method is possible	Speed control by rotor resistance method is not possible
Slip ring induction motor are used where high starting torque is required *i.e.* in hoists, cranes, elevator etc	Squirrel cage induction motor is used in lathes, drilling machine, fan, blower printing machines etc.

LINEAR INDUCTION MOTOR

A **linear induction motor** (LIM) is an AC asynchronous linear motor that works by the same general principles as other induction motors but is very typically designed to directly produce motion in a straight line. Characteristically, linear induction motors have a finite length primary, which generates end-effects, whereas with a conventional induction motor the primary is arranged in an endless loop.

Despite their name, not all linear induction motors produce linear motion; some linear induction motors are employed for generating rotations of large diameters where the use of a continuous primary would be very expensive.

As with rotary motors, linear motors frequently run on a 3 phase power supply and can support very high speeds. However, there are end-effects which reduce the force, and it's often not possible to fit a gearbox to trade off force and speed. Linear induction motors are thus frequently less energy efficient than normal rotary motors for any given required force output.

LIMs are often used where contactless force is required, where low maintenance is desirable, or where the duty cycle is low. Their practical uses include magnetic levitation, linear propulsion, and linear actuators. They have also been used for pumping liquid metals.

History

The history of linear electric motors can be traced back at least as far as the 1840s, to the work of Charles Wheatstone at King's College in London, but Wheatstone's model was too inefficient to be practical. A feasible linear induction motor is described in the US patent 782312 (1905 - inventor Alfred Zehden of Frankfurt-am-Main), for driving trains or lifts. The German engineer Hermann Kemper built a working model in 1935. In the late 1940s, professor Eric Laithwaite of Imperial College in London developed the first full-size working model.

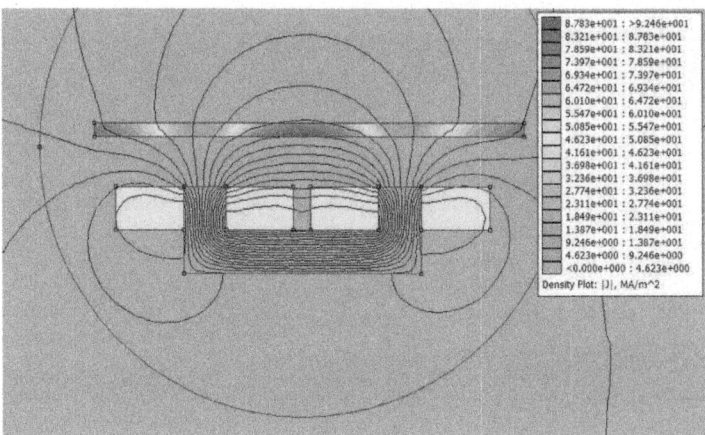

Fig. : FEMM simulation of a Cross-section of Magnetic River, coloured by electric current density.

In a single sided version the magnetic field can create repulsion forces that push the conductor away from the stator, levitating it, and carrying it along in the direction of the moving magnetic field. Laithewaite called the later versions of it magnetic river. These versions of the linear induction motor use a principle called *transverse flux* where two opposite poles are placed side by side. This permits very long poles to be used, which permits high speed and efficiency.

Construction

A linear electric motor's primary typically consists of a flat magnetic core (generally laminated) with transverse slots which are often straight cut with coils laid into the slots, with each phase giving an alternating polarity and so that the different phases physically overlap.

The secondary is frequently a sheet of aluminum, often with an iron backing plate. Some LIMs are double sided, with one primary either side of the secondary, and in this case no iron backing is needed.

Two sorts of linear motor exist, *short primary*, where the coils are truncated shorter than the secondary, and a *short secondary* where the conductive plate is smaller. Short secondary LIMs are often wound as parallel connections between coils of the same phase, whereas short primaries are usually wound in series.

The primaries of transverse flux LIMs have a series of twin poles lying transversely side-by-side, with opposite winding directions. These are typically made either with a suitably cut laminated backing plate or a series of transverse U-cores.

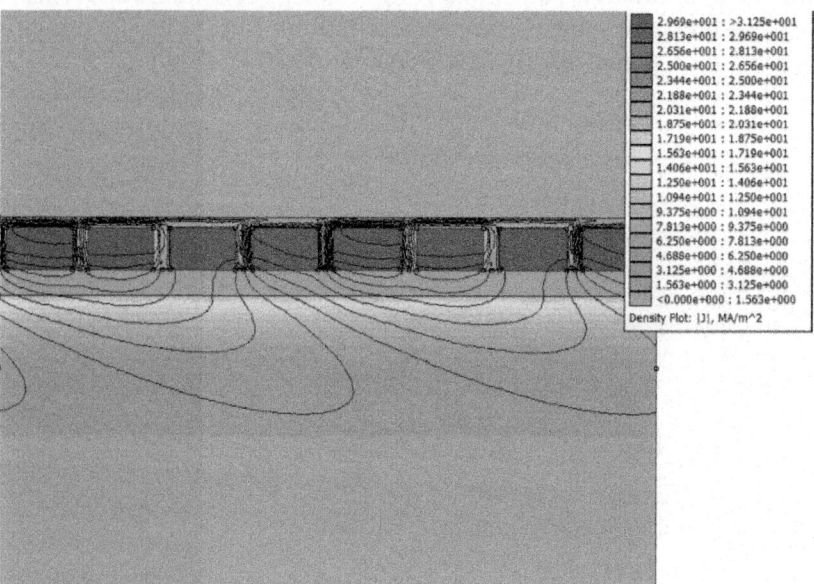

Fig. : The magnetic field of a linear motor sweeping to the left, past an aluminium block. Coloured by induced electric current.

Principles

In this design of electric motor, the force is produced by a linearly moving magnetic field acting on conductors in the field. Any conductor, be it a loop, a coil or simply a piece of plate metal, that is placed in this field will have eddy currents induced in it thus creating an opposing magnetic field, in accordance with Lenz's law. The two opposing fields will repel each other, thus creating motion as the magnetic field sweeps through the metal.

$$n_s = 2f_s/p$$

where, f_s is supply frequency in Hz, p is the number of poles, and n_s is the synchronous speed of the magnetic field in revolutions per second.

The travelling field pattern has a velocity of :

$$v_s = 2t\,f_s$$

v_s is velocity of the linear travelling field in m/s, t is the pole pitch.

For a slip of s, the speed of the secondary in a linear motor is given by

$$v_r = (1 - S)v_s$$

Forces

Thrust

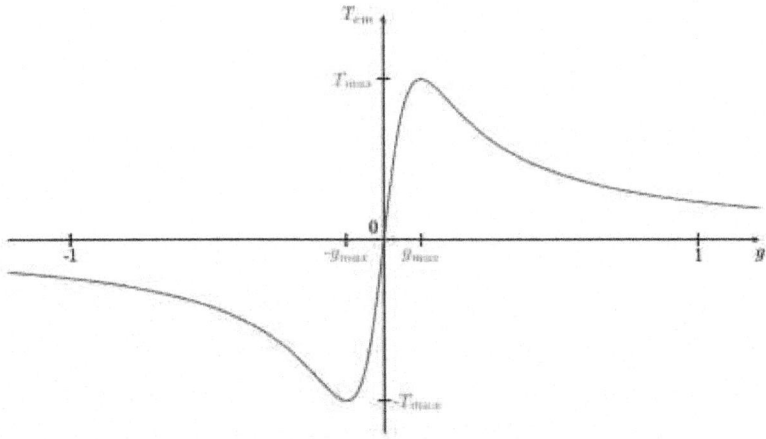

Fig. : Thrust generated as a function of slip.

The drive generated by linear induction motors is somewhat similar to conventional induction motors; the drive forces show a roughly similar characteristic shape relative to slip, albeit modulated by end effects.

End Effect

Unlike a circular induction motor, a linear induction motor shows 'end effects'. These end effects include losses in performance and efficiency that are believed to be caused by magnetic energy being carried away and lost at the end of the primary by the relative movement of the primary and secondary.

With a short secondary, the behaviour is almost identical to a rotary machine, provided it is at least two poles long, but with a short primary reduction in thrust occurs at low slip (below about 0.3) until it is eight poles or longer.

However, because of end effect, linear motors cannot 'run light'- normal induction motors are able to run the motor with a near synchronous field under low load conditions. Due to end effect this creates much more significant losses with linear motors.

Levitation

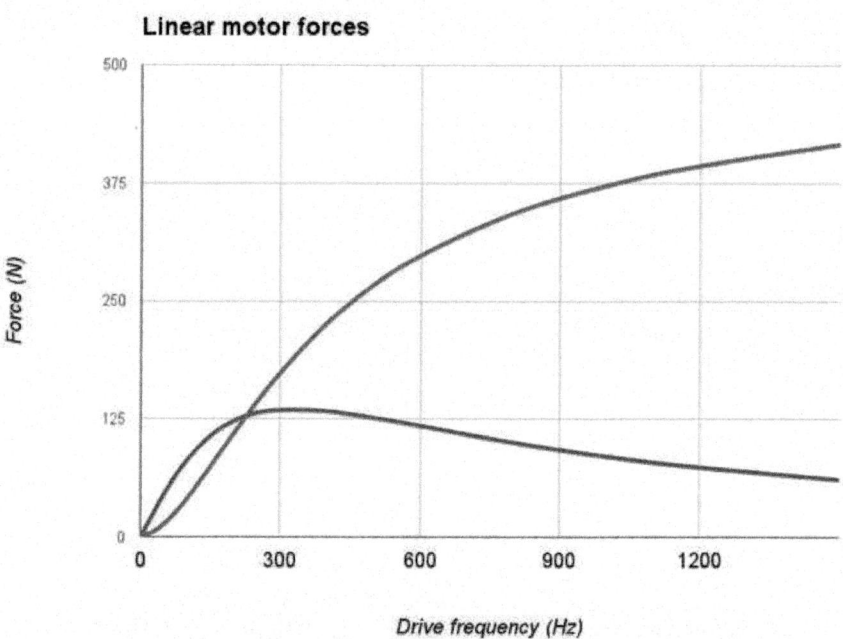

Fig. : Levitation and thrust force curves of a linear motor.

In addition, unlike a rotary motor, an electrodynamic levitation force is shown, this is zero at zero slip, and gives a roughly constant amount of force/gap as slip increases in either direction. This occurs in single sided motors, and levitation will not usually occur when an iron backing plate is used on the secondary, since this causes an attraction that overwhelms the lifting force.

Performance

Linear induction motors are often less efficient than conventional rotary induction motors; the end effects and the relatively large air gap that is often present will typically reduce the forces produced for the same electrical power. However, linear induction motors can avoid the need for gearboxes and similar drivetrains, and these have their own losses; and in any case power use is not always the most important consideration. For example, in many cases linear induction motors have far fewer moving parts, and have very low maintenance.

Uses

Because of these properties, linear motors are often used in maglev propulsion, as in the Japanese Linimo magnetic levitation train line near Nagoya.

The world's first commercial automated maglev system was a low-speed maglev shuttle that ran from the airport terminal of Birmingham International Airport to the nearby Birmingham International railway station between 1984-1995. The length of the track was 600 metres (2,000 ft), and trains "flew" at an altitude of 15 millimetres (0.59 in), levitated by electromagnets, and propelled with linear induction motors. It was in operation for nearly eleven years, but obsolescence problems with the electronic systems made it unreliable in its later years. One of the original cars is now on display at Railworld in Peterborough, together with the RTV31 hover train vehicle.

However, linear motors have been used independently of magnetic levitation, as in Bombardier's Advanced Rapid Transit systems worldwide and a number of modern Japanese subways, including Tokyo's Toei Oedo Line.

Linear induction motor technology is also used in some roller coasters. At present it is still impractical on street running trams, although this, in theory, could be done by burying it in a slotted conduit.

Outside of public transportation, vertical linear motors have been proposed as lifting mechanisms in deep mines, and the use of linear motors is growing in motion control applications. They are also often used on sliding doors, such as those of low floor trams such as the Citadis and the Eurotram. Dual axis linear motors also exist. These specialized devices have been used to provide direct X-Y motion for precision laser cutting of cloth and sheet metal, automated drafting, and cable forming. Most linear motors in use are LIM (linear induction motors) or LSM (linear synchronous motors). Linear DC motors are not used as it includes more cost and linear SRM suffers from poor thrust. So for long run in traction LIM is mostly preferred and for short run LSM is mostly preferred.

Linear induction motors have also been used for launching aircraft, the Westinghouse Electropult system in 1945 was an early example and the Electromagnetic Aircraft Launch System (EMALS) was due to be delivered in 2010.

Linear induction motors are also used in looms, magnetic levitation permit allows bobbins to float between the fibers without direct contact.

Chapter 9

ELECTROMAGNETISM

Electromagnetism, or the **electromagnetic force** is one of the four fundamental interactions in nature, the other three being the strong interaction, the weak interaction, and gravitation. This force is described by electromagnetic fields, and has innumerable physical instances including the interaction of electrically charged particles and the interaction of uncharged magnetic force fields with electrical conductors.

The word *electromagnetism* is a compound form of two Greek terms, "amber", and *magnetic*, from "magnítis líthos", which means "magnesian stone", a type of iron ore. The science of electromagnetic phenomena is defined in terms of the electromagnetic force, sometimes called the Lorentz force, which includes both electricity and magnetism as elements of one phenomenon.

During the quark epoch, the electroweak force split into the electromagnetic and weak force. The electromagnetic force plays a major role in determining the internal properties of most objects encountered in daily life. Ordinary matter takes its form as a result of intermolecular forces between individual molecules in matter. Electrons are bound by electromagnetic wave mechanics into orbitals around atomic nuclei to form atoms, which are the building blocks of molecules. This governs the processes involved in chemistry, which arise from interactions between the electrons of neighboring atoms, which are in turn determined by the interaction between electromagnetic force and the momentum of the electrons.

There are numerous mathematical descriptions of the electromagnetic field. In classical electrodynamics, electric fields are described as electric potential and electric current in Ohm's law, magnetic fields are associated with electromagnetic induction and magnetism, and Maxwell's equations describe how electric and magnetic fields are generated and altered by each other and by charges and currents.

The theoretical implications of electromagnetism, in particular the establishment of the speed of light based on properties of the "medium" of propagation (permeability and permittivity), led to the development of special relativity by Albert Einstein in 1905.

History of the Theory

Originally electricity and magnetism were thought of as two separate forces. This view changed, however, with the publication of James Clerk Maxwell's 1873 *Treatise on Electricity and Magnetism* in which the interactions of positive and negative charges were shown to be regulated by one force. There are four main effects resulting from these interactions, all of which have been clearly demonstrated by experiments :

1. Electric charges attract or repel one another with a force inversely proportional to the square of the distance between them : unlike charges attract, like ones repel.
2. Magnetic poles (or states of polarization at individual points) attract or repel one another in a similar way and always come in pairs : every north pole is yoked to a south pole.
3. An electric current in a wire creates a circular magnetic field around the wire, its direction (clockwise or counter-clockwise) depending on that of the current.
4. A current is induced in a loop of wire when it is moved towards or away from a magnetic field, or a magnet is moved towards or away from it, the direction of current depending on that of the movement.

While preparing for an evening lecture on 21 April, 1820, Hans Christian Ørsted made a surprising observation. As he was setting up his materials, he noticed a compass needle deflected from magnetic north when the electric current from the battery he was using was switched on and off. This deflection convinced him that magnetic fields radiate from all sides of a wire carrying an electric current, just as light and heat do, and that it confirmed a direct relationship between electricity and magnetism.

At the time of discovery, Ørsted did not suggest any satisfactory explanation of the phenomenon, nor did he try to represent the phenomenon in a mathematical framework. However, three months later he began more intensive investigations. Soon thereafter he published his findings, proving that an electric current produces a magnetic field as it flows through a wire. The CGS unit of magnetic induction (oersted) is named in honor of his contributions to the field of electromagnetism.

His findings resulted in intensive research throughout the scientific community in electrodynamics. They influenced French physicist André-Marie Ampère's developments of a single mathematical form to represent the magnetic forces between current-carrying conductors. Ørsted's discovery also represented a major step toward a unified concept of energy.

This unification, which was observed by Michael Faraday, extended by James Clerk Maxwell, and partially reformulated by Oliver Heaviside and Heinrich Hertz, is one of the key accomplishments of 19th century mathematical physics.

It had far-reaching consequences, one of which was the understanding of the nature of light. Unlike what was proposed in Electromagnetism, light and other electromagnetic waves are at the present seen as taking the form of quantized, self-propagating oscillatory electromagnetic field disturbances which have been called photons. Different frequencies of oscillation give rise to the different forms of electromagnetic radiation, from radio waves at the lowest frequencies, to visible light at intermediate frequencies, to gamma rays at the highest frequencies.

Ørsted was not the only person to examine the relation between electricity and magnetism. In 1802 Gian Domenico Romagnosi, an Italian legal scholar, deflected a magnetic needle by electrostatic charges. Actually, no galvanic current existed in the setup and hence no electromagnetism was present. An account of the discovery was published in 1802 in an Italian newspaper, but it was largely overlooked by the contemporary scientific community.

Overview

The electromagnetic force is one of the four known fundamental forces. The other fundamental forces are :

- the weak nuclear force, which binds to all known particles in the Standard Model, and causes certain forms of radioactive decay. (In particle physics though, the electroweak interaction is the unified description of two of the four known fundamental interactions of nature : electromagnetism and the weak interaction);

- the strong nuclear force, which binds quarks to form nucleons, and binds nucleons to form nuclei

- the gravitational force.

All other forces (*e.g.*, friction) are ultimately derived from these fundamental forces and momentum carried by the movement of particles.

The electromagnetic force is the one responsible for practically all the phenomena one encounters in daily life above the nuclear scale, with the exception of gravity. Roughly speaking, all the forces involved in interactions between atoms can be explained by the electromagnetic force acting on the electrically charged atomic nuclei and electrons inside and around the atoms, together with how these particles carry momentum by their movement. This includes the forces we experience in "pushing" or "pulling" ordinary material objects, which come from the intermolecular forces between the individual molecules in our bodies and those in the objects. It also includes all forms of chemical phenomena.

A necessary part of understanding the intra-atomic to intermolecular forces is the effective force generated by the momentum of the electrons' movement, and that electrons move between interacting atoms, carrying momentum with them. As a collection of electrons becomes more confined, their minimum momentum necessarily increases due to the Pauli exclusion principle. The behaviour of matter at the molecular scale including its density is determined by the balance between the electromagnetic force and the force generated by the exchange of momentum carried by the electrons themselves.

CLASSICAL ELECTROMAGNETISM

Classical electromagnetism (or **classical electrodynamics**) is a branch of theoretical physics that studies the interactions between electric charges and currents using an extension of the classical Newtonian model. The theory provides an excellent description of electromagnetic phenomena whenever the relevant length scales and field strengths are large enough that quantum mechanical effects are negligible. For small distances and low field strengths, such interactions are better described by quantum electrodynamics.

Fundamental physical aspects of classical electrodynamics are presented in many texts, such as those by by Feynman, Leighton and Sands, Panofsky and Phillips, and Jackson.

History

The physical phenomena that electromagnetism describes have been studied as separate fields since antiquity. For example, there were many advances in the field of optics centuries before light was understood to be an electromagnetic wave. However, the theory of electromagnetism, as it is currently understood, emerged as a unified field over the course of the 19th century, most prominently in a set of equations systemized by James Clerk Maxwell. For a detailed historical account, consult Pauli, Whittaker, and Pais.

Lorentz Force

In physics, particularly electromagnetism, the **Lorentz force** is the combination of electric and magnetic force on a point charge due to electromagnetic fields. If a particle of charge q moves with velocity $\mathbf{v}$ in the presence of an electric field $\mathbf{E}$ and a magnetic field $\mathbf{B}$, then it will experience a force. For any produced force there will be an opposite reactive force. In the case of the magnetic field, the reactive force may be obscure, but it must be accounted for.

$$\mathbf{F} = q(\mathbf{E} + \mathbf{v} \times \mathbf{B})$$

(in SI units). Variations on this basic formula describe the magnetic force on a current-carrying wire (sometimes called *Laplace force*), the electromotive force in a wire loop moving through a magnetic field (an aspect of Faraday's law of induction), and the force on a charged particle which might be traveling near the speed of light (relativistic form of the Lorentz force).

The first derivation of the Lorentz force is commonly attributed to Oliver Heaviside in 1889, although other historians suggest an earlier origin in an 1865 paper by James Clerk Maxwell. Hendrik Lorentz derived it a few years after Heaviside.

hide]Derivation of Lorentz force from classical Lagrangian (SI units)

For an **A** field, a particle moving with velocity $\mathbf{v} = \dot{\mathbf{r}}$ has potential momentum $q\mathbf{A}(\mathbf{r},t)$, so its potential energy is $q\mathbf{A}(\mathbf{r},t) \cdot \dot{\mathbf{r}}$. For a ϕ field, the particle's potential energy is $q\phi(\mathbf{r},t)$.

The total potential energy is then :

$$V = q\phi - q\mathbf{A} \cdot \dot{\mathbf{r}}$$

and the kinetic energy is :

$$T = \frac{m}{2}\dot{\mathbf{r}} \cdot \dot{\mathbf{r}}$$

hence the Lagrangian :

$$T = T - V \frac{m}{2}\dot{\mathbf{r}} \cdot \dot{\mathbf{r}} + q\mathbf{A} \cdot \dot{\mathbf{r}} - q\phi$$

$$T = \frac{m}{2}\left(\dot{x}^2 + \dot{y}^2 + \dot{z}^2\right) + q\left(\dot{x}A_x + \dot{y}A_y + \dot{z}A_z\right) - q\phi$$

Lagrange's equations are

$$\frac{d}{dt}\frac{\partial L}{\partial \dot{x}} = \frac{\partial L}{\partial x}$$

(same for y and z). So calculating the partial derivatives :

$$\frac{d}{dt}\frac{\partial L}{\partial \dot{x}} = m\ddot{x} + q\frac{dA_x}{dt}$$

$$= m\ddot{x} + \frac{q}{dt}\left(\frac{\partial A_x}{\partial t}dt + \frac{\partial A_x}{\partial x}dx + \frac{\partial A_x}{\partial y}dy + \frac{\partial A_x}{\partial z}dz\right)$$

$$= m\ddot{x} + q\left(\frac{\partial A_x}{\partial t} + \frac{\partial A_x}{\partial x}\dot{x} + \frac{\partial A_x}{\partial y}\dot{y} + \frac{\partial A_x}{\partial z}\dot{z}\right)$$

$$\frac{\partial L}{\partial x} = -q\frac{\partial \phi}{\partial x} + q\left(\frac{\partial A_x}{\partial x}\dot{x} + \frac{\partial A_y}{\partial x}\dot{y} + \frac{\partial A_z}{\partial x}\dot{z}\right)$$

equating and simplifying :

$$m\ddot{x} + q\left(\frac{\partial A_x}{\partial t} + \frac{\partial A_x}{\partial x}\dot{x} + \frac{\partial A_x}{\partial y}\dot{y} + \frac{\partial A_x}{\partial y}\dot{z}\right) = -q\frac{\partial \phi}{\partial x} + q\left(\frac{\partial \phi}{\partial x}\dot{x} + \frac{\partial A_y}{\partial x}\dot{y} + \frac{\partial A_z}{\partial x}\dot{z}\right)$$

$$F_x = -q\left(\frac{\partial \phi}{\partial x} + \frac{\partial A_x}{\partial t}\right) + q\left[\dot{y}\left(\frac{\partial A_y}{\partial x} - \frac{\partial A_x}{\partial y}\right) + \dot{z}\left(\frac{\partial A_z}{\partial x} - \frac{\partial A_x}{\partial z}\right)\right]$$

$$= qE_x + q\left[\dot{y}\,(\nabla \times \mathbf{A})_z - \dot{z}\,(\nabla \times \mathbf{A})_y\right]$$

$$= qE_x + q[\dot{\mathbf{r}} \times (\nabla \times \mathbf{A})]_x$$

$$= qEx + q(\dot{\mathbf{r}} \times \mathbf{B})_x$$

and similarly for the y and z directions. Hence the force equation is :

$F = q(E + \dot{\mathbf{r}} \times \mathbf{B})$ The potential energy depends on the velocity of the particle, so the force is velocity dependent, so it is not conservative.

The relativistic Lagrangian is

$$L = -m\sqrt{1 - \left(\frac{\dot{\mathbf{r}}}{c}\right)^2} + e\mathbf{A}\,(\mathbf{r}) \cdot \dot{\mathbf{r}} \_ e\phi(\mathbf{r})$$

The action is the relativistic arclength of the path of the particle in space time, minus the potential energy contribution, plus an extra contribution which quantum mechanically is an extra phase a charged particle gets when it is moving along a vector potential.

[hide]Derivation of Lorentz force from relativistic Lagrangian (SI units)

The equations of motion derived by extremizing the action :

$$\frac{d\mathbf{P}}{dt} = \frac{\partial L}{\partial \mathbf{r}} = e\frac{\partial \mathbf{A}}{\partial \mathbf{r}} \cdot \dot{\mathbf{r}} - e\frac{\partial \phi}{\partial \mathbf{r}}$$

$$\mathbf{P} - e\mathbf{A} = \frac{m\dot{\mathbf{r}}}{\sqrt{1 - \left(\frac{\dot{\mathbf{r}}}{c}\right)^2}}$$

are the same as Hamilton's equations of motion :

$$\frac{d\mathbf{r}}{dt} = \frac{\partial}{\partial_\mathbf{P}}\left(\sqrt{(\mathbf{P} - e\mathbf{A})^2 + (mc^2)^2} + e\phi\right)$$

$$\frac{d\mathbf{p}}{dt} = -\frac{\partial}{\partial \mathbf{r}}\left(\sqrt{(\mathbf{P} - e\mathbf{A})^2 + (mc^2)^2} + e\phi\right)$$

both are equivalent to the non-canonical form :

$$\frac{d}{dt}\left(\frac{m\dot{\mathbf{r}}}{\sqrt{1-\left(\frac{\dot{\mathbf{r}}}{c}\right)^2}}\right) = e\,(\mathbf{E} + \mathbf{v} \times \mathbf{B}).$$

This formula is the Lorentz force, representing the rate at which the EM field adds relativistic momentum to the particle.

Equation (cgs Units)

The above-mentioned formulae use SI units which are the most common among experimentalists, technicians, and engineers. In cgs-Gaussian units, which are somewhat more common among theoretical physicists, one has instead

$$\mathbf{F} = q_{cgs}\left(E_{cgs} + \frac{\mathbf{v}}{c} \times \mathbf{B}_{cgs}\right).$$

where c is the speed of light. Although this equation looks slightly different, it is completely equivalent, since one has the following relations :

$$q_{cgs} = \frac{q_{SI}}{\sqrt{4\pi\,\epsilon_0}},\quad \mathbf{E}_{cgs} = \sqrt{4\pi\,\epsilon_0}\,\mathbf{E}_{SI},\quad \mathbf{B}_{cgs} = \sqrt{4\pi/\mu_0}\,\mathbf{B}_{SI}$$

where ϵ_0 is the vacuum permittivity and μ_0 the vacuum permeability. In practice, the subscripts "cgs" and "SI" are always omitted, and the unit system has to be assessed from context.

Relativistic Form of the Lorentz Force

Covariant form of the Lorentz Force

Field Tensor

Using the metric signature (-1,1,1,1), The Lorentz force for a charge q can be written in covariant form :

$$\frac{dp^\alpha}{d\tau} = qU_\beta F^{\alpha\beta}$$

where p^α is the four-momentum, defined as :

$$p^\alpha = (p_0,\,p_1,\,p_2,\,p_3) = (\gamma mc,\,p_x,\,p_y,\,p_z)$$

the proper time of the particle, $F^{\alpha\beta}$ the contravariant electromagnetic tensor

$$F^{\alpha\beta} = \begin{pmatrix} 0 & E_x/c & E_y/c & E_z/c \\ -E_x/c & 0 & B_z & -B_y \\ -E_y/c & -B_z & 0 & -B_x \\ -E_z/c & By & -B_x & 0 \end{pmatrix}$$

and U is the covariant 4-velocity of the particle, defined as :

$$U_\beta = (U_0, U_1, U_2, U_3) = \gamma(-c, u_x, u_y, u_z),$$

where γ is the Lorentz factor defined above.

The fields are transformed to a frame moving with constant relative velocity by :

$$F'^{\mu\nu} = \Lambda^\mu{}_\alpha \Lambda^\mu F^{\alpha\beta},$$

where $\Lambda^\mu{}_a$ is the Lorentz transformation tensor.

Translation to Vector Notation

The $\alpha = 1$ component (x-component) of the force is

$$\frac{d p^1}{d\tau} = qU_\beta F^{1\beta} = q(U_0 F^{10} + U_1 F^{11} + U_2 F^{12} + U_3 F^{13}).$$

Substituting the components of the covariant electromagnetic tensor F yields

$$\frac{d p^1}{d\tau} = \left[U_0\left(\frac{-E_x}{c}\right) + U_2\,(B_z) + U_3\,(-B_y) \right].$$

Using the components of covariant four-velocity yields

$$\frac{d p^1}{d\tau} = q\gamma\left[-c\left(\frac{-E_x}{c}\right) + u_y B_z + u_z\,(-B_y) \right]$$

$$= q\gamma(E_x + u_y B_z - u_z B_y)$$

$$= q\gamma\,[E_x + (\mathbf{u} \times \mathbf{B})_x]$$

The calculation for $\alpha = 2, 3$ (force components in the y and z directions) yields similar results, so collecting the 3 equations into one :

$$\frac{d\mathbf{p}}{d\tau} = q\gamma(\mathbf{E} + \mathbf{u} \times \mathbf{B}),$$

which is the Lorentz force.

STA form of the Lorentz Force

The electric and magnetic fields are dependent on the velocity of an observer, so the relativistic form of the Lorentz force law can best be exhibited starting from a co-ordinate-independent expression for the electromagnetic and magnetic fields, F, and an arbitrary time-direction, γ_0, where

$$\mathbf{E} = (F \cdot \gamma_0)\gamma_0$$

and

$$i\mathbf{B} = (F \wedge \gamma_0)\,\gamma_0$$

F is a space-time bivector (an oriented plane segment, just like a vector is an oriented line segment), which has six degrees of freedom corresponding to boosts (rotations in space-time planes) and rotations (rotations in space-space planes). The dot product with the vector γ_0 pulls a vector (in the space algebra) from the translational part, while the wedge-product creates a trivector (in the space algebra) who is dual to a vector which is the usual magnetic field vector. The relativistic velocity is given by the (time-like) changes in a time-position vector $v = \dot{x}$, where

$$v^2 = 1,$$

(which shows our choice for the metric) and the velocity is

$$\mathbf{v} = cv \wedge \gamma_0 / (v \cdot \gamma_0).$$

The proper (invariant is an inadequate term because no transformation has been defined) form of the Lorentz force law is simply

$$F = qF \cdot v$$

Note that the order is important because between a bivector and a vector the dot product is anti-symmetric. Upon a space time split like one can obtain the velocity, and fields as above yielding the usual expression.

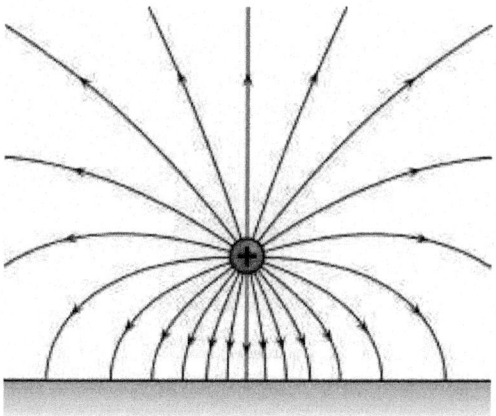

Fig. : Electric field lines emanating from a point positive electric charge suspended over an infinite sheet of conducting material.

ELECTRIC FIELD

An **electric field** is generated by electrically charged particles and time-varying magnetic fields. The electric field describes the electric force experienced by a motionless positively charged test particle at any point in space relative to the source(s) of the field. The concept of an electric field was introduced by Michael Faraday.

Qualitative Description

The electric field is a vector field. The field vector at a given point is defined as the force vector per unit charge that would be exerted on a stationary test charge at that point. An electric field is generated by electric charge, as well as by a time-varying magnetic field. Electric fields contain electrical energy with energy density proportional to the square of the field amplitude. The electric field is to charge as gravitational acceleration is to mass. The SI units of the field are newtons per coulomb ($N \cdot C^{-1}$) or, equivalently, volts per metre ($V \cdot m^{-1}$), which in terms of SI base units are $kg \cdot m \cdot s^{-3} \cdot A^{-1}$.

An electric field that changes with time, such as due to the motion of charged particles producing the field, influences the local magnetic field. That is : the electric and magnetic fields are not separate phenomena; what one observer perceives as an electric field, another observer in a different frame of reference perceives as a mixture of electric and magnetic fields. For this reason, one speaks of "electromagnetism" or "electromagnetic fields". In quantum electrodynamics, disturbances in the electromagnetic fields are called photons.

Definition

Electric Field

Consider a point charge q with position (x,y,z). Now suppose the charge is subject to a force $F_{on\,q}$ due to other charges. Since this force varies with the position of the charge and by Coulomb's Law it is defined at all points in space, $F_{on\,q}$ is a continuous function of the charge's position. This suggests that there is some property of the space that causes the force which is exerted on the charge q. This property is called the electric field and it is defined by

$$E(x, y, x) \equiv \frac{F_{on\,q}(x, y, z)}{q}$$

Notice that the magnitude of the electric field has dimensions of Force/ Charge. Mathematically, the E field can be thought of as a function that associates a vector with every point in space. Each such vector's magnitude is proportional to how much force a charge at that point would "feel" if it were present and this force would have the same direction as the electric field vector at that point. It is also important to note that the electric field defined above is caused by a configuration of *other* electric charges. This means that the charge q in the equation above is not the charge that is *creating* the electric field, but rather, being acted upon *by* it. This definition does not give a means of computing the electric field caused by a group of charges.

Superposition

Array of Discrete Point Charges

Electric fields satisfy the superposition principle. If more than one charge is present, the total electric field at any point is equal to the vector sum of the separate electric fields that each point charge would create in the absence of the others. That is,

$$E = \sum_i E_i = E_1 + E_2 + E_3 + \cdots$$

where E_i is the electric field created by the i-th point charge.

At any point of interest, the total E-field due to N point charges is simply the superposition of the E-fields due to each point charge, given by

$$E = \sum_{i=1}^{N} E_i = \frac{1}{4\pi\varepsilon_0} \sum_{i=1}^{N} \frac{Q_i}{r_i^2} \hat{r}_i .$$

where Q_i is the electric charge of the i-th point charge, $\hat{r}_i$ the corresponding unit vector of r_i, which is the position of charge Q_i with respect to the point of interest.

Continuum of Charges

It holds for an infinite number of infinitesimally small elements of charges – $i.e.$ a continuous distribution of charge. By taking the limit as N approaches infinity in the previous equation, the electric field for a continuum of charges can be given by the integral :

$$E = \int_V dE = \frac{1}{4\pi\varepsilon_0} \int_V \frac{\rho}{r^2} \hat{r} \, dV - \frac{1}{4\pi\varepsilon_0} \int_V \frac{\rho}{r^3} r \, dV$$

where ρ is the charge density (the amount of charge per unit volume), ε_0 the permittivity of free space, and dV is the differential volume element. This integral is a volume integral over the region of the charge distribution.

The equations above express the electric field of point charges as derived from Coulomb's law, which is a special case of Gauss's Law. While Coulomb's law is only true for stationary point charges, Gauss's law is true for all charges either in static form or in motion. Gauss's Law establishes a more fundamental relationship between the distribution of electric charge in space and the resulting electric field. It is one of Maxwell's equations governing electromagnetism.

Gauss's law allows the E-field to be calculated in terms of a continuous distribution of charge density. In differential form, it can be stated as

$$\nabla \cdot E = \frac{\rho}{\varepsilon_0}$$

where $\nabla\cdot$ is the divergence operator, ρ is the total charge density, including free and bound charge, in other words all the charge present in the system (per unit volume).

Electrostatic Fields

Electrostatic fields are E-fields which do not change with time, which happens when the charges are stationary.

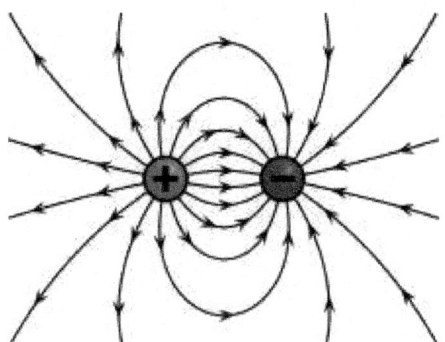

Fig. : Illustration of the electric field surrounding a positive (red) and a negative (blue) charge.

The electric field **E** at a point **r**, that is, **E(r)**, is equal to the negative gradient of the electric potential Φ(**r**), a scalar field at the same point :

$$\mathbf{E} = -\nabla\Phi$$

where ∇ is the gradient operator. This is equivalent to the force definition above, since electric potential Φ is defined by the electric potential energy U per unit (test) positive charge :

$$\Phi = \frac{U}{q}$$

and force is the negative of potential energy gradient :

$$\mathbf{F} = -\nabla U$$

If several spatially distributed charges generate such an electric potential, *e.g.* in a solid, an electric field gradient may also be defined.

Uniform Fields

A uniform field is one in which the electric field is constant at every point. It can be approximated by placing two conducting plates parallel to each other and maintaining a voltage (potential difference) between them; it is only an approximation because of edge effects. Ignoring such effects, the equation for the magnitude of the electric field E is :

$$E = -\frac{\Delta\phi}{d}$$

where $\Delta\phi$ is the potential difference between the plates and d is the distance separating the plates. The negative sign arises as positive charges repel, so a positive charge will experience a force away from the positively charged plate, in the opposite direction to that in which the voltage increases. In micro- and nanoapplications, for instance in relation to semi-conductors, a typical magnitude of an electric field is in the order of 1 volt/µm achieved by applying a voltage of the order of 1 volt between conductors spaced 1 µm apart.

Parallels between Electrostatic and Gravitational Fields

Coulomb's law, which describes the interaction of electric charges :

$$F = q \left(\frac{Q}{4\pi\varepsilon_0} \frac{\hat{\mathbf{r}}}{|\mathbf{r}|^2} \right) = q\mathbf{E}$$

is similar to Newton's law of universal gravitation :

$$F = m \left(-GM \frac{\hat{\mathbf{r}}}{|\mathbf{r}|^2} \right) = m\mathbf{g} .$$

This suggests similarities between the electric field $\mathbf{E}$ and the gravitational field $\mathbf{g}$, so sometimes mass is called "gravitational charge".

Similarities between electrostatic and gravitational forces :

1. Both act in a vacuum.
2. Both are central and conservative.
3. Both obey an inverse-square law (both are inversely proportional to square of r).

Differences between electrostatic and gravitational forces :

1. Electrostatic forces are much greater than gravitational forces for natural values of charge and mass. For instance, the ratio of the electrostatic force to the gravitational force between two electrons is about 10^{42}.
2. Gravitational forces are attractive for like charges, whereas electrostatic forces are repulsive for like charges.
3. There are not negative gravitational charges (no negative mass) while there are both positive and negative electric charges. This difference, combined with the previous two, implies that gravitational forces are always attractive, while electrostatic forces may be either attractive or repulsive.

Electrodynamic Fields

Electrodynamic fields are E-fields which do change with time, when charges are in motion.

An electric field can be produced not only by a static charge, but also by a changing magnetic field (in which case it is a non-conservative field). The electric field is then given by :

$$\mathbf{E} = -\Delta\varphi - \frac{\partial\mathbf{A}}{\partial t}$$

in which $\mathbf{B}$ satisfies

$$\mathbf{B} = \nabla \times \mathbf{A}$$

and $\nabla \times$ denotes the curl. The vector field **B** is the magnetic flux density and the vector **A** is the magnetic vector potential. Taking the curl of the electric field equation we obtain,

$$\nabla \times \mathbf{E} = -\frac{\partial \mathbf{B}}{\partial t}$$

which is Faraday's law of induction, another one of Maxwell's equations.

Energy in the Electric Field

The electrostatic field stores energy. The energy density u (energy per unit volume) is given by

$$u = \frac{1}{2}\varepsilon \,|\, \mathbf{E}\,|^2 ,$$

where ε is the permittivity of the medium in which the field exists, and **E** is the electric field vector (in newtons per coulomb).

The total energy U stored in the electric field in a given volume V is therefore

$$U = \frac{1}{2}\varepsilon \int_V |\,\mathbf{E}\,|^2 \mathrm{d}V ,$$

Further Extensions

Definitive Equation of Vector Fields

In the presence of matter, it is helpful in electromagnetism to extend the notion of the electric field into three vector fields, rather than just one :

$$\mathbf{D} = \varepsilon_0 \, \mathbf{E} + \mathbf{P}$$

where **P** is the electric polarization – the volume density of electric dipole moments, and **D** is the electric displacement field. Since **E** and **P** are defined separately, this equation can be used to define **D**. The physical interpretation of **D** is not as clear as **E** (effectively the field applied to the material) or **P** (induced field due to the dipoles in the material), but still serves as a convenient mathematical simplification, since Maxwell's equations can be simplified in terms of free charges and currents.

Constitutive Relation

The **E** and **D** fields are related by the permittivity of the material, ε.

For linear, homogeneous, isotropic materials **E** and **D** are proportional and constant throughout the region, there is no position dependence : For inhomogeneous materials, there is a position dependence throughout the material :

$$\mathbf{D}(\mathbf{r}) = \varepsilon \mathbf{E}(\mathbf{r})$$

For anisotropic materials the **E** and **D** fields are not parallel, and so **E** and **D** are related by the permittivity tensor (a 2nd order tensor field), in component form :

$$D_i = \varepsilon_{ij} E_j$$

For non-linear media, **E** and **D** are not proportional. Materials can have varying extents of linearity, homogeneity and isotropy.

Electromagnetic Waves

A changing electromagnetic field propagates away from its origin in the form of a wave. These waves travel in vacuum at the speed of light and exist in a wide spectrum of wavelengths. Examples of the dynamic fields of electromagnetic radiation (in order of increasing frequency) : radio waves, microwaves, light (infrared, visible light and ultra-violet), x-rays and gamma rays. In the field of particle physics this electromagnetic radiation is the manifestation of the electromagnetic interaction between charged particles.

General Field Equations

As simple and satisfying as Coulomb's equation may be, it is not entirely correct in the context of classical electromagnetism. Problems arise because changes in charge distributions require a non-zero amount of time to be "felt" elsewhere (required by special relativity).

For the fields of general charge distributions, the retarded potentials can be computed and differentiated accordingly to yield Jefimenko's Equations.

Retarded potentials can also be derived for point charges, and the equations are known as the Liénard–Wiechert potentials. The scalar potential is :

$$\varphi = \frac{1}{4\pi\varepsilon_0} \frac{q}{|\mathbf{r} - \mathbf{r}_q(t_{ret})| - \dfrac{v_q(t_{ret})}{c} \cdot (\mathbf{r} - \mathbf{r}_q(t_{ret}))}$$

where q is the point charge's charge and **r** is the position. $\mathbf{r}_q$ and $\mathbf{v}_q$ are the position and velocity of the charge, respectively, as a function of retarded time. The vector potential is similar :

$$\mathbf{A} = \frac{\mu_0}{4\pi} \frac{q\mathbf{v}_q(t_{ret})}{|\mathbf{r} - \mathbf{r}_q(t_{ret})| - \dfrac{v_q(t_{ret})}{c} \cdot (\mathbf{r} - \mathbf{r}_q(t_{ret}))}$$

These can then be differentiated accordingly to obtain the complete field equations for a moving point particle.

Models

Branches of classical electromagnetism such as optics, electrical and electronic engineering consist of a collection of relevant mathematical models of different degrees of simplification and idealization to enhance the understanding of specific electrodynamics phenomena, cf. An electrodynamics phenomenon is determined by the particular fields, specific densities of electric charges and currents, and the particular transmission medium. Since there are infinitely many of them, in modelling there is a need for some typical, representative

(a) electrical charges and currents, *e.g.* moving pointlike charges and electric and magnetic dipoles, electric currents in a conductor etc.;

(b) electromagnetic fields, *e.g.* voltages, the Liénard–Wiechert potentials, the monochromatic plane waves, optical rays; radio waves, microwaves, infrared radiation, visible light, ultra-violet radiation, X-rays, gamma rays etc.;

(c) transmission media, *e.g.* electronic components, antennas, electromagnetic waveguides, flat mirrors, mirrors with curved surfaces convex lenses, concave lenses; resistors, inductors, capacitors, switches; wires, electric and optical cables, transmission lines, integrated circuits etc.;

all of which have only few variable characteristics.

PHOTOELECTRIC EFFECT

The **photoelectric effect** is the observation that many metals emit electrons when light shines upon them. Electrons emitted in this manner may be called *photoelectrons*.

According to classical electromagnetic theory, this effect can be attributed to the transfer of energy from the light to an electron in the metal. From this perspective, an alteration in either the amplitude or wavelength of light would induce changes in the rate of emission of electrons from the metal. Furthermore, according to this theory, a sufficiently dim light would be expected to show a lag time between the initial shining of its light and the subsequent emission of an electron. However, the experimental results did not correlate with either of the two predictions made by this theory.

Instead, as it turns out, electrons are only dislodged by the photoelectric effect if light reaches or exceeds a threshold frequency, below which no electrons can be emitted from the metal regardless of the amplitude and temporal length of exposure of light. To make sense of the fact that light can eject electrons even if its intensity is low, Albert Einstein proposed that a beam of light is not a wave propagating through space, but rather a collection of discrete wave packets (photons), each with energy hf. This shed light on Max Planck's previous discovery of the Planck relation ($E = hf$) linking energy (E) and frequency (f) as arising from quantization of energy. The factor h is known as the Planck constant.

In 1887, Heinrich Hertz discovered that electrodes illuminated with ultra-violet light create electric sparks more easily. In 1905 Albert Einstein published a paper that explained experimental data from the photoelectric effect as being the result of light energy being carried in discrete quantized packets. This discovery led to the quantum revolution. Einstein was awarded the Nobel Prize in 1921 for "his discovery of the law of the photoelectric effect".

The photoelectric effect requires photons with energies from a few electron-volts to over 1 MeV in elements with a high atomic number. Study of the photoelectric effect led to important steps in understanding the quantum nature of light and electrons and influenced the formation of the concept of wave–particle duality. Other phenomena where light affects the movement of electric charges include the photoconductive effect (also known as photoconductivity or photoresistivity), the photovoltaic effect, and the photoelectrochemical effect.

Emission Mechanism

The photons of a light beam have a characteristic energy proportional to the frequency of the light. In the photoemission process, if an electron within some material absorbs the energy of one photon and acquires more energy than the work function (the electron binding energy) of the material, it is ejected. If the photon energy is too low, the electron is unable to escape the material. Since an increase in the intensity of low-frequency light will only increase the number of low-energy photons sent over a given interval of time, this change in intensity will not create any single photon with enough energy to dislodge an electron. Thus, the energy of the emitted electrons does not depend on the intensity of the incoming light, but only on the energy (equivalently frequency) of the individual photons. It is an interaction between the incident photon and the outermost electrons.

Electrons can absorb energy from photons when irradiated, but they usually follow an "all or nothing" principle. All of the energy from one photon must be absorbed and used to liberate one electron from atomic binding, or else the energy is re-emitted. If the photon energy is absorbed, some of the energy liberates the electron from the atom, and the rest contributes to the electron's kinetic energy as a free particle.

Experimental Observations of Photoelectric Emission

The theory of the photoelectric effect must explain the experimental observations of the emission of electrons from an illuminated metal surface.

For a given metal, there exists a certain minimum frequency of incident radiation below which no photoelectrons are emitted. This frequency is called the *threshold frequency*. Increasing the frequency of the incident beam, keeping the number of incident photons fixed (this would result in a proportionate increase in energy) increases the maximum kinetic energy of the photoelectrons emitted. Thus the stopping voltage increases. The number of electrons also changes because the probability that each photon results in an emitted electron is a function of photon energy. If the intensity of the incident radiation is increased, there is no effect on the kinetic energies of the photoelectrons.

Above the threshold frequency, the maximum kinetic energy of the emitted photoelectron depends on the frequency of the incident light, but is independent of the intensity of the incident light so long as the latter is not too high.

For a given metal and frequency of incident radiation, the rate at which photoelectrons are ejected is directly proportional to the intensity of the incident light. Increase in intensity of incident beam (keeping the frequency fixed) increases the magnitude of the photoelectric current, though stopping voltage remains the same.

The time lag between the incidence of radiation and the emission of a photoelectron is very small, less than 10^{-9} second.

The direction of distribution of emitted electrons peaks in the direction of polarization (the direction of the electric field) of the incident light, if it is linearly polarized.

Mathematical Description

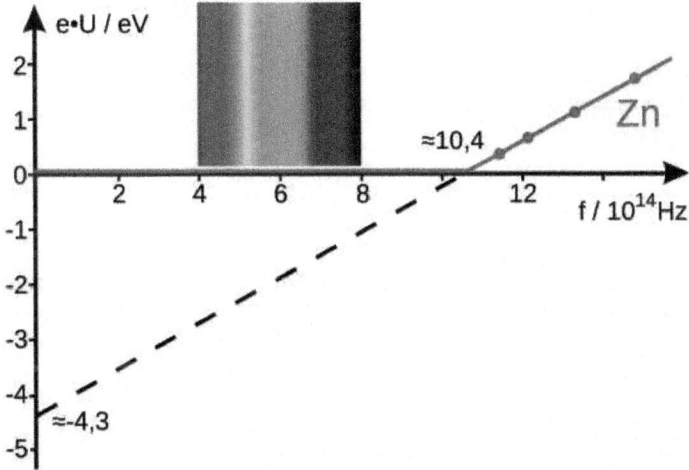

Fig. : Diagram of the maximum kinetic energy as a function of the frequency of light on zinc.

The maximum kinetic energy K_{max} of an ejected electron is given by

$$K_{max} = hf - \varphi,$$

where h is the Planck constant and f is the frequency of the incident photon. The term φ is the work function (sometimes denoted W, or ϕ), which gives the minimum energy required to remove a delocalised electron from the surface of the metal. The work function satisfies

$$\varphi = h f_0,$$

where f_0 is the threshold frequency for the metal. The maximum kinetic energy of an ejected electron is then

$$K_{max} = h(f - f_0).$$

Kinetic energy is positive, so we must have $f > f_0$ for the photoelectric effect to occur.

Stopping Potential

The relation between current and applied voltage illustrates the nature of the photoelectric effect. For discussion, a light source illuminates a plate P, and another plate electrode Q collects any emitted electrons. We vary the potential between P and Q and measure the current flowing in the external circuit between the two plates.

If the frequency and the intensity of the incident radiation are fixed, the photoelectric current increases gradually with an increase in positive potential on collector electrode until all the photoelectrons emitted are collected. The photoelectric current attains a saturation value and does not increase further for any increase in the positive potential. The saturation current depends on the intensity of illumination, but not its wavelength.

If we apply a negative potential to plate Q with respect to plate P and gradually increase it, the photoelectric current decreases until it is zero, at a certain negative potential on plate Q. The minimum negative potential given to plate Q at which the photoelectric current becomes zero is called stopping potential or cut off potential

i.   For the given frequency of incident radiation, the stopping potential is independent of its intensity.

ii. For a given frequency of the incident radiation, the stopping potential Vo is related to the maximum kinetic energy of the photoelectron that is just stopped from reaching plate Q. If m is the mass and v_{max} is the maximum velocity of photoelectron emitted, then

$$K_{max} = \frac{1}{2} m v_{max}^2$$

If q_e is the charge on the electron and V_0 is the stopping potential, then the work done by the retarding potential in stopping the electron $= q_e V_0$, which gives

$$\frac{1}{2} m v_{max}^2 = q_e V_0$$

The above relation shows that the maximum velocity of the emitted photoelectron is independent of the intensity of the incident light. Hence,

$$K_{max} = q_e V_0$$

The stopping voltage varies linearly with frequency of light, but depends on the type of material. For any particular material, there is a threshold frequency that must be exceeded, independent of light intensity, to observe any electron emission.

Three-step Model

In the X-ray regime, the photoelectric effect in crystalline material is often decomposed into three steps :

1. Inner photoelectric effect. The hole left behind can give rise to auger effect, which is visible even when the electron does not leave the material. In molecular solids phonons are excited in this step and may be visible as lines in the final electron energy. The inner photoeffect has to be dipole allowed. The transition rules for atoms translate via the tight-binding model onto the crystal. They are similar in geometry to plasma oscillations in that they have to be transversal.

2. Ballistic transport of half of the electrons to the surface. Some electrons are scattered.

3. Electrons escape from the material at the surface.

In the three-step model, an electron can take multiple paths through these three steps. All paths can interfere in the sense of the path integral formulation. For surface states and molecules the three-step model does still make some sense as even most atoms have multiple electrons which can scatter the one electron leaving.

History

When a surface is exposed to electromagnetic radiation above a certain threshold frequency (typically visible light for alkali metals, near ultra-violet for other metals, and extreme ultra-violet for non-metals), the radiation is absorbed and electrons are emitted. Light, and especially ultra-violet light, discharges negatively electrified bodies with the production of rays of the same nature as cathode rays. Under certain circumstances it can directly ionize gases. The first of these phenomena was discovered by Hertz and Hallwachs in 1887. The second was announced first by Philipp Lenard in 1900.

The ultra-violet light to produce these effects may be obtained from an arc lamp, or by burning magnesium, or by sparking with an induction coil between zinc or cadmium terminals, the light from which is very rich in ultra-violet rays. Sunlight is not rich in ultra-violet rays, as these have been absorbed by the atmosphere, and it does not produce nearly so large an effect as the arc-light. Many substances besides metals discharge negative electricity under the action of ultra-violet light : lists of these substances will be found in papers by G. C. Schmidt and O. Knoblauch.

19th Century

In 1839, Alexandre Edmond Becquerel discovered the photovoltaic effect while studying the effect of light on electrolytic cells. Though not equivalent to the photoelectric effect, his work on photovoltaics was instrumental in showing a strong relationship between light and electronic properties of materials. In 1873, Willoughby Smith discovered photoconductivity in selenium while testing the metal for its high resistance properties in conjunction with his work involving submarine telegraph cables.

Johann Elster (1854–1920) and Hans Geitel (1855–1923), students in Heidelberg, developed the first practical photoelectric cells that could be used to measure the intensity of light. Elster and Geitel had investigated with great success the effects produced by light on electrified bodies.

In 1887, Heinrich Hertz observed the photoelectric effect and the production and reception of electromagnetic waves. He published these observations in the journal Annalen der Physik. His receiver consisted of a coil with a spark gap, where a spark would be seen upon detection of electromagnetic waves. He placed the apparatus in a darkened box to see the spark better. However, he noticed that the maximum spark length was reduced when in the box. A glass panel placed between the source of electromagnetic waves and the receiver absorbed ultra-violet radiation that assisted the electrons in jumping across the gap. When removed, the spark length would increase. He observed no decrease in spark length when he replaced glass with quartz, as quartz does not absorb UV radiation. Hertz concluded his months of investigation and reported the results obtained. He did not further pursue investigation of this effect.

The discovery by Hertz in 1887 that the incidence of ultra-violet light on a spark gap facilitated the passage of the spark, led immediately to a series of investigations by Hallwachs, Hoor, Righi and Stoletow. on the effect of light, and especially of ultra-violet light, on charged bodies. It was proved by these investigations that a newly cleaned surface of zinc, if charged with negative electricity, rapidly loses this charge however small it may be when ultra-violet light falls upon the surface; while if the surface is uncharged to begin with, it acquires a positive charge when exposed to the light, the negative electrification going out into the gas by which the metal is surrounded; this positive electrification can be much increased by directing a strong airblast against the surface. If however the zinc surface is positively electrified it suffers no loss of charge when exposed to the light : this result has been questioned, but a very careful examination of the phenomenon by Elster and Geitel has shown that the loss observed under certain circumstances is due to the discharge by the light reflected from the zinc surface of negative electrification on neighbouring conductors induced by the positive charge, the negative electricity under the influence of the electric field moving up to the positively electrified surface.

With regard to the *Hertz effect*, the researches from the start showed a great complexity of the phenomenon of photoelectric fatigue — that is, the progressive diminution of the effect observed upon fresh metallic surfaces. According to an important research by Wilhelm Hallwachs, ozone played an important part in the phenomenon. However, other elements enter such as oxidation, the humidity, the mode of polish of the surface, etc. It was at the time not even sure that the fatigue is absent in a vacuum.

In the period from February 1888 and until 1891, a detailed analysis of photoeffect was performed by Aleksandr Stoletov with results published in 6 works; four of them in *Comptes Rendus*, one review in *Physikalische Revue* (translated from Russian), and the last work in *Journal de Physique*. First, in these works Stoletov invented a new experimental setup which was more suitable for a quantitative analysis of photoeffect. Using this setup, he discovered the direct proportionality between the intensity of light and the induced photo-electric current (the first law of photoeffect or Stoletov's law). One of his other findings resulted from measurements of the dependence of the intensity of the electric photo current on the gas pressure, where he found the existence of an optimal gas pressure P_m corresponding to a maximum photocurrent; this property was used for a creation of solar cells.

In 1899, J. J. Thomson investigated ultra-violet light in Crookes tubes. Thomson deduced that the ejected particles were the same as those previously found in the cathode ray, later called electrons, which he called "corpuscles". In the research, Thomson enclosed a metal plate (a cathode) in a vacuum tube, and exposed it to high frequency radiation. It was thought that the oscillating electromagnetic fields caused the atoms' field to resonate and, after reaching a certain amplitude, caused a subatomic "corpuscle" to be emitted, and current to be detected. The amount of this current varied with the intensity and colour of the radiation. Larger radiation intensity or frequency would produce more current.

20th Century

The discovery of the ionization of gases by ultra-violet light was made by Philipp Lenard in 1900. As the effect was produced across several centimeters of air and made very great positive and small negative ions, it was natural to interpret the phenomenon, as did J. J. Thomson, as a *Hertz effect* upon the solid or liquid particles present in the gas.

In 1902, Lenard observed that the energy of individual emitted electrons increased with the frequency (which is related to the colour) of the light.

This appeared to be at odds with Maxwell's wave theory of light, which predicted that the electron energy would be proportional to the intensity of the radiation.

Lenard observed the variation in electron energy with light frequency using a powerful electric arc lamp which enabled him to investigate large changes in intensity, and that had sufficient power to enable him to investigate the variation of potential with light frequency. His experiment directly measured potentials, not electron kinetic energy : he found the electron energy by relating it to the maximum stopping potential (voltage) in a phototube. He found that the calculated maximum electron kinetic energy is determined by the frequency of the light. For example, an increase in frequency results in an increase in the maximum kinetic energy calculated for an electron upon liberation – ultra-violet radiation would require a higher applied stopping potential to stop current in a phototube than blue light. However Lenard's results were qualitative rather than quantitative because of the difficulty in performing the experiments : the experiments needed to be done on freshly cut metal so that the pure metal was observed, but it oxidised in a matter of minutes even in the partial vacuums he used. The current emitted by the surface was determined by the light's intensity, or brightness : doubling the intensity of the light doubled the number of electrons emitted from the surface.

The researches of Langevin and those of Eugene Bloch have shown that the greater part of the Lenard effect is certainly due to this 'Hertz effect'. The Lenard effect upon the gas itself nevertheless does exist. Refound by J. J. Thomson and then more decisively by Frederic Palmer, Jr., it was studied and showed very different characteristics than those at first attributed to it by Lenard.

In 1905, Albert Einstein solved this apparent paradox by describing light as composed of discrete quanta, now called photons, rather than continuous waves. Based upon Max Planck's theory of black-body radiation, Einstein theorized that the energy in each quantum of light was equal to the frequency multiplied by a constant, later called Planck's constant. A photon above a threshold frequency has the required energy to eject a single electron, creating the observed effect. This discovery led to the quantum revolution in physics and earned Einstein the Nobel Prize in Physics in 1921. By wave-particle duality the effect can be analyzed purely in terms of waves though not as conveniently.

Albert Einstein's mathematical description of how the photoelectric effect was caused by absorption of quanta of light was in one of his 1905 papers, named *"On a Heuristic Viewpoint Concerning the Production and Transformation of Light"*. This paper proposed the simple description of "light quanta", or photons, and showed how they explained such phenomena as the photoelectric effect. His simple explanation in terms of absorption of discrete quanta of light explained the features of the phenomenon and the characteristic frequency.

The idea of light quanta began with Max Planck's published law of black-body radiation (*"On the Law of Distribution of Energy in the Normal Spectrum"*) by assuming that Hertzian oscillators could only exist at energies E proportional to the frequency f of the oscillator by $E = hf$, where h is Planck's constant. By assuming that light actually consisted of discrete energy packets, Einstein wrote an equation for the photoelectric effect that agreed with experimental results. It explained why the energy of photoelectrons was dependent only on the *frequency* of the incident light and not on its *intensity* : a low-intensity, high-frequency source could supply a few high energy photons, whereas a high-intensity, low-frequency source would supply no photons of sufficient individual energy to dislodge any electrons. This was an enormous theoretical leap, but the concept was strongly resisted at first because it contradicted the wave theory of light that followed naturally from James Clerk Maxwell's equations for electromagnetic behaviour, and more generally, the assumption of infinite divisibility of energy in physical systems. Even after experiments showed that Einstein's equations for the photoelectric effect were accurate, resistance to the idea of photons continued, since it appeared to contradict Maxwell's equations, which were well-understood and verified.

Einstein's work predicted that the energy of individual ejected electrons increases linearly with the frequency of the light. Perhaps surprisingly, the precise relationship had not at that time been tested. By 1905 it was known that the energy of photoelectrons increases with increasing *frequency* of incident light and is independent of the *intensity* of the light. However, the manner of the increase was not experimentally determined until 1914 when Robert Andrews Millikan showed that Einstein's prediction was correct.

The photoelectric effect helped to propel the then-emerging concept of wave–particle duality in the nature of light. Light simultaneously possesses the characteristics of both waves and particles, each being manifested according to the circumstances. The effect was impossible to understand in terms of the classical wave description of light, as the energy of the emitted electrons did not depend on the intensity of the incident radiation. Classical theory predicted that the electrons would 'gather up' energy over a period of time, and then be emitted.

Uses and Effects

Photomultipliers

These are extremely light-sensitive vacuum tubes with a photocathode coated onto part (an end or side) of the inside of the envelope. The photocathode contains combinations of materials such as caesium, rubidium and antimony specially selected to provide a low work function, so when illuminated even by very low levels of light, the photocathode readily releases electrons. By means of a series of electrodes (dynodes) at ever-higher potentials, these electrons are accelerated and substantially increased in number through secondary emission to provide a readily detectable output current. Photomultipliers are still commonly used wherever low levels of light must be detected.

Image sensors

Video camera tubes in the early days of television used the photoelectric effect, for example, Philo Farnsworth's "Image dissector" used a screen charged by the photoelectric effect to transform an optical image into a scanned electronic signal.

Gold-leaf Electroscope

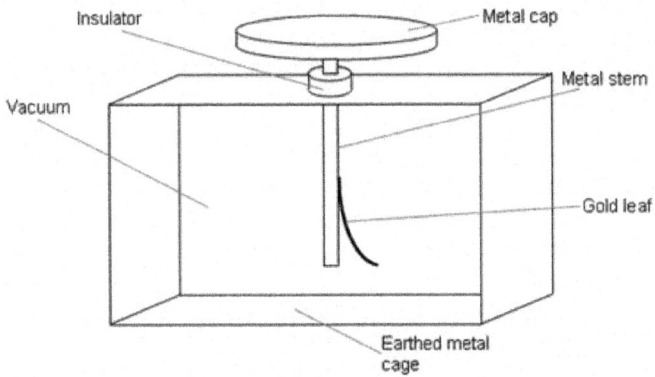

Fig. : The gold leaf electroscope.

Gold-leaf electroscopes are designed to detect static electricity. Charge placed on the metal cap spreads to the stem and the gold leaf of the electroscope. Because they then have the same charge, the stem and leaf repel each other. This will cause the leaf to bend away from the stem.

The electroscope is an important tool in illustrating the photoelectric effect. For example, if the electroscope is negatively charged throughout, there is an excess of electrons and the leaf is separated from the stem. If high-frequency light shines on the cap, the electroscope discharges and the leaf will fall limp. This is because the frequency of the light shining on the cap is above the cap's threshold frequency. The photons in the light have enough energy to liberate electrons from the cap, reducing its negative charge. This will discharge a negatively charged electroscope and further charge a positive electroscope. However, if the electromagnetic radiation hitting the metal cap does not have a high enough frequency (its frequency is below the threshold value for the cap), then the leaf will never discharge, no matter how long one shines the low-frequency light at the cap.

Photoelectron Spectroscopy

Since the energy of the photoelectrons emitted is exactly the energy of the incident photon minus the material's work function or binding energy, the work function of a sample can be determined by bombarding it with a monochromatic X-ray source or UV source, and measuring the kinetic energy distribution of the electrons emitted.

Photoelectron spectroscopy is usually done in a high-vacuum environment, since the electrons would be scattered by gas molecules if they were present. However, some companies are now selling products that allow photoemission in air. The light source can be a laser, a discharge tube, or a synchrotron radiation source.

The concentric hemispherical analyser (CHA) is a typical electron energy analyzer, and uses an electric field to change the directions of incident electrons, depending on their kinetic energies. For every element and core (atomic orbital) there will be a different binding energy. The many electrons created from each of these combinations will show up as spikes in the analyzer output, and these can be used to determine the elemental composition of the sample.

Spacecraft

The photoelectric effect will cause spacecraft exposed to sunlight to develop a positive charge. This can be a major problem, as other parts of the spacecraft in shadow develop a negative charge from nearby plasma, and the imbalance can discharge through delicate electrical components. The static charge created by the photoelectric effect is self-limiting, though, because a more highly charged object gives up its electrons less easily.

Moon Dust

Light from the sun hitting lunar dust causes it to become charged through the photoelectric effect. The charged dust then repels itself and lifts off the surface of the Moon by electrostatic levitation. This manifests itself almost like an "atmosphere of dust", visible as a thin haze and blurring of distant features, and visible as a dim glow after the sun has set. This was first photographed by the Surveyor program probes in the 1960s. It is thought that the smallest particles are repelled up to kilometers high, and that the particles move in "fountains" as they charge and discharge.

Night Vision Devices

Photons hitting a thin film of alkali metal or semi-conductor material such as gallium arsenide in an image intensifier tube cause the ejection of photoelectrons due to the photoelectric effect. These are accelerated by an electrostatic field where they strike a phosphor coated screen, converting the electrons back into photons. Intensification of the signal is achieved either through acceleration of the electrons or by increasing the number of electrons through secondary emissions, such as with a Micro-channel plate. Sometimes a combination of both methods is used. Additional kinetic energy is required to move an electron out of the conduction band and into the vacuum level. This is known as the electron affinity of the photocathode and is another barrier to photoemission other than the forbidden band, explained by the band gap model. Some materials such as Gallium Arsenide have an effective electron affinity that is below the level of the conduction band. In these materials, electrons that move to the conduction band are all of sufficient energy to be emitted from the material and as such, the film that absorbs photons can be quite thick. These materials are known as negative electron affinity materials.

Cross-section

The photoelectric effect is one interaction mechanism between photons and atoms. It is one of 12 theoretically possible interactions.

At the high photon energies comparable to the electron rest energy of 511 keV, Compton scattering, another process, may take place. Above twice this (1.022 MeV) pair production may take place. Compton scattering and pair production are examples of two other competing mechanisms.

Indeed, even if the photoelectric effect is the favoured reaction for a particular single-photon bound-electron interaction, the result is also subject to statistical processes and is not guaranteed, albeit the photon has certainly disappeared and a bound electron has been excited (usually K or L shell electrons at gamma ray energies). The probability of the photoelectric effect occurring is measured by the cross-section of interaction, σ. This has been found to be a function of the atomic number of the target atom and photon energy. A crude approximation, for photon energies above the highest atomic binding energy, is given by :

$$\sigma = \text{constant} \cdot \frac{Z^n}{E^3}$$

Here Z is atomic number and n is a number which varies between 4 and 5. (At lower photon energies a characteristic structure with edges appears, K edge, L edges, M edges, etc.) The obvious interpretation follows that the photoelectric effect rapidly decreases in significance, in the gamma ray region of the spectrum, with increasing photon energy, and that photoelectric effect increases steeply with atomic number. The corollary is that high-Z materials make good gamma-ray shields, which is the principal reason that lead (Z = 82) is a preferred and ubiquitous gamma radiation shield.

QUANTITIES AND UNITS

Electromagnetic units are part of a system of electrical units based primarily upon the magnetic properties of electric currents, the fundamental SI unit being the ampere. The units are :

- ampere (electric current)
- coulomb (electric charge)
- farad (capacitance)
- henry (inductance)
- ohm (resistance)
- tesla (magnetic flux density)
- volt (electric potential)
- watt (power)
- weber (magnetic flux).

In the electromagnetic cgs system, electric current is a fundamental quantity defined via Ampère's law and takes the permeability as a dimensionless quantity (relative permeability) whose value in a vacuum is unity. As a consequence, the square of the speed of light appears explicitly in some of the equations interrelating quantities in this system.

Formulas for physical laws of electromagnetism (such as Maxwell's equations) need to be adjusted depending on what system of units one uses. This is because there is no one-to-one correspondence between electromagnetic units in SI and those in CGS, as is the case for mechanical units. Furthermore, within CGS, there are several plausible choices of electromagnetic units, leading to different unit "sub-systems", including Gaussian, "ESU", "EMU", and Heaviside–Lorentz. Among these choices, Gaussian units are the most common today, and in fact the phrase "CGS units" is often used to refer specifically to CGS-Gaussian units.

ELECTROMAGNETIC PHENOMENA

With the exception of gravitation, electromagnetic phenomena as described by quantum electrodynamics (which includes classical electrodynamics as a limiting case) account for almost all physical phenomena observable to the unaided human senses, including light and other electromagnetic radiation, all of chemistry, most of mechanics (excepting gravitation), and, of course, magnetism and electricity. Magnetic monopoles (and "Gilbert" dipoles) are not strictly electromagnetic phenomena, since in standard electromagnetism, magnetic fields are generated not by true "magnetic charge" but by currents. There are, however, condensed matter analogs of magnetic monopoles in exotic materials (spin ice) created in the laboratory.

ELECTROMAGNETIC INDUCTION

Electromagnetic Induction is the Induction of an electromotive force in a circuit by varying the magnetic flux linked with the circuit. The phenomenon was first investigated in 1830-31 by Joseph Henry and Michael Faraday, who discovered that when the magnetic field around an electromagnet was increased and decreased, an electric current should be detected by nearby conductor. A current can also be induced by constantly moving a permanent magnet in and out of a coil of wire, or by constantly moving a conductor near a stationary permanent magnet. The induced electromotive force is proportional to the rate of change of the magnetic flux cutting across the circuit.

COMPUTATIONAL ELECTROMAGNETICS

Computational electromagnetics, computational electrodynamics or **electromagnetic modelling** is the process of modelling the interaction of electromagnetic fields with physical objects and the environment.

It typically involves using computationally efficient approximations to Maxwell's equations and is used to calculate antenna performance, electromagnetic compatibility, radar cross-section and electromagnetic wave propagation when not in free space.

A specific part of computational electromagnetics deals with electromagnetic radiation scattered and absorbed by small particles.

Background

Several real-world electromagnetic problems like electromagnetic scattering, electromagnetic radiation, modelling of waveguides etc., are not analytically calculable, for the multitude of irregular geometries found in actual devices. Computational numerical techniques can overcome the inability to derive closed form solutions of Maxwell's equations under various constitutive relations of media, and boundary conditions. This makes *computational electromagnetics* (CEM), important to the design, and modelling of antenna, radar, satellite and other communication systems, nanophotonic devices and high speed silicon electronics, medical imaging, cell-phone antenna design, among other applications.

CEM typically solves the problem of computing the E (Electric), and H (Magnetic) fields across the problem domain (*e.g.*, to calculate antenna radiation pattern for an arbitrarily shaped antenna structure). Also calculating power flow direction (Poynting vector), a waveguide's normal modes, media-generated wave dispersion, and scattering can be computed from the E and H fields. CEM models may or may not assume symmetry, simplifying real world structures to idealized cylinders, spheres, and other regular geometrical objects. CEM models extensively make use of symmetry, and solve for reduced dimensionality from 3 spatial dimensions to 2D and even 1D.

An eigenvalue problem formulation of CEM allows us to calculate steady state normal modes in a structure. Transient response and impulse field effects are more accurately modelled by CEM in time domain, by FDTD. Curved geometrical objects are treated more accurately as finite elements FEM, or non-orthogonal grids. Beam propagation method can solve for the power flow in waveguides. CEM is application specific, even if different techniques converge to the same field and power distributions in the modelled domain.

Overview of Methods

One approach is to discretize the space in terms of grids (both orthogonal, and non-orthogonal) and solving Maxwell's equations at each point in the grid. Discretization consumes computer memory, and solving the equations takes significant time. Large scale CEM problems face memory and CPU limitations. As of 2007, CEM problems require supercomputers, high performance clusters, vector processors and/or parallel computer. Typical formulations involve either time-stepping through the equations over the whole domain for each time instant; or through banded matrix inversion to calculate the weights of basis functions, when modelled by finite element methods; or matrix products when using transfer matrix methods; or calculating integrals when using method of moments (MoM); or using fast fourier transforms, and time iterations when calculating by the split-step method or by BPM.

Choice of Methods

Choosing the right technique for solving a problem is important, as choosing the wrong one can either result in incorrect results, or results which take excessively long to compute. However, the name of a technique does not always tell one how it is implemented, especially for commercial tools, which will often have more than one solver.

Davidson gives two tables comparing the FEM, MoM and FDTD techniques in the way they are normally implemented.

Maxwell's Equations in Hyperbolic PDE form

Maxwell's equations can be formulated as a hyperbolic system of partial differential equations. This gives access to powerful techniques for numerical solutions.

It is assumed that the waves propagate in the (x,y)-plane and restrict the direction of the magnetic field to be parallel to the z-axis and thus the electric field to be parallel to the (x,y) plane. The wave is called a transverse electric (TE) wave. In 2D and no polarization terms present, Maxwell's equations can then be formulated as :

$$\frac{\partial}{\partial t}\bar{u} + A\frac{\partial}{\partial x}\bar{u} + B\frac{\partial}{\partial y}\bar{u} + C\bar{u} = g$$

where u, A, B, and C are defined as

$$u = \begin{pmatrix} E_x \\ E_y \\ H_z \end{pmatrix},$$

$$A = \begin{pmatrix} 0 & 0 & 0 \\ 0 & 0 & \frac{1}{\epsilon} \\ 0 & \frac{1}{\mu} & 0 \end{pmatrix},$$

$$B = \begin{pmatrix} 0 & 0 & \frac{-1}{\epsilon} \\ 0 & 0 & 0 \\ \frac{-1}{\mu} & 0 & 0 \end{pmatrix},$$

$$C = \begin{pmatrix} \frac{\sigma}{\epsilon} & 0 & 0 \\ 0 & \frac{\sigma}{\epsilon} & 0 \\ 0 & 0 & 0 \end{pmatrix}.$$

Integral Equation Solvers

The Discrete Dipole Approximation

The discrete dipole approximation is a flexible technique for computing scattering and absorption by targets of arbitrary geometry. The formulation is based on integral form of Maxwell equations. The DDA is an approximation of the continuum target by a finite array of polarizable points. The points acquire dipole moments in response to the local electric field. The dipoles of course interact with one another via their electric fields, so the DDA is also sometimes referred to as the coupled dipole approximation. The resulting linear system of equations is commonly solved using conjugate gradient iterations. The discretization matrix has symmetries (the integral form of Maxwell equations has form of convolution) enabling Fast Fourier Transform to multiply matrix times vector during conjugate gradient iterations.

Method of Moments (MoM) or Boundary Element Method (BEM)

The **method of moments (MoM)** or **boundary element method (BEM)** is a numerical computational method of solving linear partial differential equations which have been formulated as integral equations (*i.e.* in *boundary integral* form). It can be applied in many areas of engineering and science including fluid mechanics, acoustics, electromagnetics, fracture mechanics, and plasticity.

MoM has become more popular since the 1980s. Because it requires calculating only boundary values, rather than values throughout the space, it is significantly more efficient in terms of computational resources for problems with a small surface/volume ratio. Conceptually, it works by constructing a "mesh" over the modelled surface. However, for many problems, BEM are significantly less efficient than volume-discretization methods (finite element method, finite difference method, finite volume method). Boundary element formulations typically give rise to fully populated matrices. This means that the storage requirements and computational time will tend to grow according to the square of the problem size. By contrast, finite element matrices are typically banded (elements are only locally connected) and the storage requirements for the system matrices typically grow linearly with the problem size. Compression techniques (*e.g.* multipole expansions or adaptive cross-approximation/hierarchical matrices) can be used to ameliorate these problems, though at the cost of added complexity and with a success-rate that depends heavily on the nature and geometry of the problem.

BEM is applicable to problems for which Green's functions can be calculated. These usually involve fields in linear homogeneous media. This places considerable restrictions on the range and generality of problems suitable for boundary elements. Non-linearities can be included in the formulation, although they generally introduce volume integrals which require the volume to be discretized before solution, removing an oft-cited advantage of BEM.

Fast Multipole Method (FMM)

The **fast multipole method (FMM)** is an alternative to MoM or Ewald summation. It is an accurate simulation technique and requires less memory and processor power than MoM. The FMM was first introduced by Greengard and Rokhlin and is based on the multipole expansion technique. The first application of the FMM in computational electromagnetics was by Engheta *et. al.* (1992). FMM can also be used to accelerate MoM.

Partial Element Equivalent Circuit (PEEC) Method

The **partial element equivalent circuit (PEEC)** is a 3D full-wave modelling method suitable for combined electromagnetic and circuit analysis. Unlike MoM, PEEC is a full spectrum method valid from dc to the maximum frequency determined by the meshing. In the PEEC method, the integral equation is interpreted as Kirchhoff's voltage law applied to a basic PEEC cell which results in a complete circuit solution for 3D geometries. The equivalent circuit formulation allows for additional SPICE type circuit elements to be easily included. Further, the models and the analysis apply to both the time and the frequency domains. The circuit equations resulting from the PEEC model are easily constructed using a modified loop analysis (MLA) or modified nodal analysis (MNA) formulation. Besides providing a direct current solution, it has several other advantages over a MoM analysis for this class of problems since any type of circuit element can be included in a straightforward way with appropriate matrix stamps. The PEEC method has recently been extended to include non-orthogonal geometries. This model extension, which is consistent with the classical orthogonal formulation, includes the Manhattan representation of the geometries in addition to the more general quadrilateral and hexahedral elements. This helps in keeping the number of unknowns at a minimum and thus reduces computational time for non-orthogonal geometries.

Differential Equation Solvers

Finite-difference Time-domain (FDTD)

Finite-difference time-domain (FDTD) is a popular CEM technique. It is easy to understand. It has an exceptionally simple implementation for a full wave solver. It is at least an order of magnitude less work to implement a basic FDTD solver than either an FEM or MoM solver. FDTD is the only technique where one person can realistically implement oneself in a reasonable time frame, but even then, this will be for a quite specific problem. Since it is a time-domain method, solutions can cover a wide frequency range with a single simulation run, provided the time step is small enough to satisfy the Nyquist–Shannon sampling theorem for the desired highest frequency.

FDTD belongs in the general class of grid-based differential time-domain numerical modelling methods. Maxwell's equations (in partial differential form) are modified to central-difference equations, discretized, and implemented in software. The equations are solved in a cyclic manner : the electric field is solved at a given instant in time, then the magnetic field is solved at the next instant in time, and the process is repeated over and over again.

The basic FDTD algorithm traces back to a seminal 1966 paper by Kane Yee in IEEE Transactions on Antennas and Propagation. Allen Taflove originated the descriptor "Finite-difference time-domain" and its corresponding "FDTD" acronym in a 1980 paper in IEEE Transactions on Electromagnetic Compatibility. Since about 1990, FDTD techniques have emerged as the primary means to model many scientific and engineering problems addressing electromagnetic wave interactions with material structures. An effective technique based on a time-domain finite-volume discretization procedure was introduced by Mohammadian *et. al.* in 1991. Current FDTD modelling applications range from near-DC (ultralow-frequency geophysics involving the entire Earth-ionosphere waveguide) through microwaves (radar signature technology, antennas, wireless communications devices, digital interconnects, biomedical imaging/treatment) to visible light (photonic crystals, nanoplasmonics, solitons, and biophotonics). Approximately 30 commercial and university-developed software suites are available.

Multiresolution Time-domain (MRTD)

MRTD is an adaptive alternative to the finite difference time domain method (FDTD) based on wavelet analysis.

Finite Element Method (FEM)

The **finite element method (FEM)** is used to find approximate solution of partial differential equations (PDE) and integral equations. The solution approach is based either on eliminating the time derivatives completely (steady state problems), or rendering the PDE into an equivalent ordinary differential equation, which is then solved using standard techniques such as finite differences, etc.

In solving partial differential equations, the primary challenge is to create an equation which approximates the equation to be studied, but which is numerically stable, meaning that errors in the input data and intermediate calculations do not accumulate and destroy the meaning of the resulting output. There are many ways of doing this, with various advantages and disadvantages. The Finite Element Method is a good choice for solving partial differential equations over complex domains or when the desired precision varies over the entire domain.

Finite Integration Technique (FIT)

The **finite integration technique (FIT)** is a spatial discretization scheme to numerically solve electromagnetic field problems in time and frequency domain. It preserves basic topological properties of the continuous equations such as conservation of charge and energy. FIT was proposed in 1977 by Thomas Weiland and has been enhanced continually over the years. This method covers the full range of electromagnetics (from static up to high frequency) and optic applications and is the basis for commercial simulation tools.

The basic idea of this approach is to apply the Maxwell equations in integral form to a set of staggered grids. This method stands out due to high flexibility in geometric modelling and boundary handling as well as incorporation of arbitrary material distributions and material properties such as anisotropy, non-linearity and dispersion. Furthermore, the use of a consistent dual orthogonal grid (*e.g.* Cartesian grid) in conjunction with an explicit time integration scheme (*e.g.* leap-frog-scheme) leads to compute and memory-efficient algorithms, which are especially adapted for transient field analysis in radio frequency (RF) applications.

Pseudospectral Time Domain (PSTD)

This class of marching-in-time computational techniques for Maxwell's equations uses either discrete Fourier or Chebyshev transforms to calculate the spatial derivatives of the electric and magnetic field vector components that are arranged in either a 2-D grid or 3-D lattice of unit cells. PSTD causes negligible numerical phase velocity anisotropy errors relative to FDTD, and therefore allows problems of much greater electrical size to be modelled.

Pseudo-spectral Spatial Domain (PSSD)

PSSD solves Maxwell's equations by propagating them forward in a chosen spatial direction. The fields are therefore held as a function of time, and (possibly) any transverse spatial dimensions. The method is pseudo-spectral because temporal derivatives are calculated in the frequency domain with the aid of FFTs. Because the fields are held as functions of time, this enables arbitrary dispersion in the propagation medium to be rapidly and accurately modelled with minimal effort. However, the choice to propagate forward in space (rather than in time) brings with it some subtleties, particularly if reflections are important.

Transmission Line Matrix (TLM)

Transmission line matrix (TLM) can be formulated in several means as a direct set of lumped elements solvable directly by a circuit solver (ala SPICE, HSPICE, *et. al.*), as a custom network of elements or via a scattering matrix approach. TLM is a very flexible analysis strategy akin to FDTD in capabilities, though more codes tend to be available with FDTD engines.

Locally-One-Dimensional FDTD (LOD-FDTD)

This is an implicit method. In this method, in two-dimensional case, Maxwell equations are computed in two steps, whereas in three-dimensional case Maxwell equations are divided into three spatial co-ordinate directions. Stability and dispersion analysis of the three-dimensional LOD-FDTD method have been discussed in detail.;;

Other methods

Eigen Mode Expansion (EME)

Eigen mode expansion (EME) is a rigorous bi-directional technique to simulate electromagnetic propagation which relies on the decomposition of the electromagnetic fields into a basis set of local eigenmodes. The eigenmodes are found by solving Maxwell's equations in each local cross-section. Eigenmode expansion can solve Maxwell's equations in 2D and 3D and can provide a fully vectorial solution provided that the mode solvers are vectorial. It offers very strong benefits compared with the FDTD method for the modelling of optical waveguides, and it is a popular tool for the modelling of fiber optics and silicon photonics devices.

Physical Optics (PO)

Physical optics (PO) is the name of a high frequency approximation (short-wavelength approximation) commonly used in optics, electrical engineering and applied physics. It is an intermediate method between geometric optics, which ignores wave effects, and full wave electromagnetism, which is a precise theory. The word "physical" means that it is more physical than geometrical optics and not that it is an exact physical theory.

The approximation consists of using ray optics to estimate the field on a surface and then integrating that field over the surface to calculate the transmitted or scattered field. This resembles the Born approximation, in that the details of the problem are treated as a perturbation.

Uniform Theory of Diffraction (UTD)

The **uniform theory of diffraction (UTD)** is a high frequency method for solving electromagnetic scattering problems from electrically small discontinuities or discontinuities in more than one dimension at the same point.

The uniform theory of diffraction approximates near field electromagnetic fields as quasi-optical and uses ray diffraction to determine diffraction coefficients for each diffracting object-source combination. These coefficients are then used to calculate the field strength and phase for each direction away from the diffracting point. These fields are then added to the incident fields and reflected fields to obtain a total solution.

Validation

Validation is one of the key issues facing electromagnetic simulation users. The user must understand and master the validity domain of its simulation. The measure is, "how far from the reality are the results?"

Answering this question involves three steps :

- **Comparison between simulation results and analytical formulation** — For example, assessing the value of the radar cross-section of a plate with the analytical formula :

$$\text{RCS}_{\text{Plate}} = \frac{4\pi A^2}{\lambda^2},$$

where **A** is the surface of the plate and λ is the wavelength. The next curve presenting the RCS of a plate computed at 35 GHz can be used as reference example.

- **Cross-comparison between codes** — One example is the cross-comparison of results from method of moments and asymptotic methods in their validity domains.

- **Comparison of simulation results with measurement** — The final validation step is made by comparison between measurements and simulation. For example, the RCS calculation and the measurement of a complex metallic object at 35 GHz. The computation implements GO, PO and PTD for the edges.

Validation processes can clearly reveal that some differences can be explained by the differences between the experimental setup and its reproduction in the simulation environment.

ELECTROMAGNETIC WAVE EQUATION

The **electromagnetic wave equation** is a second-order partial differential equation that describes the propagation of electromagnetic waves through a medium or in a vacuum. It is a three-dimensional form of the wave equation. The homogeneous form of the equation, written in terms of either the electric field **E** or the magnetic field **B**, takes the form :

$$\left(\nabla^2 - \mu \in \frac{\partial^2}{\partial t^2} \right) \mathbf{E} = 0$$

$$\left(\nabla^2 - \mu \in \frac{\partial^2}{\partial t^2} \right) \mathbf{B} = 0$$

where

$$c = \frac{1}{\sqrt{\mu \in}}$$

is the speed of light in a medium with permeability (μ), and permittivity ($\in$), and ∇^2 is the Laplace operator. In a vacuum, $c = c_0 = 299{,}792{,}458$ meters per second, which is the speed of light in free space. The electromagnetic wave equation derives from Maxwell's equations. It should also be noted that in most older literature, **B** is called the *magnetic flux density* or *magnetic induction*.

The Origin of the Electromagnetic Wave Equation

In his 1864 paper titled A Dynamical Theory of the Electromagnetic Field, Maxwell utilized the correction to Ampère's circuital law that he had made in part III of his 1861 paper On Physical Lines of Force. In *Part VI* of his 1864 paper titled *Electromagnetic Theory of Light*, Maxwell combined displacement current with some of the other equations of electromagnetism and he obtained a wave equation with a speed equal to the speed of light. He commented :

The agreement of the results seems to show that light and magnetism are affections of the same substance, and that light is an electromagnetic disturbance propagated through the field according to electromagnetic laws.

Maxwell's derivation of the electromagnetic wave equation has been replaced in modern physics education by a much less cumbersome method involving combining the corrected version of Ampère's circuital law with Faraday's law of induction.

To obtain the electromagnetic wave equation in a vacuum using the modern method, we begin with the modern 'Heaviside' form of Maxwell's equations. In a vacuum- and charge-free space, these equations are :

$$\nabla \cdot \mathbf{E} = 0$$

$$\nabla \cdot \mathbf{E} = -\frac{\partial \mathbf{B}}{\partial t}$$

$$\nabla \cdot \mathbf{B} = 0$$

$$\nabla \cdot \mathbf{B} = \mu_0 \varepsilon_0 \frac{\partial \mathbf{E}}{\partial t}$$

where $\rho = 0$ because there's no charge density in free space.

Taking the curl of the curl equations gives :

$$\nabla \times (\nabla \times \mathbf{E}) = -\frac{\partial}{\partial t} \nabla \times \mathbf{B} = -\mu_0 \varepsilon_0 \frac{\partial^2 \mathbf{E}}{\partial t^2}$$

$$\nabla \times (\nabla \times \mathbf{B}) = \mu_0 \varepsilon_0 \frac{\partial}{\partial t} \nabla \times \mathbf{E} = -\mu_0 \varepsilon_0 \frac{\partial^2 \mathbf{B}}{\partial t^2}$$

We can use the vector identity

$$\nabla \times (\nabla \times \mathbf{V}) = \nabla (\nabla \cdot \mathbf{V}) - \nabla^2 \mathbf{V}$$

where $\mathbf{V}$ is any vector function of space. And

$$\nabla^2 \mathbf{V} = \nabla \cdot (\nabla \mathbf{V})$$

where $\nabla \mathbf{V}$ is a dyadic which when operated on by the divergence operator $\nabla \cdot$ yields a vector. Since

$$\nabla \cdot \mathbf{E} = 0$$

$$\nabla \cdot \mathbf{B} = 0$$

then the first term on the right in the identity vanishes and we obtain the wave equations :

$$\frac{\partial^2 \mathbf{E}}{\partial t^2} - c_0^2 \cdot \nabla^2 \mathbf{E} = 0$$

$$\frac{\partial^2 \mathbf{B}}{\partial t^2} - c_0^2 \cdot \nabla^2 \mathbf{B} = 0$$

where

$$c_0 = \frac{1}{\sqrt{\mu_0 \varepsilon_0}} \ 2.99792458 \times 10^8 \ \text{m/s}$$

is the speed of light in free space.

Covariant form of the Homogeneous Wave Equation

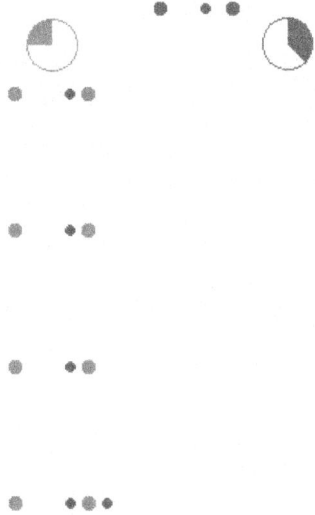

Fig. : Time dilation in transversal motion. The requirement that the speed of light is constant in every inertial reference frame leads to the theory of Special Relativity.

These relativistic equations can be written in contravariant form as

$$\Box A^\mu = 0$$

where the electromagnetic four-potential is

$$A^\mu = (\phi/c, \mathbf{A})$$

with the Lorenz gauge condition :

$$\partial_\mu A^\mu = 0,$$

where

$$-\Box = \nabla^2 - \frac{1}{c^2}\frac{\partial^2}{\partial t^2}$$

is the d'Alembertian operator. (The square box is not a typographical error; it is the correct symbol for this operator.)

Homogeneous Wave Equation in Curved Spacetime

The electromagnetic wave equation is modified in two ways, the derivative is replaced with the covariant derivative and a new term that depends on the curvature appears.

$$- A^{\alpha,\beta}{}_{;\beta} + R^\alpha{}_\beta A^\beta = 0$$

where $R^\alpha{}_\beta$ is the Ricci curvature tensor and the semicolon indicates covariant differentiation.

The generalization of the Lorenz gauge condition in curved spacetime is assumed :

$$A^{\mu}{}_{;\mu} = 0.$$

Inhomogeneous Electromagnetic Wave Equation

Localized time-varying charge and current densities can act as sources of electromagnetic waves in a vacuum. Maxwell's equations can be written in the form of a wave equation with sources. The addition of sources to the wave equations makes the partial differential equations inhomogeneous.

Solutions to the Homogeneous Electromagnetic Wave Equation

The general solution to the electromagnetic wave equation is a linear superposition of waves of the form

$$\mathbf{E}(\mathbf{r},t) = g(\phi(\mathbf{r},t)) = g\,(\omega t - \mathbf{k} \cdot \mathbf{r})$$

$$\mathbf{B}(\mathbf{r},\,t) = g(\phi(\mathbf{r},\,t)) = g(\omega t - \mathbf{k} \cdot \mathbf{r})$$

for virtually *any* well-behaved function g of dimensionless argument φ, where ω is the angular frequency (in radians per second), and $\mathbf{k} = (k_x, k_y, k_z)$ is the wave vector (in radians per meter).

Although the function g can be and often is a monochromatic sine wave, it does not have to be sinusoidal, or even periodic. In practice, g cannot have infinite periodicity because any real electromagnetic wave must always have a finite extent in time and space. As a result, and based on the theory of Fourier decomposition, a real wave must consist of the superposition of an infinite set of sinusoidal frequencies.

In addition, for a valid solution, the wave vector and the angular frequency are not independent; they must adhere to the dispersion relation :

$$k = |k| = \frac{w}{c} = \frac{2\pi}{\lambda}$$

where k is the wave number and λ is the wavelength. The variable c can only be used in this equation when the electromagnetic wave is in a vacuum.

Monochromatic, Sinusoidal Steady-state

The simplest set of solutions to the wave equation result from assuming sinusoidal waveforms of a single frequency in separable form :

$$\mathbf{E}(\mathbf{r},t) = \text{Re}\{\mathbf{E}(\mathbf{r})e^{i\omega t}\}$$

where :

- i is the imaginary unit,
- $\omega = 2\pi f$ is the angular frequency in radians per second,
- f is the frequency in hertz, and

- $e^{i\omega t} = \cos(\omega t) + i \sin(\omega t)$ is Euler's formula.

Plane Wave Solutions

Consider a plane defined by a unit normal vector

$$n = \frac{k}{k}.$$

Then planar traveling wave solutions of the wave equations are

$$E(r) = E_0 e^{-ik \cdot r}$$

and

$$B(r) = B_0 e^{-ik \cdot r}$$

where $r = (x, y, z)$ is the position vector (in meters).

These solutions represent planar waves travelling in the direction of the normal vector n. If we define the z direction as the direction of n. and the x direction as the direction of E., then by Faraday's Law the magnetic field lies in the y direction and is related to the electric field by the relation $c^2 \dfrac{\partial B}{\partial z} = \dfrac{\partial E}{\partial t}$. Because the divergence of the electric and magnetic fields are zero, there are no fields in the direction of propagation.

This solution is the linearly polarized solution of the wave equations. There are also circularly polarized solutions in which the fields rotate about the normal vector.

Spectral Decomposition

Because of the linearity of Maxwell's equations in a vacuum, solutions can be decomposed into a superposition of sinusoids. This is the basis for the Fourier transform method for the solution of differential equations. The sinusoidal solution to the electromagnetic wave equation takes the form

$$E(r, t) = E_0 \cos(\omega t - k \cdot r + \phi_0)$$

and

$$B(r, t) = B_0 \cos(\omega t - k \cdot r + \phi_0)$$

where

> t is time (in seconds),
> ω is the angular frequency (in radians per second),
> $k = (k_x, k_y, k_z)$ is the wave vector (in radians per meter), and
> ϕ_0 is the phase angle (in radians).

The wave vector is related to the angular frequency by

$$k = |k| = \frac{\omega}{c} = \frac{2\pi}{\lambda}$$

where k is the wavenumber and λ is the wavelength.

The electromagnetic spectrum is a plot of the field magnitudes (or energies) as a function of wavelength.

Multipole Expansion

Assuming monochromatic fields varying in time as $e^{-i\omega t}$, if one uses Maxwell's Equations to eliminate **B**, the electromagnetic wave equation reduces to the Helmholtz Equation for **E** :

$$(\nabla^2 + k^2)\mathbf{E} = 0, \mathbf{B} = -\frac{i}{k}\nabla \times \mathbf{E},$$

with $k = \omega/c$ as given above. Alternatively, one can eliminate **E** in favor of **B** to obtain :

$$(\nabla^2 + k^2)\mathbf{B} = 0, \mathbf{E} = -\frac{i}{k}\nabla \times \mathbf{B}.$$

A generic electromagnetic field with frequency ω can be written as a sum of solutions to these two equations. The three-dimensional solutions of the Helmholtz Equation can be expressed as expansions in spherical harmonics with coefficients proportional to the spherical Bessel functions. However, applying this expansion to each vector component of **E** or **B** will give solutions that are not generically divergence-free ($\nabla \cdot \mathbf{E} = \nabla \cdot \mathbf{B} = 0$), and therefore require additional restrictions on the coefficients.

The multipole expansion circumvents this difficulty by expanding not **E** or **B**, but $\mathbf{r} \cdot \mathbf{E}$ or $\mathbf{r} \cdot \mathbf{B}$ into spherical harmonics. These expansions still solve the original Helmholtz equations for **E** and **B** because for a divergence-free field **F**, $\nabla^2 (\mathbf{r} \cdot \mathbf{F}) = \mathbf{r} \cdot (\nabla^2 \mathbf{F})$. The resulting expressions for a generic electromagnetic field are :

$$\mathbf{E} = e^{-i\omega t} \sum_{l,m} \sqrt{l(l+1)} \left[a_E(l,m)\, \mathbf{E}_{l,m}^{(E)} + a_M(l,m)\, \mathbf{E}_{l,m}^{(M)} \right]$$

$$\mathbf{B} = e^{-i\omega t} \sum_{l,m} \sqrt{l(l+1)} \left[a_E(l,m)\, \hat{\mathbf{A}}_{l,m}^{(E)} + a_M(l,m)\mathbf{B}_{l,m}^{(M)} \right]$$

where $\mathbf{E}_{l,m}^{(E)}$ and $\mathbf{B}_{l,m}^{(E)}$ are the *electric multipole fields of order* (l, m), and $\mathbf{E}_{l,m}^{(M)}$ and $\mathbf{B}_{l,m}^{(M)}$ are the corresponding *magnetic multipole fields*, and $a_E(l,m)$ and $a_M(l,m)$ are the coefficients of the expansion. The multipole fields are given by

$$\mathbf{E}_{l,m}^{(E)} = \sqrt{l(l+1)} \left[B_l^{(1)}\, h_l^{(1)}(kr) + B_l^{(2)}\, h_l^{(2)}(kr) \right] \Phi_{l,m}$$

$$\mathbf{E}_{l,m}^{(E)} = \frac{i}{k}\nabla \times \mathbf{B}_{l,m}^{(E)}$$

$$\mathbf{E}_{l,m}^{(M)} = \sqrt{l(l+1)} \left[E_l^{(1)}\, h_l^{(1)}(kr) + E_l^{(2)}\, h_l^{(2)}(kr) \right] \Phi_{l,m}$$

$$\mathbf{B}_{l,m}^{(M)} = -\frac{i}{k}\nabla \times \mathbf{E}_{l,m}^{(M)},$$

where $h_l^{(1,2)}(x)$ are the spherical Hankel functions, $E_l^{(1,2)}$ and $B_l^{(1,2)}$ are determined by boundary conditions, and $\Phi_{l,m} = \dfrac{1}{\sqrt{l(l+1)}}(\mathbf{r} \times \nabla)\, Y_{l,m}$ are vector spherical harmonics normalized so that

$$\int \Phi_{l,m}^{*} \cdot \Phi_{l,m} d\Omega \ d\Omega = \delta_{l,l'}\, \delta m, m'.$$

The multipole expansion of the electromagnetic field finds application in a number of problems involving spherical symmetry, for example antennae radiation patterns, or nuclear gamma decay. In these applications, one is often interested in the power radiated in the far-field. In this regions, the **E** and **B** fields asymptote to

$$\mathbf{B} \approx \frac{e^{i(kr-\omega t)}}{kr} \sum_{l,m} (-i)^{l+1} [a_E(l,m)\Phi_{l,m} + a_M(l,m)\,\hat{\mathbf{r}} \times \Phi_{l,m}]$$

$$\mathbf{B} \approx \mathbf{B} \times \mathbf{r}.$$

The angular distribution of the time-averaged radiated power is then given by

$$\frac{dP}{d\Omega} \approx \frac{1}{2k^2} \left| \sum_{l,m} (-i)^{l+1} [a_E(l,m)\Phi_{l,m} \times \hat{\mathbf{r}} + a_M(l,m)\Phi_{l,m}] \right|^2 .$$

Other Solutions

Other spherically and cylindrically symmetric analytic solutions to the electromagnetic wave equations are also possible.

In spherical co-ordinates the solutions to the wave equation can be written as follows :

$$\mathbf{E}(\mathbf{r}, t) = \frac{1}{r}\mathbf{E}_0 \cos(\omega t - \mathbf{k}\cdot\mathbf{r} + \phi_0),$$

$$\mathbf{E}(\mathbf{r}, t) = \frac{1}{r}\mathbf{E}_0 \cos(\omega t - \mathbf{k}\cdot\mathbf{r} + \phi_0)$$

and

$$\mathbf{B}(\mathbf{r}, t) = \frac{1}{r}\mathbf{B}_0 \cos(\omega t - \mathbf{k}\cdot\mathbf{r} + \phi_0),$$

$$\mathbf{B}(\mathbf{r}, t) = \frac{1}{r}\mathbf{B}_0 \sin(\omega t - \mathbf{k}\cdot\mathbf{r} + \phi_0)$$

These can be rewritten in terms of the spherical Bessel function.

In cylindrical co-ordinates, the solutions to the wave equation are the ordinary Bessel function of integer order.

ELECTROMECHANICS

In engineering, **electromechanics** combines electrical and mechanical processes and procedures drawn from electrical engineering and mechanical engineering. Electrical engineering in this context also encompasses electronics engineering.

Devices which carry out electrical operations by using moving parts are known as electromechanical. Strictly speaking, a manually operated switch is an electromechanical component, but the term is usually understood to refer to devices such as relays, which allow a voltage or current to control other, isolated voltages and currents by mechanically switching sets of contacts, solenoids, by which a voltage can actuate a moving linkage, vibrators, which convert DC to AC with vibrating sets of contacts, etc.

Before the development of modern electronics, electromechanical devices were widely used in complicated systems sub-systems, including electric typewriters, teletypes, very early television systems, and the very early electromechanical digital computers.

History of Electromechanics

Relays originated with telegraphy as electromechanical devices used to regenerate telegraph signals. In 1885, Michael Pupin at Columbia University taught mathematical physics and electromechanics until 1931.

The Strowger switch, the Panel switch, and similar ones were widely used in early automated telephone exchanges. Crossbar switches were first widely installed in the middle 20th century in Sweden, the United States, Canada, and Great Britain, and these quickly spread to the rest of the world - especially to Japan. The electromechanical television systems of the late 19th century were less successful.

Electric typewriters developed, up to the 1980s, as "power-assisted typewriters". They contained a single electrical component, the motor. Where the keystroke had previously moved a typebar directly, now it engaged mechanical linkages that directed mechanical power from the motor into the typebar. This was also true of the later IBM Selectric. At Bell Labs, in the 1940s, the Bell Model V computer was developed. It was an electromechanical relay-based device; cycles took seconds. In 1968 electromechanical systems were still under serious consideration for an aircraft flight control computer, until a device based on large scale integration electronics was adopted in the Central Air Data Computer.

Modern Practice

Beginning in the last third of the century, much equipment which for most of the 20th century would have used electromechanical devices for control, has come to use less expensive and more reliable integrated microcontroller circuits containing ultimately a few million transistors, and a program to carry out the same task through logic, with electromechanical components only where moving parts, such as mechanical electric actuators, are a requirement. Such chips have replaced most electromechanical devices, are used in most simple feedback control systems, and appear in huge numbers in everything from traffic lights to washing machines.

As of 2010, approximately 16,400 people work as electro-mechanical technicians in the US, about 1 out of every 9000 workers. Their median annual wage is about 50% more than the median annual wage over all occupations.

ELECTROMAGNET

An **electromagnet** is a type of magnet in which the magnetic field is produced by electric current. The magnetic field disappears when the current is turned off. Electromagnets are widely used as components of other electrical devices, such as motors, generators, relays, loudspeakers, hard disks, MRI machines, scientific instruments, and magnetic separation equipment, as well as being employed as industrial lifting electromagnets for picking up and moving heavy iron objects like scrap iron.

An electric current flowing in a wire creates a magnetic field around the wire, due to Ampere's law. To concentrate the magnetic field, in an electromagnet the wire is wound into a coil with many turns of wire lying side by side. The magnetic field of all the turns of wire passes through the center of the coil, creating a strong magnetic field there. A coil forming the shape of a straight tube (a helix) is called a solenoid. Much stronger magnetic fields can be produced if a "core" of ferromagnetic material, such as soft iron, is placed inside the coil. The ferromagnetic core increases the magnetic field to thousands of times the strength of the field of the coil alone, due to the high magnetic permeability μ of the ferromagnetic material. This is called a ferromagnetic-core or iron-core electromagnet.

The direction of the magnetic field through a coil of wire can be found from a form of the right-hand rule. If the fingers of the right hand are curled around the coil in the direction of current flow (conventional current, flow of positive charge) through the windings, the thumb points in the direction of the field inside the coil. The side of the magnet that the field lines emerge from is defined to be the *north pole*.

The main advantage of an electromagnet over a permanent magnet is that the magnetic field can be rapidly manipulated over a wide range by controlling the amount of electric current. However, a continuous supply of electrical energy is required to maintain the field.

How the Iron Core Works

The material of the core of the magnet (usually iron) is composed of small regions called magnetic domains that act like tiny magnets. Before the current in the electromagnet is turned on, the domains in the iron core point in random directions, so their tiny magnetic fields cancel each other out, and the iron has no large scale magnetic field. When a current is passed through the wire wrapped around the iron, its magnetic field penetrates the iron, and causes the domains to turn, aligning parallel to the magnetic field, so their tiny magnetic fields add to the wire's field, creating a large magnetic field that extends into the space around the magnet. The larger the current passed through the wire coil, the more the domains align, and the stronger the magnetic field is. Finally all the domains are lined up, and further increases in current only cause slight increases in the magnetic field : this phenomenon is called saturation.

When the current in the coil is turned off, most of the domains lose alignment and return to a random state and the field disappears. However some of the alignment persists, because the domains have difficulty turning their direction of magnetization, leaving the core a weak permanent magnet. This phenomenon is called hysteresis and the remaining magnetic field is called remanent magnetism. The residual magnetization of the core can be removed by degaussing.

History

Danish scientist Hans Christian Ørsted discovered in 1820 that electric currents create magnetic fields. British scientist William Sturgeon invented the electromagnet in 1824. His first electromagnet was a horseshoe-shaped piece of iron that was wrapped with about 18 turns of bare copper wire (insulated wire didn't exist yet). The iron was varnished to insulate it from the windings. When a current was passed through the coil, the iron became magnetized and attracted other pieces of iron; when the current was stopped, it lost magnetization. Sturgeon displayed its power by showing that although it only weighed seven ounces (roughly 200 grams), it could lift nine pounds (roughly 4 kilos) when the current of a single-cell battery was applied. However, Sturgeon's magnets were weak because the uninsulated wire he used could only be wrapped in a single spaced out layer around the core, limiting the number of turns.

Beginning in 1827, US scientist Joseph Henry systematically improved and popularized the electromagnet. By using wire insulated by silk thread he was able to wind multiple layers of wire on cores, creating powerful magnets with thousands of turns of wire, including one that could support 2,063 lb (936 kg). The first major use for electromagnets was in telegraph sounders.

The magnetic domain theory of how ferromagnetic cores work was first proposed in 1906 by French physicist Pierre-Ernest Weiss, and the detailed modern quantum mechanical theory of ferromagnetism was worked out in the 1920s by Werner Heisenberg, Lev Landau, Felix Bloch and others.

Uses of Electromagnets

Electromagnets are very widely used in electric and electromechanical devices, including :

- Motors and generators
- Transformers

- Relays, including reed relays originally used in telephone exchanges
- Electric bells and buzzers
- Loudspeakers and earphones
- Actuators
- Magnetic recording and data storage equipment : tape recorders, VCRs, hard disks
- Scientific instruments such as MRI machines and mass spectrometers
- Particle accelerators
- Magnetic locks
- Magnetic separation equipment, used for separating magnetic from non-magnetic material, for example separating ferrous metal from other material in scrap.
- Industrial lifting magnets
- Electromagnetic suspension used for MAGLEV trains.

Analysis of Ferromagnetic Electromagnets

The magnetic field of electromagnets in the general case is given by Ampere's Law :

$$\int J \cdot d\mathbf{A} = \oint \mathbf{H} \cdot d\mathbf{l}$$

which says that the integral of the magnetizing field H around any closed loop of the field is equal to the sum of the current flowing through the loop. Another equation used, that gives the magnetic field due to each small segment of current, is the Biot–Savart law. Computing the magnetic field and force exerted by ferromagnetic materials is difficult for two reasons. First, because the strength of the field varies from point to point in a complicated way, particularly outside the core and in air gaps, where *fringing fields* and *leakage flux* must be considered. Second, because the magnetic field B and force are non-linear functions of the current, depending on the non-linear relation between B and H for the particular core material used. For precise calculations, computer programs that can produce a model of the magnetic field using the finite element method are employed.

Magnetic Circuit – the Constant B Field Approximation

In many practical applications of electromagnets, such as motors, generators, transformers, lifting magnets, and loudspeakers, the iron core is in the form of a loop or magnetic circuit, possibly broken by a few narrow air gaps. This is because the magnetic field lines are in the form of closed loops. Iron presents much less "resistance" (reluctance) to the magnetic field than air, so a stronger field can be obtained if most of the magnetic field's path is within the core.

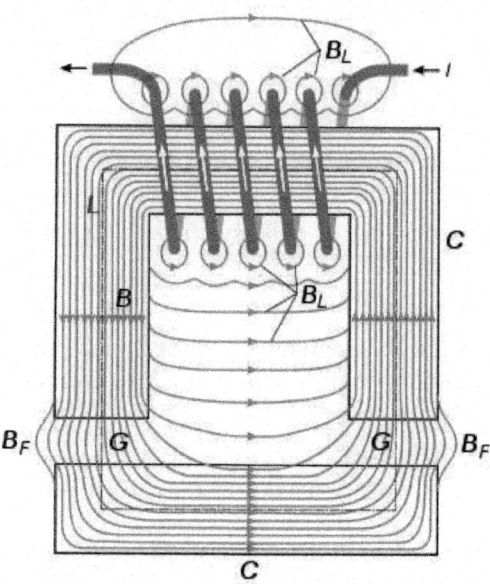

Fig. : Magnetic field *(green)* of a typical electromagnet, with the iron core C forming a closed loop with two air gaps G in it.

B – magnetic field in the core

B_F – "fringing fields". In the gaps G the magnetic field lines "bulge" out, so the field strength is less than in the core : $B_F < B$

B_L – leakage flux; magnetic field lines which don't follow complete magnetic circuit

L – average length of the magnetic circuit used in eq. 1 below. It is the sum of the length L_{core} in the iron core pieces and the length L_{gap} in the air gaps G. Both the leakage flux and the fringing fields get larger as the gaps are increased, reducing the force exerted by the magnet.

Since most of the magnetic field is confined within the outlines of the core loop, this allows a simplification of the mathematical analysis. A common simplifying assumption satisfied by many electromagnets, which will be used in this section, is that the magnetic field strength B is constant around the magnetic circuit and zero outside it. Most of the magnetic field will be concentrated in the core material (C). Within the core the magnetic field (B) will be approximately uniform across any cross-section, so if in addition the core has roughly constant area throughout its length, the field in the core will be constant. This just leaves the air gaps (G), if any, between core sections. In the gaps the magnetic field lines are no longer confined by the core, so they 'bulge' out beyond the outlines of the core before curving back to enter the next piece of core material, reducing the field strength in the gap. The bulges *(B_F)* are called *fringing fields*. However, as long as the length of the gap is smaller than the cross-section dimensions of the core, the field in the gap will be approximately the same as in the core. In addition, some of the magnetic field lines (B_L) will take 'short cuts' and not pass through the entire core circuit, and thus will not contribute to the force exerted by the magnet. This also includes field lines that encircle the wire windings but do not enter the core. This is called *leakage flux*. Therefore the equations in this section are valid for electromagnets for which :

1. the magnetic circuit is a single loop of core material, possibly broken by a few air gaps
2. the core has roughly the same cross-sectional area throughout its length.
3. any air gaps between sections of core material are not large compared with the cross-sectional dimensions of the core.
4. there is negligible leakage flux

The main non-linear feature of ferromagnetic materials is that the B field saturates at a certain value, which is around 1.6 teslas (T) for most high permeability core steels. The B field increases quickly with increasing current up to that value, but above that value the field levels off and becomes almost constant, regardless of how much current is sent through the windings. So the strength of the magnetic field possible from an iron core electromagnet is limited to around 1.6 to 2 T.

Magnetic Field Created by a Current

The magnetic field created by an electromagnet is proportional to both the number of turns in the winding, N, and the current in the wire, I, hence this product, NI, in ampere-turns, is given the name magnetomotive force. For an electromagnet with a single magnetic circuit, of which length L_{core} of the magnetic field path is in the core material and length L_{gap} is in air gaps, Ampere's Law reduces to :

$$NI = H_{core} L_{core} + H_{gap} L_{gap}$$

$$NI = B\left(\frac{L_{core}}{\mu} + \frac{L_{gap}}{\mu_0} \right) \tag{1}$$

where $\mu = B/H$

$\mu_0 = 4\pi (10^{-7})$ **N** · **A**$^{-2}$ is the permeability of free space (or air); note that **A** in this definition is amperes.

This is a non-linear equation, because the permeability of the core, μ, varies with the magnetic field B. For an exact solution, the value of μ at the B value used must be obtained from the core material hysteresis curve. If B is unknown, the equation must be solved by numerical methods. However, if the magnetomotive force is well above saturation, so the core material is in saturation, the magnetic field will be approximately the saturation value B_{sat} for the material, and won't vary much with changes in NI. For a closed magnetic circuit (no air gap) most core materials saturate at a magnetomotive force of roughly 800 ampere-turns per meter of flux path.

For most core materials, $\mu_r \approx \mu/\mu_0 = 2000 - 6000$. So in equation (1) above, the second term dominates. Therefore, in magnetic circuits with an air gap, the strength of the magnetic field B depends strongly on the length of the air gap, and the length of the flux path in the core doesn't matter much.

Force Exerted by Magnetic Field

The force exerted by an electromagnet on a section of core material is :

$$F = \frac{B^2 A}{2\mu_0} \tag{2}$$

The 1.6 T limit on the field mentioned above sets a limit on the maximum force per unit core area, or pressure, an iron-core electromagnet can exert; roughly :

$$\frac{F}{A} = \frac{B_{sat}^2}{2\mu_0} \approx 1000 \text{ kPa} = 10^6 \text{ N/m}^2 = 145 \text{ ibf} \cdot \text{in}^{-2}$$

In more intuitive units it's useful to remember that at 1T the magnetic pressure is approximately 4 atmospheres, or kg/cm^2.

Given a core geometry, the B field needed for a given force can be calculated from (2); if it comes out to much more than 1.6 T, a larger core must be used.

Closed Magnetic Circuit

For a closed magnetic circuit (no air gap), such as would be found in an electromagnet lifting a piece of iron bridged across its poles, equation (1) becomes :

$$B = \frac{NI\mu}{L} \tag{3}$$

Substituting into (2), the force is :

$$F = \frac{\mu^2 N^2 I^2 A}{2\mu_0 L^2} \tag{4}$$

It can be seen that to maximize the force, a core with a short flux path L and a wide cross-sectional area A is preferred. To achieve this, in applications like lifting magnets and loudspeakers (although loudspeaker magnets need an airgap) a flat cylindrical design is often used. The winding is wrapped around a short wide cylindrical core that forms one pole, and a thick metal housing that wraps around the outside of the windings forms the other part of the magnetic circuit, bringing the magnetic field to the front to form the other pole.

Force Between Electromagnets

The above methods are inapplicable when most of the magnetic field path is outside the core. For electromagnets (or permanent magnets) with well defined 'poles' where the field lines emerge from the core, the force between two electromagnets can be found using the 'Gilbert model' which assumes the magnetic field is produced by fictitious 'magnetic charges' on the surface of the poles, with pole strength m and units of Ampere-turn meter. Magnetic pole strength of electromagnets can be found from :

$$m = \frac{NIA}{L}$$

The force between two poles is :

$$F = \frac{\mu_0 m_1 m_2}{4\pi r^2}$$

This model doesn't give the correct magnetic field inside the core, and thus gives incorrect results if the pole of one magnet gets too close to another magnet.

Side Effects in Large Electromagnets

There are several side effects which become important in large electromagnets and must be provided for in their design :

Ohmic Heating

The only power consumed in a DC electromagnet is due to the resistance of the windings, and is dissipated as heat. Some large electromagnets require cooling water circulating through pipes in the windings to carry off the waste heat.

Since the magnetic field is proportional to the product NI, the number of turns in the windings N and the current I can be chosen to minimize heat losses, as long as their product is constant. Since the power dissipation, $P = I^2R$, increases with the square of the current but only increases approximately linearly with the number of windings, the power lost in the windings can be minimized by reducing I and increasing the number of turns N proportionally, or using thicker wire to reduce the resistance. For example halving I and doubling N halves the power loss, as does doubling the area of the wire. In either case, increasing the amount of wire reduces the ohmic losses. For this reason, electromagnets often have a significant thickness of windings.

However, the limit to increasing N or lowering the resistance is that the windings take up more room between the magnet's core pieces. If the area available for the windings is filled up, more turns require going to a smaller diameter of wire, which has higher resistance, which cancels the advantage of using more turns. So in large magnets there is a minimum amount of heat loss that can't be reduced. This increases with the square of the magnetic flux B^2.

Inductive Voltage Spikes

An electromagnet is a large inductor, and resists changes in the current through its windings. Any sudden changes in the winding current cause large voltage spikes across the windings. This is because when the current through the magnet is increased, such as when it is turned on, energy from the circuit must be stored in the magnetic field. When it is turned off the energy in the field is returned to the circuit.

If an ordinary switch is used to control the winding current, this can cause sparks at the terminals of the switch. This doesn't occur when the magnet is switched on, because the voltage is limited to the power supply voltage. But when it is switched off, the energy in the magnetic field is suddenly returned to the circuit, causing a large voltage spike and an arc across the switch contacts, which can damage them. With small electromagnets a capacitor is often used across the contacts, which reduces arcing by temporarily storing the current. More often a diode is used to prevent voltage spikes by providing a path for the current to recirculate through the winding until the energy is dissipated as heat. The diode is connected across the winding, oriented so it is reverse-biased during steady state operation and doesn't conduct. When the supply voltage is removed, the voltage spike forward-biases the diode and the reactive current continues to flow through the winding, through the diode and back into the winding. A diode used in this way is called a flyback diode.

Large electromagnets are usually powered by variable current electronic power supplies, controlled by a microprocessor, which prevent voltage spikes by accomplishing current changes slowly, in gentle ramps. It may take several minutes to energize or deenergize a large magnet.

Lorentz Forces

In powerful electromagnets, the magnetic field exerts a force on each turn of the windings, due to the Lorentz force $q\mathbf{v} \times \mathbf{B}$ acting on the moving charges within the wire. The Lorentz force is perpendicular to both the axis of the wire and the magnetic field. It can be visualized as a pressure between the magnetic field lines, pushing them apart. It has two effects on an electromagnet's windings :

- The field lines within the axis of the coil exert a radial force on each turn of the windings, tending to push them outward in all directions. This causes a tensile stress in the wire.
- The leakage field lines between each turn of the coil exert a repulsive force between adjacent turns, tending to push them apart.

The Lorentz forces increase with B^2. In large electromagnets the windings must be firmly clamped in place, to prevent motion on power-up and power-down from causing metal fatigue in the windings. In the Bitter design, below, used in very high field research magnets, the windings are constructed as flat disks to resist the radial forces, and clamped in an axial direction to resist the axial ones.

Core Losses

In alternating current (AC) electromagnets, used in transformers, inductors, and AC motors and generators, the magnetic field is constantly changing. This causes energy losses in their magnetic cores that are dissipated as heat in the core. The losses stem from two processes :

- *Eddy currents* : From Faraday's law of induction, the changing magnetic field induces circulating electric currents inside nearby conductors, called eddy currents. The energy in these currents is dissipated as heat in the electrical resistance of the conductor, so they are a cause of energy loss. Since the magnet's iron core is conductive, and most of the magnetic field is concentrated there, eddy currents in the core are the major problem. Eddy currents are closed loops of current that flow in planes perpendicular to the magnetic field. The energy dissipated is proportional to the area enclosed by the loop. To prevent them, the cores of AC electromagnets are made of stacks of thin steel sheets, or laminations, oriented parallel to the magnetic field, with an insulating coating on the surface. The insulation layers prevent eddy current from flowing between the sheets. Any remaining eddy currents must flow within the cross-section of each individual lamination, which reduces losses greatly. Another alternative is to use a ferrite core, which is a non-conductor.

- *Hysteresis losses* : Reversing the direction of magnetization of the magnetic domains in the core material each cycle causes energy loss, because of the coercivity of the material. These losses are called hysteresis. The energy lost per cycle is proportional to the area of the hysteresis loop in the *BH* graph. To minimize this loss, magnetic cores used in transformers and other AC electromagnets are made of "soft" low coercivity materials, such as silicon steel or soft ferrite.

The energy loss per cycle of the AC current is constant for each of these processes, so the power loss increases linearly with frequency.

High Field Electromagnets

Superconducting Electromagnets

When a magnetic field higher than the ferromagnetic limit of 1.6 T is needed, superconducting electromagnets can be used. Instead of using ferromagnetic materials, these use superconducting windings cooled with liquid helium, which conduct current without electrical resistance. These allow enormous currents to flow, which generate intense magnetic fields. Superconducting magnets are limited by the field strength at which the winding material ceases to be superconducting. Current designs are limited to 10–20 T, with the current (2009) record of 33.8 T. The necessary refrigeration equipment and cryostat make them much more expensive than ordinary electromagnets. However, in high power applications this can be offset by lower operating costs, since after startup no power is required for the windings, since no energy is lost to ohmic heating. They are used in particle accelerators, MRI machines, and research.

Bitter Electromagnets

Both iron-core and superconducting electromagnets have limits to the field they can produce. Therefore the most powerful man-made magnetic fields have been generated by *air-core* nonsuperconducting electromagnets of a design invented by Francis Bitter in 1933, called Bitter electromagnets. Instead of wire windings, a Bitter magnet consists of a solenoid made of a stack of conducting disks, arranged so that the current moves in a helical path through them, with a hole through the center where the maximum field is created. This design has the mechanical strength to withstand the extreme Lorentz forces of the field, which increase with B^2. The disks are pierced with holes through which cooling water passes to carry away the heat caused by the high current. The strongest continuous field achieved with a resistive magnet is currently (2008) 35 T, produced by a Bitter electromagnet. The strongest continuous magnetic field, 45 T, was achieved with a hybrid device consisting of a Bitter magnet inside a superconducting magnet.

Fig. : The most powerful electromagnet in the world, the 45 T hybrid Bitter-superconducting magnet at the US National High Magnetic Field Laboratory, Tallahassee, Florida, USA.

Exploding Electromagnets

The factor limiting the strength of electromagnets is the inability to dissipate the enormous waste heat, so more powerful fields, up to 100 T, have been obtained from resistive magnets by sending brief pulses of current through them. The most powerful manmade magnetic fields have been created by using explosives to compress the magnetic field inside an electromagnet as it is pulsed. The implosion compresses the magnetic field to values of around 1000 T for a few microseconds. While this method may seem very destructive there are methods to control the blast so that neither the experiment nor the magnetic structure are harmed, by redirecting the brunt of the force radially outwards. These devices are known as destructive pulsed electromagnets. They are used in physics and materials science research to study the properties of materials at high magnetic fields.

Definition of Terms

A	square meter	cross-sectional area of core
B	tesla	Magnetic field (Magnetic flux density)
F	newton	Force exerted by magnetic field
H	ampere per meter	Magnetizing field
I	ampere	Current in the winding wire
L	meter	Total length of the magnetic field path $L_{core} + L_{gap}$
L_{core}	meter	Length of the magnetic field path in the core material
L_{gap}	meter	Length of the magnetic field path in air gaps
m_1, m_2	ampere meter	Pole strength of the electromagnet
μ	newton per square ampere	Permeability of the electromagnet core material
μ_0	newton per square ampere	Permeability of free space (or air) = $4\pi(10^{-7})$
μ_r	-	Relative permeability of the electromagnet core material
N	-	Number of turns of wire on the electromagnet
r	meter	Distance between the poles of two electromagnets

RELATIVISTIC ELECTROMAGNETISM

Relativistic electromagnetism is a modern teaching strategy for developing electromagnetic field theory from Coulomb's law and Lorentz transformations. Though Coulomb's law expresses action at a distance, it is an easily understood *electric force* principle. The more sophisticated view of electromagnetism expressed by electromagnetic fields in spacetime can be approached by applying space time symmetries. In certain special configurations it is possible to exhibit magnetic effects due to relative charge density in various simultaneous hyperplanes. This approach to physics education and the education and training of electrical and electronics engineers can be seen in the Encyclopædia Britannica (1956), The Feynman Lectures on Physics (1964), Edward M. Purcell (1965), Jack R. Tessman (1966), W.G.V. Rosser (1968), Anthony French (1968), and Dale R. Corson & Paul Lorrain (1970). This approach provides some preparation for understanding of magnetic forces involved in the Biot–Savart law, Ampère's law, and Maxwell's equations.

Einstein's Motivation

In 1953 Albert Einstein wrote to the Cleveland Physics Society on the occasion of a commemoration of the Michelson–Morley experiment. In that letter he wrote :

> What led me more or less directly to the special theory of relativity was the conviction that the electromotive force acting on a body in motion in a magnetic field was nothing else but an electric field.

This statement by Einstein reveals that he investigated spacetime symmetries to determine the complementarity of electric and magnetic forces.

Introduction

Purcell argued that the question of an electric field in one inertial frame of reference, and how it looks from a different reference frame moving with respect to the first, is *crucial* to understand fields created by moving sources. In the special case, the sources that create the field are at rest with respect to one of the reference frames. Given the electric field in the frame where the sources are at rest, Purcell asked : what is the electric field in some other frame?

He stated that the fundamental assumption is that, knowing the electric field at some point (in space and time) in the *rest* frame of the sources, and knowing the relative velocity of the two frames provided all the information needed to calculate the electric field at the same point in the other frame. In other words, the electric field in the other frame does not depend on the particular distribution of the source charges, only on the local value of the electric field in the first frame at that point. He assumed that the electric field is a **complete** representation of the influence of the far-away charges.

Alternatively, introductory treatments of magnetism introduce the Biot–Savart law, which describes the magnetic field associated with an electric current. An observer at rest with respect to a system of static, free charges will see no magnetic field. However, a moving observer looking at the same set of charges does perceive a current, and thus a magnetic field.

Uniform Electric Field — Simple Analysis

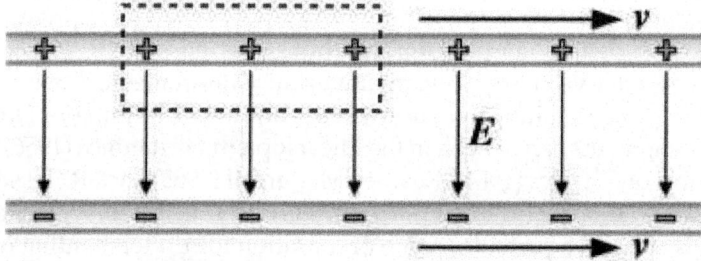

Fig. : Two oppositely charged plates produce uniform electric field even when moving. The electric field is shown as 'flowing' from top to bottom plate. The Gaussian pill box (at rest) can be used to find the strength of the field.

Consider the very simple situation of a charged parallel-plate capacitor, whose electric field (in its rest frame) is uniform (neglecting edge effects) between the plates and zero outside.

To calculate the electric field of this charge distribution in a reference frame where it is in motion, suppose that the motion is in a direction parallel to the plates as shown in figure. The plates will then be shorter by a factor of :

$$\sqrt{1 - v^2 / c^2}$$

than they are in their rest frame, but the distance between them will be the same. Since charge is independent of the frame in which it is measured, the total charge on each plate is also the same. So the charge per unit area on the plates is therefore larger than in the rest frame by a factor of :

$$\frac{1}{\sqrt{1 - v^2 / c^2}}$$

The field between the plates is therefore stronger by this factor.

More Rigorous Analysis

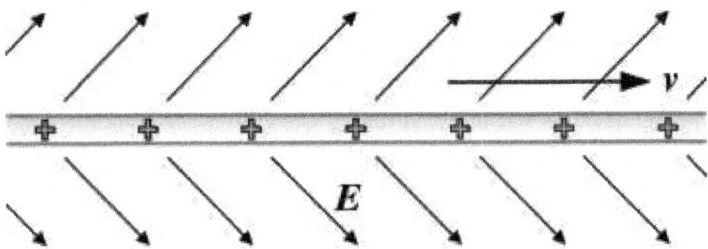

Fig. a : The electric field lines are shown flowing outward from the positive plate.

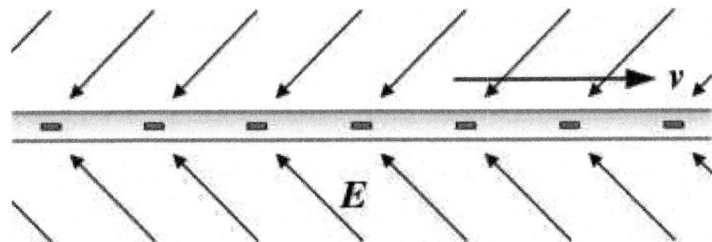

Fig. b : The electric field lines flow inward toward the negative plate.

Consider the electric field of a single, infinite plate of positive charge, moving parallel to itself. The field must be uniform both above and below the plate, since it is uniform in its rest frame. We also assume that knowing the field in one frame is sufficient for calculating it in the other frame.

The plate however *could* have a non-zero component of electric field in the direction of motion as in Fig a. Even in this case, the field of the infinite plane of negative charge must be equal and opposite to that of the positive plate (as in Fig b), since the combination of plates is neutral and cannot therefore produce any net fields. When the plates are separated, the horizontal components still cancel, and the resultant is a uniform vertical field as shown in Fig.

If Gauss's law is applied to pillbox as shown in Fig, it can be shown that the magnitude of the electric field between the plates is given by :

$$|E'| = \frac{\sigma'}{\epsilon_0}$$

where the prime (') indicates the value measured in the frame in which the plates are moving. σ represents the surface charge density of the positive plate. Since the plates are contracted in length by the factor

$$\sqrt{1 - v^2 / c^2}$$

then the surface charge density in the primed frame is related to the value in the rest frame of the plates by :

$$\sigma' = \frac{\sigma}{\sqrt{1 - v^2 / c^2}}$$

But the electric field in the rest frame has value σ / ϵ_0 and the field points in the same direction on both of the frames, so

$$E' = \frac{E}{\sqrt{1 - v^2 / c^2}}$$

The E field in the primed frame is therefore stronger than in the unprimed frame. If the direction of motion is perpendicular to the *plates*, length contraction of the plates does not occur, but the distance between them is reduced. This closer spacing however does not affect the strength of the electric field. So for motion parallel to the electric field E,

$$E' = E$$

In the general case where motion is in a diagonal direction relative to the field the field is merely a superposition of the perpendicular and parallel fields, each generated by a set of plates at right angles to each other as shown in Fig. Since both sets of plates are length contracted, the two components of the E field are

$$E'_y = \frac{E_y}{\sqrt{1 - v^2 / c^2}}$$

and

$$E'_x = E_x$$

where the y subscript denotes perpendicular, and the x subscript, parallel.

These transformation equations only apply if the source of the field is at rest in the unprimed frame.

The Field of a Moving Point Charge

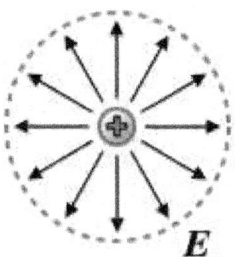

Fig. : A point charge at rest, surrounded by an imaginary sphere.

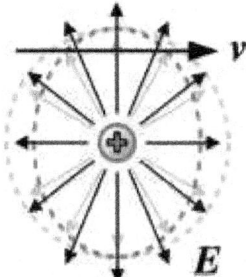

Fig. : A view of the electric field of a point charge moving at constant velocity.

A very important application of the electric field transformation equations is to the field of a single point charge moving with constant velocity. In its rest frame, the electric field of a positive point charge has the same strength in all directions and points directly away from the charge. In some other reference frame the field will appear differently.

In applying the transformation equations to a non-uniform electric field, it is important to record not only the value of the field, but also at what point in space it has this value.

In the rest frame of the particle, the point charge can be imagined to be surrounded by a spherical shell which is also at rest. In *our* reference frame, however, both the particle and its sphere are moving.

Consider the value of the electric field at any point on the surface of the sphere. Let x and y be the components of the displacement (in the rest frame of the charge), from the charge to a point on the sphere, measured parallel and perpendicular to the direction of motion as shown in the figure. Because the field in the rest frame of the charge points directly away from the charge, its components are in the same ratio as the components of the displacement :

$$\frac{E_y}{E_x} = \frac{y}{x}$$

In our reference frame, where the charge is moving, the displacement x' in the direction of motion is length-contracted :

$$x' = x\sqrt{1 - v^2 / c^2}$$

The electric field at any point on the sphere points directly away from the charge. (b) In a reference frame where the charge and the sphere are moving to the right, the sphere is length-contracted but the vertical component of the field is stronger. These two effects combine to make the field again point directly away from the current location of the charge. (While the y component of the displacement is the same in both frames).

However, according to the above results, the y component of the field is enhanced by a similar factor :

$$E'_y = \frac{E_y}{\sqrt{1 - v^2 / c^2}}$$

whilst the x component of the field is the same in both frames. The ratio of the field components is therefore

$$\frac{E'_y}{E'_x} = \frac{E_y}{E_x\sqrt{1 - v^2 / c^2}} = \frac{y'}{x'}$$

So, the field in the primed frame points directly away from the charge, just as in the unprimed frame. A view of the electric field of a point charge moving at constant velocity is shown in figure. The faster the charge is moving, the more noticeable the enhancement of the perpendicular component of the field becomes. If the speed of the charge is much less than the speed of light, this enhancement is often negligible. But under certain circumstances, it is crucially important even at low velocities.

The Origin of Magnetic Forces

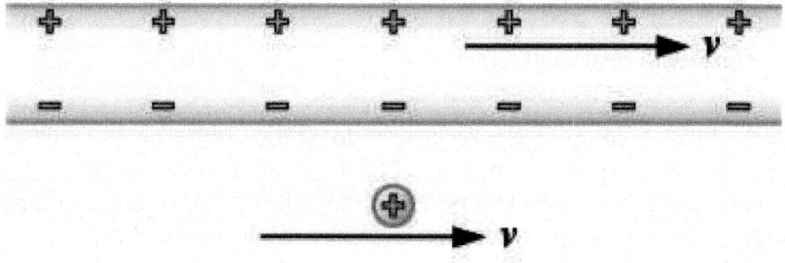

Fig., lab frame : A horizontal wire carrying a current, represented by evenly spaced positive charges moving to the right whilst an equal number of negative charges remain at rest, with a positively charged particle outside the wire and traveling in a direction parallel to the current.

In the simple model of events in a wire stretched out horizontally, a current can be represented by the evenly spaced positive charges, moving to the right, whilst an equal number of negative charges remain at rest. If the wire is electrostatically neutral, the distance between adjacent positive charges must be the same as the distance between adjacent negative charges.

Assume that in our 'lab frame' (Figure), we have a positive test charge, Q, outside the wire, traveling parallel to the current, at the speed, v, which is equal to the speed of the moving charges in the wire. It should experience a magnetic force, which can be easily confirmed by experiment.

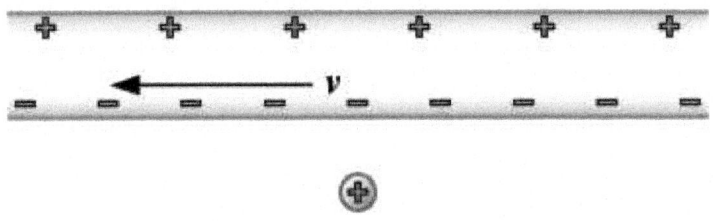

Fig., test charge frame : The same situation as in fig., but viewed from the reference frame in which positive charges are at rest. The negative charges flow to the left. The distance between the negative charges is length-contracted relative to the lab frame, while the distance between the positive charges is expanded, so the wire carries a net negative charge.

Inside 'test charge frame'(Fig.), the only possible force is the electrostatic force $F_e = Q * E$ because, although the magnetic field is the same, the test charge is at rest and, therefore, cannot feel it. In this frame, the negative charge density has Lorentz-contracted with respect to what we had in lab frame because of the increased speed. This means that spacing between charges has reduced by the Lorentz factor with respect to the lab frame spacing, l :

$$l_- = l\sqrt{1 - v^2/c^2}$$

Thus, positive charges have Lorentz-expanded (because their speed has dropped) :

$$l_+ = l/\sqrt{1 - v^2/c^2}$$

Both of these effects combine to give the wire a net negative charge in the test charge frame. Since the negatively charged wire exerts an attractive force on a positively charged particle, the test charge will therefore be attracted and will move toward the wire.

For $v \ll c$, we can concretely compute both, the magnetic force sensed in the lab frame

$$F_m = \frac{Qvl}{2\pi \epsilon_0 c^2 R}$$

and electrostatic force, sensed in the test charge frame, where we first compute the charge density with respect to the lab frame length, l :

$$\lambda = \frac{q}{l_+} - \frac{q}{l_-} = \frac{q}{l}\left(\sqrt{1-v^2/c^2} - 1/\sqrt{1-v^2/c^2}\right) \approx \frac{q}{l}\left(1-0.5\left(\frac{v^2}{c^2}\right)-1-0.5\left(\frac{v^2}{c^2}\right)\right) = -\frac{qv^2}{lc^2}$$

and, keeping in mind that current $I = \frac{q}{t} = q\frac{v}{l}$, resulting electrostatic force

$$F_e = Q_E = Q\frac{\lambda}{2\pi\,\epsilon_0\,R} = \frac{Qqv^2}{2\pi\,\epsilon_0\,c^2 Rl} = \frac{QvI}{2\pi\,\epsilon_0\,c^2 R}$$

which comes out exactly equal to the magnetic force sensed in the lab frame, $F_e = F_m$.

If the currents are in opposite directions, consider the charge moving to the left. No charges are now at rest in the reference frame of the test charge. The negative charges are moving with speed v in the test charge frame so their spacing is again :

$$l_{(-)} = l\sqrt{1-v^2/c^2}$$

The distance between positive charges is more difficult to calculate. The relative velocity should be less than $2v$ due to special relativity. For simplicity, assume it is $2v$. The positive charge spacing contraction is then :

$$\sqrt{1-(2v/c)^2}$$

relative to its value in their rest frame. Now its value in their rest frame was found to be

$$l_{(+)} = \frac{l}{\sqrt{1-v^2/c^2}}$$

So the final spacing of positive charges is :

$$l_{(+)} = \frac{l}{\sqrt{1-v^2/c^2}}\sqrt{1-(2v/c)^2}$$

To determine whether $l_{(+)}$ or $l_{(-)}$ is larger we assume that $v \ll c$ and use the binomial approximation that

$$(1+x)^p \approx 1 + px \quad \text{when } x \ll 1$$

After some algebraic calculation it is found that $l_{(+)} < l_{(-)}$, and so the wire is positively charged in the frame of the test charge.

One may think that the picture, presented here, is artificial because electrons, which accelerated in fact, must condense in the lab frame, making the wire charged. Naturally, however, all electrons feel the same accelerating force and, therefore, identically to the Bell's spaceships, the distance between them does not change in the lab frame (*i.e.* expands in their proper moving frame). Rigid bodies, like trains, don't expand however in their proper frame, and, therefore, really contract, when observed from the stationary frame.

Calculating the Magnetic Field

The Lorentz Force Law

A moving test charge near a wire carrying current will experience a magnetic force dependent on the velocity of the moving charges in the wire. If the current is flowing to the right, and a positive test charge is moving below the wire, then there is a force in a direction 90° counterclockwise from the direction of motion.

The Magnetic Field of a Wire

Calculation of the magnitude of the force exerted by a current-carrying wire on a moving charge is equivalent to calculating the magnetic field produced by the wire. Consider again the situation shown in figures. The latter figure, showing the situation in the reference frame of the test charge, is reproduced in the figure. The positive charges in the wire, each with charge q, are at rest in this frame, while the negative charges, each with charge −q, are moving to the left with speed v. The average distance between the negative charges in this frame is length-contracted to :

$$\sqrt{1 - v^2 / c^2}$$

where is the distance between them in the lab frame. Similarly, the distance between the positive charges is not length-contracted :

$$\sqrt{1 - v^2 / c^2}$$

Both of these effects give the wire a net negative charge in the test charge frame, so that it exerts an attractive force on the test charge.

Chapter 10

RESEARCH ON THE FAULT COEFFICIENT IN COMPLEX ELECTRICAL ENGINEERING

Yi Sun[1,†], Yagang Zhang [2,3,†,]* and Yinding Wang [3,†]

[1]   Hebei Electric Power Research Institute, Shijiazhuang, Hebei 050022, China; E-Mail: yisunsjz@163.com

[2]   State Key Laboratory of Alternate Electrical Power System with Renewable Energy Sources, North China Electric Power University, Beijing 102206, China

[3]   Interdisciplinary Mathematics Institute, University of South Carolina, Columbia, SC 29208, USA; E-Mail: dnagct@gmail.com

†   These authors contributed equally to this work.

*   Author to whom correspondence should be addressed; E-Mail: yagangzhang@ncepu.edu.cn; Tel.: +1-803-7771-731.

Academic Editor: Hung-Yu Wang

ABSTRACT

Fault detection and isolation in a complex system are research hotspots and frontier problems in the reliability engineering field. Fault identification can be regarded as a procedure of excavating key characteristics from massive failure data, then classifying and identifying fault samples. In this paper, based on the fundamental of feature extraction about the fault coefficient, we will discuss the fault coefficient feature in complex electrical engineering in detail. For general fault types in a complex power system, even if there is a strong white Gaussian stochastic interference, the fault coefficient feature is still accurate and reliable. The results about comparative analysis of noise influence will also demonstrate the strong anti-interference ability and great redundancy of the fault coefficient feature in complex electrical engineering.

Keywords

Fault coefficient; Gaussian interference; BPA; noise influence; PMU

1. INTRODUCTION

Fault detection is always one of the core problems in the complex electrical engineering field. Generally speaking, fault detection can be divided into three categories based on analytical models, signal processing, and knowledge [1-5]. Among these fault detection technologies, the analytical models based fault detection technology was the earliest development and the most systematical research. It can construct a system model based mainly on the connections between components in a composition system and can be roughly divided into state estimation method, parameter estimation method and equivalence space method [6-8]. The fault detection based on signal processing can avoid establishing an object's mathematical model, and it can directly analyze measurable information by signal model, such as correlation function, frequency spectrum, high order statistics, autoregressive moving average process and so on. The fault can be detected by extracting amplitude, variance, frequency and other characteristic values [9,10]. The fault detection based on signal processing mainly includes principle component analysis method, absolute value testing, tendency testing, Kullback detection, self-adaptable filter detection, *etc.* Similar to signal processing based fault detection, the fault detection based on knowledge is mainly applied to a nonlinear system. At the knowledge level, on the basis of knowledge processing technologies, the dialectical logic and mathematical logic will be integrated, and the symbol processing and numerical processing will be unified. It will also mainly include an expert system, fuzzy reasoning, neural network and rough set, *etc.* [11-14].

In fact, whichever fault detection technology we will adopt, the effective extraction of the fault feature is always the key problem. According to complex electrical engineering, we have carried out large numbers of basic research [15-18]. In this paper, based on the fundamental of feature extraction about the fault coefficient, we will discuss the fault coefficient in complex electrical engineering.

The paper is organized as follows. In Section 2, the fundamental of feature extraction about fault coefficient are introduced. In Section 3, the Phasor Measurement Unit and Wide Area Measurement System are described, and they have provided a real-time data platform for improving the security and stabilization of power system. In Section 4, for general fault types in complex electrical engineering, the feature extraction process of fault coefficient is clarified in detail. Finally, the paper is concluded in Section 5.

2. THE FUNDAMENTAL OF FEATURE EXTRACTION ABOUT THE FAULT COEFFICIENT

For general, fault factor model [19,20],

$$x_i = \mu_i + a_{i1}f_1 + a_{i2}f_2 + \dots + a_{im}f_m + \varepsilon_i,$$

$$(i = 1,2,\dots,n) \tag{1}$$

where $f_1, f_2, \dots, f_m$ are m fault factors (common factor), a_{ij} is the loading of x_i on f_j, μ_i is mean value of x_i, and ε_i is specific factor of x_i.

If x is a random vector that each component has been standardized, then the correlation coefficient between x_i and f_j is

$$\rho(x_i, f_j) = \frac{Cov(x_i, f_j)}{\sqrt{V(x_i)}\sqrt{V(f_j)}} = Cov(x_i, f_j) = a_{ij} \tag{2}$$

At this point, a_{ij} is just the correlation coefficient between x_i and f_j. Calculating variance on both sides of the fault factor model,

$$V(x_i) = a_{i1}^2 V(f_1) + a_{i2}^2 V(f_2) + \cdots + a_{im}^2 V(f_m) + V(\varepsilon_i)$$

$$= a_{i1}^2 + a_{i2}^2 + \cdots + a_{im}^2 + \sigma_i^2 \tag{3}$$

$$(i = 1, 2, \cdots, p)$$

let

$$h_i^2 = \sum_{j=1}^{m} a_{ij}^2, \quad (i = 1, 2, \cdots, p) \tag{4}$$

so

$$\sigma_{ii} = h_i^{\,2} + \sigma_i^2, \ (i = 1, 2, \ldots, p) \tag{5}$$

h_i^2 is communality of $f_1, f_2, \ldots, f_m$ on x, and σ_i^2 is specific variance of ε_i on x_i.

For $V(x_i) = a_{i1}^2 V(f_1) + a_{i2}^2 V(f_2) + \cdots + a_{im}^2 V(f_m) + V(\varepsilon_i)$, then

$$\sum_{i=1}^{p} V(x_i) = \sum_{i=1}^{p} a_{i1}^2 V(f_1) + \sum_{i=1}^{p} a_{i2}^2 V(f_2) + \cdots + \sum_{i=1}^{p} a_{im}^2 V(f_m) + \sum_{i=1}^{p} V(\varepsilon_i)$$

$$= g_1^2 + g_2^2 + \cdots + g_m^2 + \sum_{i=1}^{p} \sigma_i^2 \tag{6}$$

where

$$g_j^2 = \sum_{i=1}^{p} a_{ij}^2, \quad (j = 1, 2, \cdots, m) \tag{7}$$

g_j^2 can be considered to be the total variance contribution of f_j on $x_1, x_2, \ldots, x_p$.

Suppose $x_1, x_2, \ldots, x_n$ are a group of p-dimensional samples, then the estimations of the sample mean value μ and the sample covariance matrix Σ are respectively,

$$\bar{x} = \frac{1}{n} \sum_{i=1}^{n} x_i$$

$$S = \frac{1}{n-1} \sum_{i=1}^{n} (x_i - \bar{x})(x_i - \bar{x})' \tag{8}$$

In order to construct a fault factor model, one needs to estimate factor loading matrix $A = (a_{ij})_{p \times m}$ and special variance matrix $D = diag\,(\sigma_1^2, \sigma_2^2, \ldots, \sigma_p^2)$. The parameter estimation methods most commonly used include: principal component method, principal factor method and the maximum likelihood method. Suppose the characteristic values of sample covariance matrix S are $\hat{\lambda}_1 \geq \hat{\lambda}_2 \geq \cdots \geq \hat{\lambda}_p \geq 0$ in turn, and the corresponding orthogonal unit characteristic vectors are $\hat{i}_1, \hat{i}_2, \cdots, \hat{i}_p$. One should select relatively small m, which can make the cumulative variance contri-

bution rate $\dfrac{\sum\limits_{i=1}^{m}\hat{\lambda}_i}{\sum\limits_{i=1}^{p}\hat{\lambda}_i}$ reach a higher percentage, then S can be approximately decom-

posed as follows,

$$S = \hat{\lambda}_1\hat{t}_1\hat{t}_1' + \cdots + \hat{\lambda}_m\hat{t}_m\hat{t}_m' + \hat{\lambda}_{m+1}\hat{t}_{m+1}\hat{t}_{m+1}' + \cdots + \hat{\lambda}_p\hat{t}_p\hat{t}_p'$$
$$\approx \hat{\lambda}_1\hat{t}_1\hat{t}_1' + \cdots + \hat{\lambda}_m\hat{t}_m\hat{t}_m' + \hat{D} \qquad (9)$$
$$= \hat{A}\hat{A}' + \hat{D}$$

where $\hat{A} = (\sqrt{\hat{\lambda}_1}\hat{t}_1, \cdots, \sqrt{\hat{\lambda}_m}\hat{t}_m) = (\hat{a}_{ij})_{p \times m}$, $\hat{D} = diag(\hat{\sigma}_1^2, \hat{\sigma}_2^2, \cdots, \hat{\sigma}_p^2)$.

Suppose each component of original vector x have been standardized, if random vector x satisfies fault factor model, then

$$R = AA' + D \qquad (10)$$

where R is the correlation matrix of x. Let

$$R^* = R - D = AA' \qquad (11)$$

then R^* is called reduced correlation matrix of x. In fact, R^* is a nonnegative definite matrix.

Suppose $\hat{\sigma}_i^2$ is an appropriate initial estimation of special variance σ_i^2, then the reduced correlation matrix can be estimated as

$$\hat{R}^* = \hat{R} - \hat{D} = \begin{pmatrix} \hat{h}_1^2 & r_{12} & \cdots & r_{1p} \\ r_{21} & \hat{h}_2^2 & \cdots & r_{2p} \\ \vdots & \vdots & \vdots & \vdots \\ r_{p1} & r_{p2} & \cdots & \hat{h}_p^2 \end{pmatrix} \qquad (12)$$

where $\hat{R} = (r_{ij})$, $\hat{D} = diag(\hat{\sigma}_1^2, \hat{\sigma}_2^2, \cdots, \hat{\sigma}_p^2)$. And $\hat{h}_i^2 = 1 - \hat{\sigma}_i^2$ of is the initial estimation of h_i^2. Suppose the preceding m characteristic values of $\hat{R}^*$ are $\hat{\lambda}_1^* \geq \hat{\lambda}_2^* \geq \cdots \geq \hat{\lambda}_m^* \geq 0$ in turn, corresponding orthogonal unit characteristic vectors are $\hat{t}_1^*, \hat{t}_2^*, \cdots, \hat{t}_m^*$, then the principal factor solution of A is

$$\hat{A} = (\sqrt{\hat{\lambda}_1^*}\hat{t}_1^*, \sqrt{\hat{\lambda}_2^*}\hat{t}_2^*, \cdots, \sqrt{\hat{\lambda}_m^*}\hat{t}_m^*) \qquad (13)$$

Therefore the ultimate estimation of σ_i^2 can be expressed as

$$\hat{\sigma}_i^2 = 1 - \hat{h}_i^2 = 1 - \sum_{j=1}^{m} \hat{a}_{ij}^2, \quad (i = 1, 2, \cdots, p) \qquad (14)$$

3. PHASOR MEASUREMENT UNIT AND WIDE AREA MEASUREMENT SYSTEM

With the construction of long distance heavily stressed transmission system and the actualization of large scale interconnected power grids, the safety and stability problems of modern power system has become acute, which has also put forward a high requirement for the safety monitoring technology of power network. Based on mature Global Positioning System (GPS) technique and communication technique, Phasor Measurement Unit (PMU) has high stability and reliability, high precision, strong processing, calculating, memorizing and communication capabilities, friendly man-machine interface and openness, it has provided the foundations for dynamic monitoring in electric power system. Under unified time scales, PMU can directly afford the variation curve of voltage, current, phase angle, power, *etc.* in transient process of power grid, which has also created the conditions for state estimation, on-line security monitoring and dynamic supervising of power grid. On the basis of PMU, Wide Area Measurement System (WAMS) can reflect the dynamic changes of the whole power network in real time. As a new technology and important means which can realize real-time dynamic monitoring in power grid, WAMS has provided a real-time data platform for improving the security and stabilization of power system.

Table 1. The primary technical specifications of Phasor Measurement Unit.

Items	Technical Specifications
Sampling for analog input	Sampling frequency 4800 Hz, at least 36 channels ,extensible by multiple
GPS timing accuracy	$1 \mu S$
Error limit for angle	0.01 degree
Error limit for amplitude	0.2%(Relative error)
Error limit for power	0.5%(Relative error)
Error limit for frequency	0.001 Hz, measurement range 45–55 Hz
Dynamic data retained time	25 frames/sec, 50 frames/sec, 100 frames/sec
Dynamic data retained time	≥ 14 days
Dynamic data output rate	25 frames/sec, 50 frames/sec, 100 frames/sec
Time range for fault recording	-5 sec~$+15$ sec

As to the technical specification of the PMUs, a list of the items in one commercial product of power system real-time dynamic monitoring system is taken as an example and shown in Table 1. The relevant technical details of Wide Area Measurement System and Phasor Measurement Unit can refer to [15,21].

In the research of this paper, the data sources are from Wide Area Measurement System, and they can be gathered by Phasor Measurement Unit, of course, the whole process can be completed by Bonneville Power Administration (BPA) simulations. Thus, the real-time property can be guaranteed by Phasor Measurement Unit and Wide Area Measurement System. Besides, the feature extraction technology of fault coefficient advanced in this paper can be able to extract accurate fault characteristics with high efficiency, and it has strong anti-interference ability and great redundancy.

4. FAULT COEFFICIENT FEATURE EXTRACTION IN COMPLEX ELECTRICAL ENGINEERING

In order to illustrate the superior feature extraction abilities of fault coefficient feature extraction fundamental, IEEE 39-BUS New England power system will be an experimental subject. The electric diagram of IEEE 39-BUS New England power system is presented in Figure 1. In the network architecture, failures will be discussed in detail. In particular, different levels of Gaussian stochastic interference will be introduced. The anti-interference capability of the fault coefficient feature extraction technology will be deeply analyzed.

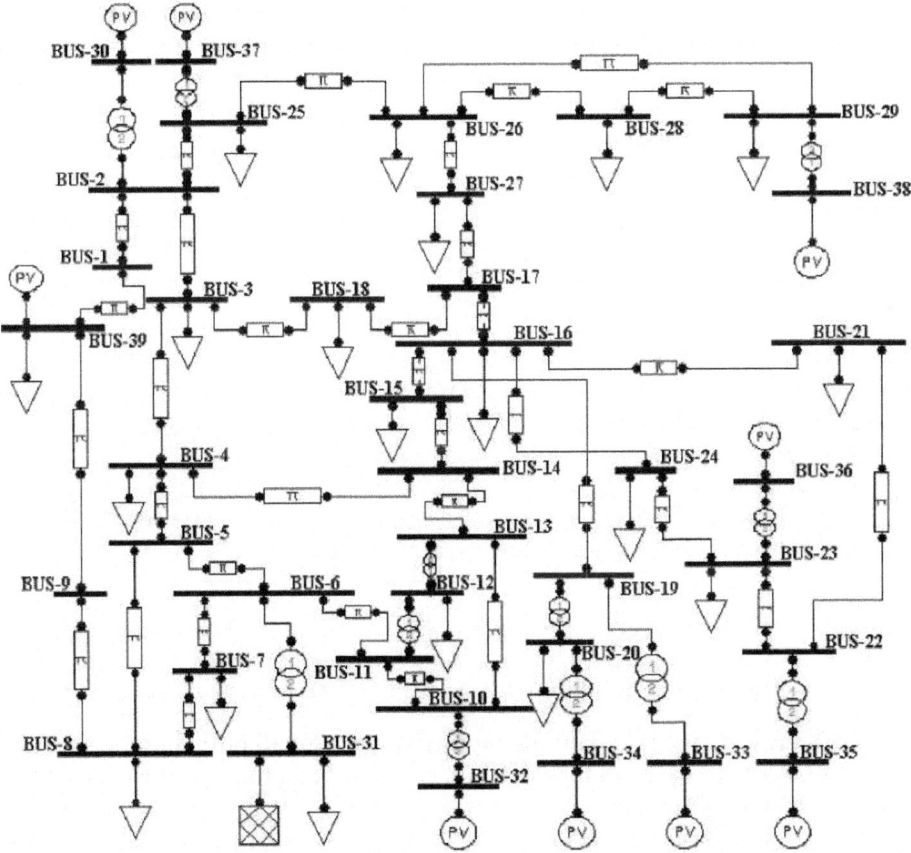

Figure 1. IEEE 39-BUS New England power system.

4.1. Fault Coefficient Feature Extraction in Asymmertical Short Circuit Fault

In the IEEE 39-BUS New England power system, an asymmertical short circuit fault breaks out suddenly, and BUS-18 is the actual fault location. By BPA simulations, one can get the node negative sequence voltages. Meanwhile, one has also introduced a white Gaussian stochastic noise $N(0,0.003^2)$, and carries out detailed analysis about these original electrical information vectors.

   According to the fundamental of feature extraction about the fault coefficient, one can calculate the initial characteristic values, initial characteristic vectors, squared loadings and rotation squared loadings. Among them, the initial characteristic value of the first fault factor is 0.01563622, the variance percentage is 0.9852, and the cumulative variance percentage is also 0.9852. For the second fault factor, the initial characteristic value of the first fault factor is 0.00008786, the variance percentage is 0.0056, and the cumulative variance percentage is 0.9908. For the third fault factor, the initial characteristic value of the first fault factor is 0.00007585, the variance percentage is 0.0048, and the cumulative variance percentage is 0.9956. For the fourth fault factor, the initial characteristic value of the first fault factor is 0.00007056, the variance percentage is 0.0044, and the cumulative variance percentage is 1.0000. Actually, only the first fault factor needs to be extracted in this simulation.

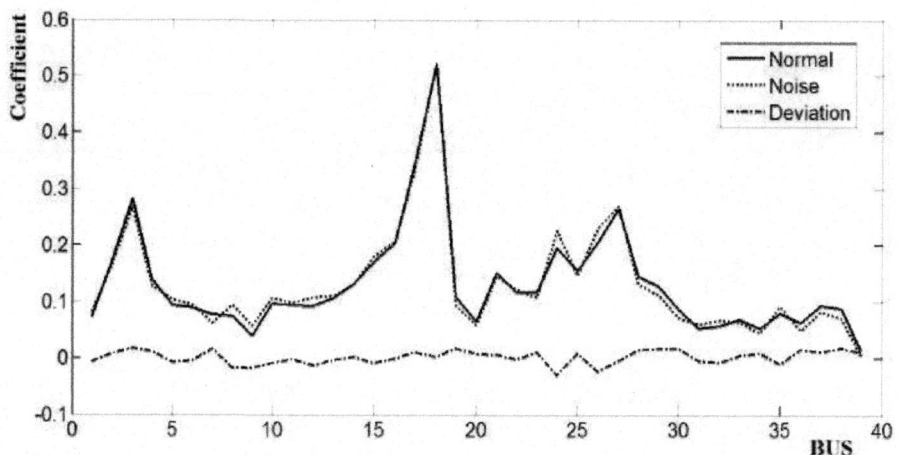

Figure 2. The noise influence in asymmertical short circuit fault.

   Furthermore, we calculate the fault coefficient of all fault factors, as shown in Table 2. Because we have focused on the first fault factor, and the general form of the first fault factor is:

Fault1 = 0.080184BUS1+0.162261BUS2+0.266219BUS3+0.127872BUS4

+ 0.101886BUS5+0.094243BUS6+0.060710BUS7+0.092570BUS8

+0.056463BUS9+0.105765BUS10+0.096482BUS11+0.106159BUS12

+0.109040BUS13+0.128760BUS14+0.178810BUS15+0.204963BUS16

+0.337853BUS17+0.520984BUS18+0.092290BUS19+0.057301BUS20

+0.144469BUS21+0.120420BUS22+0.106459BUS23+0.224937BUS24 (15)

+0.146205BUS25+0.228639BUS26+0.269287BUS27+0.130908BUS28

$$+0.110627BUS29+0.069968BUS30+0.058711BUS31 + 0.065727BUS32$$

$$+0.063605BUS33 + 0.044253BUS34+0.091074BUS35+0.047710BUS36$$

$$+0.082034BUS37+0.069996BUS38 + 0.003890BUS39$$

In the first fault factor, we will concentrate on the coefficient feature. For all of these coefficients in Fault1, the coefficient of BUS-18 is 0.520984, which is the biggest one. Consequently, we come to the conclusion that BUS-18 is just the fault BUS. In the meantime, the expression of the first fault factor without the interference of white Gaussian stochastic noise has also been obtained through the same approach, namely

Fault1' = 0.072173BUS1+0.168284BUS2+0.282617BUS3+0.138272BUS4

$$+ 0.092896BUS5+0.088251BUS6+0.076103BUS7+0.073602BUS8$$

$$+0.037873BUS9+0.095754BUS10+0.093253BUS11+0.091824BUS12$$

$$+0.105044BUS13+0.130054BUS14+0.168642BUS15+0.202942BUS16$$

$$+0.347645BUS17+0.521645BUS18+0.107902BUS19+0.063419BUS20$$

$$+0.150062BUS21+0.116477BUS22+0.115762BUS23+0.193652BUS24 \quad (16)$$

$$+0.153635BUS25+0.204371BUS26+0.263681BUS27+0.144346BUS28$$

$$+0.126838BUS29+0.086465BUS30+0.052879BUS31 + 0.056809BUS32$$

$$+0.067171BUS33 + 0.051807BUS34+0.078961BUS35+0.062526BUS36$$

$$+0.092181BUS37+0.087179BUS38 + 0.012862BUS39$$

In the above expression, the BUS-18 fault certainly has been identified. Let's further discuss the noise influence for fault coefficients shown in Figure 2. The total average deviation level is 0.00041346. Thus, despite the fact that there is stochastic influence of white Gaussian noise, the fault coefficient feature is still distinct.

Table 2. The fault coefficient in asymmertical short circuit fault.

BUS	1	2	3	4	5
BUS-1	0.080184	-0.131590	-0.101984	0.080719	-0.003175
BUS-2	0.162261	0.083017	-0.250354	-0.026744	-0.013902
BUS-3	0.266219	0.001688	-0.118200	-0.064936	-0.015069
BUS-4	0.127872	-0.312857	0.106917	-0.028192	0.007209
BUS-5	0.101886	-0.139861	0.059463	-0.079273	0.003251
BUS-6	0.094243	-0.105051	-0.019010	0.066930	-0.004484
BUS-7	0.060710	0.063427	-0.020301	0.092355	-0.011011
BUS-8	0.092570	0.178899	0.444781	-0.033685	-0.011589
BUS-9	0.056463	0.241210	-0.003291	0.021823	-0.015598
BUS-10	0.105765	-0.019603	0.252826	-0.001555	-0.005394

(Contd...)

BUS-11	0.096482	0.078763	-0.153092	0.514546	-0.033082
BUS-12	0.106159	-0.291748	-0.167308	-0.441553	0.025298
BUS-13	0.109040	-0.154105	-0.140309	0.315717	-0.014559
BUS-14	0.128760	0.029623	-0.235643	0.039650	-0.012181
BUS-15	0.178810	0.104209	-0.050374	-0.018207	-0.015830
BUS-16	0.204963	-0.218805	0.025884	0.064883	-0.006450
BUS-17	0.337853	0.222826	-0.222634	-0.000227	-0.032912
BUS-18	0.520984	0.008621	0.103751	-0.109780	-0.029657
BUS-19	0.092290	0.056933	0.056848	0.250654	-0.019615
BUS-20	0.057301	-0.102682	-0.000145	-0.105831	0.005540
BUS-21	0.144469	-0.108241	0.150710	0.117057	-0.009449
BUS-22	0.120420	-0.151639	0.326165	0.254042	-0.011526
BUS-23	0.106459	0.019505	-0.207799	0.109369	-0.013268
BUS-24	0.224937	0.025912	0.090412	0.105611	-0.020465
BUS-25	0.146205	0.161939	-0.111320	-0.011973	-0.016714
BUS-26	0.228639	-0.168502	-0.098185	-0.027499	-0.006507
BUS-27	0.269287	0.036244	-0.110314	-0.081870	-0.016063
BUS-28	0.130908	0.062338	0.239487	-0.136104	-0.004834
BUS-29	0.110627	-0.203351	0.176203	0.087581	-0.001551
BUS-30	0.069968	0.010437	-0.040059	-0.089817	-0.001214
BUS-31	0.058711	-0.027466	0.051000	-0.165169	0.004804
BUS-32	0.065727	0.044988	-0.002490	0.044075	0.995845
BUS-33	0.063605	-0.108394	0.026131	-0.216139	0.010330
BUS-34	0.044253	0.222910	0.019923	-0.096440	-0.008673
BUS-35	0.091074	0.437935	0.166688	0.058293	-0.027958
BUS-36	0.047710	0.113603	0.083594	-0.173864	-0.000377
BUS-37	0.082034	-0.157442	0.263679	0.102439	-0.002176
BUS-38	0.069996	0.287236	0.128940	-0.199584	-0.008440
BUS-39	0.003890	0.035845	-0.112979	-0.053292	0.000200

4.2. Fault Coefficient Feature Extraction in Symmetrical Short Circuit Fault

Now, let us further analyze a more complex symmetrical short circuit fault in IEEE 39-BUS New England power system. This time BUS-18 is subjected to a three-phase short circuit fault. By BPA simulations, the node positive sequence voltages have been calculated. In this simulation, let's introduce a stronger white Gaussian stochastic noise $N(0,0.06^2)$.

Based on the fundamental of feature extraction about fault coefficient, the initial characteristic values, initial characteristic vectors, squared loadings and rotation squared loadings can also be calculated. Among them, the initial characteristic value of the first fault factor is 1.83004634, the variance percentage is 0.8828, and the cumulative variance percentage is also 0.8828. For the second fault factor, the initial characteristic value of the first fault factor is 0.24305527, the variance percentage is 0.1172, and the cumulative variance percentage is 1.0000. Obviously, the first fault factor is just what we are seeking.

The fault coefficient of all fault factors can be obtained through the same approach (see Table 3). Therefore, the general form of the first fault factor can be written as:

Fault1= 0.074346BUS1+0.193963BUS2+0.195029BUS3+0.169430BUS4
   + 0.088094BUS5+0.106626BUS6+0.090224BUS7+0.139135BUS8
   +0.042615BUS9+0.130098BUS10+0.117286BUS11+0.130002BUS12
   +0.123483BUS13+0.115265BUS14+0.160039BUS15+0.156472BUS16
   +0.349494BUS17+0.380180BUS18+0.080870BUS19+0.094795BUS20
   +0.173913BUS21+0.162841BUS22+0.104185BUS23+0.239691BUS24 (17)
   +0.122016BUS25+0.234122BUS26+0.245774BUS27+0.109912BUS28
   +0.109377BUS29+0.137135BUS30+0.156919BUS31+ 0.064740BUS32
   +0.041073BUS33+ 0.061483BUS34+0.039357BUS35+0.087746BUS36
   +0.133753BUS37+0.083942BUS38+ 0.258362BUS39

Table 3. The fault coefficient in symmetrical short circuit fault.

BUS	1	2	3	4	5
BUS-1	0.074346	0.052964	-0.001593	-0.000682	-0.005143
BUS-2	0.193963	0.012428	-0.007633	-0.007139	-0.012111
BUS-3	0.195029	0.039759	-0.006921	-0.006016	-0.012461
BUS-4	0.169430	-0.011105	-0.007275	-0.007172	-0.010351
BUS-5	0.088094	0.160465	0.000814	0.003357	-0.007109
BUS-6	0.106626	0.068563	-0.002489	-0.001293	-0.007299
BUS-7	0.090224	0.073833	-0.001669	-0.000420	-0.006341
BUS-8	0.139135	-0.062473	-0.007449	-0.008164	-0.007946
BUS-9	0.042615	0.066201	0.000078	0.001137	-0.003320
BUS-10	0.130098	0.068635	-0.003453	-0.002218	-0.008750
BUS-11	0.117286	0.075571	-0.002734	-0.001416	-0.008030
BUS-12	0.130002	0.026162	-0.004623	-0.004025	-0.008303
BUS-13	0.123483	0.080095	-0.002864	-0.001468	-0.008460

(Contd...)

Table 3. The fault coefficient in symmetrical short circuit fault. (Contd...)

BUS	1	2	3	4	5
BUS-14	0.115265	0.027552	-0.003978	-0.003383	-0.007407
BUS-15	0.160039	0.012319	-0.006241	-0.005803	-0.010014
BUS-16	0.156472	0.151342	-0.002250	0.000264	-0.011238
BUS-17	0.349494	0.184110	-0.009282	-0.005971	-0.023503
BUS-18	0.380180	0.074222	-0.013582	-0.011868	-0.024257
BUS-19	0.080870	-0.058891	-0.004954	-0.005708	-0.004384
BUS-20	0.094795	0.064737	-0.002108	-0.000989	-0.006528
BUS-21	0.173913	0.086564	-0.004759	-0.003186	-0.011643
BUS-22	0.162841	-0.009310	-0.006954	-0.006835	-0.009963
BUS-23	0.104185	-0.067407	-0.006148	-0.006993	-0.005736
BUS-24	0.239691	0.013811	-0.009475	-0.008888	-0.014950
BUS-25	0.122016	0.064783	-0.003227	-0.002063	-0.008211
BUS-26	0.234122	-0.000439	-0.009640	-0.009275	-0.014458
BUS-27	0.245774	-0.016099	-0.010552	-0.010404	-0.015015
BUS-28	0.109912	0.208295	0.001239	0.004533	-0.008954
BUS-29	0.109377	-0.027594	-0.005261	-0.005501	-0.006470
BUS-30	0.137135	-0.124739	-0.009088	-0.010739	-0.007175
BUS-31	0.156919	-0.015092	-0.006870	-0.006848	-0.009537
BUS-32	0.064740	-0.166245	-0.007259	-0.009646	-0.002272
BUS-33	0.041073	-0.027614	0.998774	0.000000	0.000000
BUS-34	0.061483	0.010516	-0.002238	-0.001983	0.998048
BUS-35	0.039357	-0.042477	-0.002793	0.998318	0.000000
BUS-36	0.087746	0.191227	0.001679	0.004682	-0.007407
BUS-37	0.133753	0.032855	-0.004592	-0.003888	-0.008604
BUS-38	0.083942	0.085576	-0.001086	0.000329	-0.006075
BUS-39	0.258362	-0.843064	-0.033934	-0.046151	-0.007201

In the first fault factor, the coefficient feature is always crucial. For all of these coefficients in Fault1, the coefficient of BUS-18 is 0.380180, which is also the biggest one. As a result, we come to the conclusion that BUS-18 is the fault BUS. At the same time, the expression of the first fault factor without the interference of white Gaussian stochastic noise can be described:

Fault1 ' = 0.080303BUS1+0.171509BUS2+0.266442BUS3+0.165149BUS4

+0.124600BUS5+0.119829BUS6+0.112038BUS7+0.109471BUS8

+0.059017BUS9+0.124305BUS10+0.122919BUS11+0.125781BUS12

+0.133050BUS13+0.155540BUS14+0.186957BUS15+0.205857BUS16

+0.306195BUS17+0.425979BUS18+0.129325BUS19+0.097090BUS20

+0.165830BUS21+0.131574BUS22+0.129598BUS23+0.195589BUS24(18)

+0.161128BUS25+0.200178BUS26+0.247223BUS27+0.148679BUS28

+0.132369BUS29+0.105041BUS30+0.075669BUS31 0.079530BUS32

+0.088549BUS33 0.081280BUS34+0.096113BUS35+0.07853+1BUS36

+0.108539BUS37+0.095568BUS38 0.022421BUS39

In the former expression, the real fault location BUS-18 certainly has also been identified. In addition, the noise influence for fault coefficients is presented in Figure 3. In this simulation, the total average deviation level is 0.0016. Therefore, the fault coefficient feature is completely reliable.

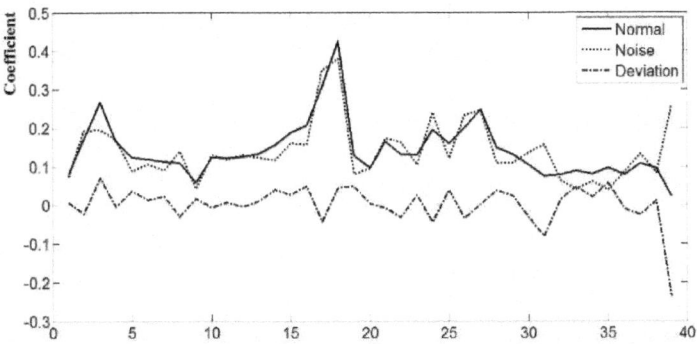

Figure 3. The noise influence in symmetrical short circuit fault.

5. CONCLUSIONS

In order to realize real time and accurate fault detection in complex electrical systems, one needs to obtain fault information of electric networks. In the process of power system operation, rich monitoring data includes the system's normal operation information. Once the fault of a system or equipment occurs, large amounts of fault feature information will be presented, such as operation sequence of protection and circuit breaker recorded by fault recorder, the switching action information, electrical quantities and measuring information provided by the Wide Area Measurement System. All of these have provided data guaranteed for accurate and reliable fault detection.

In this paper, based on the fundamental of feature extraction about the fault coefficient, we have discussed the fault coefficient in complex electrical engineering. According to the research in this paper, the system failure is corresponding to the variable with the biggest coefficient in the first fault factor. For asymmetrical short circuit fault and symmetrical short circuit fault in IEEE 39-BUS New England power system, even if there is strong white Gaussian stochastic interference, the fault coefficient feature is still accurate and reliable. The feature extraction technology about fault coefficient proposed in this paper can extract fault characteristics with high efficiency, and it can satisfy the real-time and accuracy requirements of Wide Area Measurement System and Phasor Measurement Unit. The comparative analysis results of noise influence have also demonstrated the strong anti-interference ability and great redundancy of the fault coefficient feature in complex electrical engineering.

ACKNOWLEDGMENTS

The authors thank the earnest and careful anonymous referees for their thoughtful and constructive suggestions that led to a considerable improvement of the paper. This research was supported partly by the National Key Basic Research Project (973 Program) of China (2012CB215200), the NSFC (51277193), the Specialized Research Fund for the Doctoral Program of Higher Education (20110036110003), the Fundamental Research Funds for the Central Universities (2014ZD43) and the Natural Science Foundation of Hebei Province.

Author Contributions

This paper is a result of the full collaboration of all the authors. Yagang Zhang guided the whole research process; Yi Sun wrote Case Study and Methodology; Yinding Wang performed the experiments. All authors discussed the results and commented on the manuscript.

Conflicts of Interest

The authors declare no conflict of interest.

REFERENCES

1. Ajami, A.; Daneshvar, M. Data driven approach for fault detection and diagnosis of turbine in thermal power plant using Independent Component Analysis (ICA). *Int. J. Electr. Power Energy Syst.* **2012**, *43*, 728–735.

2. Frank, P.M. Analytical and qualitative model-based fault diagnosis — A survey and some new results. *Eur. J. Control* **1996**, *2*, 6–28.

3. Chakraborty, S.; Keller, E.; Ray, A.; Mayer, J. Detection and estimation of demagnetization faults in permanent magnet synchronous motors. *Electr. Power Syst. Res.* **2013**, *96*, 225–236.

4. Patton, R.J.; Frank, P.M.; Clark, R. *Fault Diagnosis in Dynamic Systems Theory and Application*; Prentice Hall: Herfordahire, UK, 1989.

5. Frank, P.M. Fault detection in dynamic system using analytical and knowledge-based redundancy — A survey and some new result. *Automation* **1990**, *26*, 459–474.

6. Cho, Y.S.; Jang, G. New technique for enhancing the accuracy of HVDC systems in state estimation. *Int. J. Electr. Power Energy Syst.* **2014**, *54*, 658–663.

7. Korres, G.N.; Tzavellas, A.; Galinas, E. A distributed implementation of multi-area power system state estimation on a cluster of computers. *Electr. Power Syst. Res.* **2013**, *102*, 20–32.

8. Guo, Y.; Wu, W.C.; Zhang, B.M.; Sun, H.B. A method for evaluating the accuracy of power system state estimation results based on correntropy. *Int. J. Electr. Power Energy Syst.* **2014**, *60*, 45–52.

9. Geramifard, O.; Xu, J.X.; Pana, S.K. Fault detection and diagnosis in synchronous motors using hidden Markov model-based semi-nonparametric approach. *Eng. Appl. Artif. Intell.* **2013**, *26*, 1919–1929.

10. Zhang, Y.G.; Wang, Z.P.; Zhang, J.F. Fault identification based on NLPCA in complex electrical engineering. *J. Electr. Eng.* **2012**, *63*, 255–260.

11. Chai, W.; Qiao, J.F. Passive robust fault detection using RBF neural modeling based on set membership identification. *Eng. Appl. Artif. Intell.* **2014**, *28*, 1–12.

12. Zhu, D.; Gao, Q.W.; Sun, D.; Lu, Y.X.; Peng, S.L. A detection method for bearing faults using null space pursuit and S transform. *Signal Process.* **2014**, *96*, 80–89.

13. Zhou, J.; Guo, A.H.; Celler, B.; Su, S. Fault detection and identification spanning multiple processes by integrating PCA with neural network. *Appl. Soft Comput.* **2014**, *14*, 4–11.

14. Harmouche, J.; Delpha, C.; Diallo, D. Incipient fault detection and diagnosis based on Kullback–Leibler divergence using Principal Component Analysis: Part I. *Signal Process.* **2014**, *94*, 278–287.

15. Zhang, Y.G.; Wang, Z.P.; Zhang, J.F.; Ma, J. Fault localization in electrical power systems: A pattern recognition approach. *Int. J. Electr. Power Energy Syst.* **2011**, *33*, 791–798.

16. Zhang, Y.G.; Wang, Z.P.; Zhao, S.Q. BDA fault detection in complex electric power systems. *Int. Rev. Electr. Eng.* **2012**, *7*, 3638–3645.

17. Zhang, Y.G.; Wang, Z.P. New fault discrimination under the influence of Rayleigh noise. *Adv. Electr. Comput. Eng.* **2013**, *13*, 27–32.

18. Zhang, Y.G.; Zhao, Z.; Wang, Z.P. Comprehensive detection and isolation of fault in complicated electrical engineering. *Electr. Eng.* **2013**, *19*, 31–34.

19. Wand, X.M. *Applied Multivariate Analysis*; Shanghai University of Finance and Economics Press: Shanghai, China, 2009.

20. Hair, J.; Joseph, F.; Anderson, R.E.; Tatham, R.L.; Black, W.C. *Multivariate Data Analysis*; Prentice Hall: Upper Saddle River, NJ, USA, 1998.

21. Martin, K.E.; Benmouyal, G.; Adamiak, M.G.; Begovic, M. IEEE Standard for Synchrophasors for Power Systems. *IEEE Trans. Power Deliv.* **1998**, *13*, 73–77.

This page left intentionally blank.